Lecture Notes in Computer Science 1144

Edited by G. Goos, J. Hartmanis and J. van Leeuwen

Advisory Board: W. Brauer   D. Gries   J. Stoer

Springer
Berlin
Heidelberg
New York
Barcelona
Budapest
Hong Kong
London
Milan
Paris
Santa Clara
Singapore
Tokyo

Jean Ponce   Andrew Zisserman
Martial Hebert   (Eds.)

Object Representation in Computer Vision II

ECCV '96 International Workshop
Cambridge, U.K., April 13-14, 1996
Proceedings

 Springer

Series Editors

Gerhard Goos, Karlsruhe University, Germany

Juris Hartmanis, Cornell University, NY, USA

Jan van Leeuwen, Utrecht University, The Netherlands

Volume Editors

Jean Ponce
University of Illinois, Beckman Institute
Urbana, IL 61801, USA
E-mail: ponce@cs.uiuc.edu

Andrew Zisserman
Oxford University, Department of Engineering Science
Oxford OX1 3PJ, UK
E-mail: az@robots.oxford.ac.uk

Martial Hebert
Carnegie Mellon University, Robotics Institute
5000 Forbes Avenue, Pittsburgh, PA 15213, USA
E-mail: martial_hebert@ius.cs.cmu.edu

Cataloging-in-Publication data applied for

Die Deutsche Bibliothek - CIP-Einheitsaufnahme

Object representation in computer vision : international
workshop ; proceedings / ECCV '96, Cambridge, UK, April 13 -
14, 1996. Jean Ponce ... (ed.). - Berlin ; Heidelberg ; New York
; Barcelona ; Budapest ; Hong Kong ; London ; Milan ; Paris ;
Santa Clara ; Singapore ; Tokyo : Springer, 1996
  (Lecture notes in computer science ; Vol. 1144)
  ISBN 3-540-61750-7
NE: Ponce, Jean [Hrsg.]; ECCV <4, 1996, Cambridge>; GT

CR Subject Classification (1991): I.5.4, I.3, I.2.9-10, J.2, I.4

ISSN 0302-9743
ISBN 3-540-61750-7 Springer-Verlag Berlin Heidelberg New York

© Springer-Verlag Berlin Heidelberg 1996
Printed in Germany

Typesetting: Camera-ready by author
SPIN 10513712   06/3142 – 5 4 3 2 1 0   Printed on acid-free paper

Preface

To build on the momentum of the workshop on Object Representation in Computer Vision held in New York City in 1994, we decided to organize a second one in conjunction with the ECCV'96 conference. Holding this workshop in the beautiful grounds of Robinson College at the University of Cambridge, just before a major vision conference, offered us an excellent opportunity to open its participation to a wider international audience and to assess the progress accomplished in the sixteen months since the New York meeting.

The Cambridge workshop took place on April 13–14, 1996; it brought together over eighty computer vision researchers and included three invited talks by experts in the field, fifteen oral presentations selected from forty-five submissions, and three panel discussions. This book is the outcome of the workshop. It consists of a report that includes a summary of the three panels held during the workshop and a discussion of their findings, followed by the papers presented at the workshop. The report itself, along with abstracts of the contributed papers, is also available through the World Wide Web.[1]

We wish to thank our invited speakers, Professors Takeo Kanade, Jan Koenderink and Ram Nevatia, for delivering excellent, thought-provoking lectures. Many thanks also to the members of our program committee: Mike Brady, Roberto Cipolla, David Forsyth, Dan Huttenlocher, Katsushi Ikeuchi, David Lowe, Roger Mohr, Joe Mundy, Shree Nayar, Charlie Rothwell, Amnon Shashua, Kokichi Sugihara, Chris Taylor, Luc Van Gool, Alan Yuille, and Mourad Zerroug, and to our panelists, a group that included many of the above as well as David Kriegman, Jitendra Malik, Alison Noble, Steve Shafer, Richard Szeliski, and Larry Wolff. Finally, special thanks to Jennet Batten and Sharon Collins for helping us with the organization of the workshop and the preparation of this book.

June 1996

Jean Ponce
Andrew Zisserman
Martial Hebert

---

[1] http://www.ius.cs.cmu.edu/usr/users/hebert/www/eccv-workshop/report.html

Contents

3D representations and applications

Report on the 1996 International Workshop on Object Representation in Computer Vision

Jean Ponce[1], Martial Hebert[2], Andrew Zisserman[3]

[1] Beckman Institute, University of Illinois, Urbana, IL 61801, USA
[2] Robotics Institute, Carnegie-Mellon University, Pittsburgh, PA 15213, USA
[3] Department of Engineering Science, Oxford University, Oxford OX1 3PJ, UK

1 Introduction

The visual world is not obviously parsed into symbols. Of course, images can be thought of as maps defined over two-dimensional domains, and we understand the process of image formation from the physical interaction between surfaces, volumes and light, but the "vocabulary" used by the brain to parse its visual input remains largely inaccessible. This makes the task of constructing artificial vision systems different from, and perhaps more difficult than, other AI tasks such as natural language understanding, where the world structure (the syntax and semantics of natural language, for example) is imposed. This is also a fascinating side of computer vision, since we are free to invent the visual representations most appropriate for a given task.

Articulating the principles that should govern this invention process was the motivation for organizing the first workshop on object representation in computer vision. That workshop was held in New York City in December of 1994[4] and was co-sponsored by two US federal agencies: the National Science Foundation and the Advanced Research Projects Agency; its goal was to evaluate current approaches to object representation and to identify important issues and promising research directions. The success of the workshop prompted us to organize a second one in conjunction with the ECCV'96 conference held at the University of Cambridge in April 1996. Its goals were:

- to present the state of the art of research on object representation for object recognition;
- to assess the progress achieved in key areas identified during the first workshop, e.g., part decomposition and quasi invariants; and,
- to explore the representational issues involved in applications that go beyond traditional object recognition, e.g., image databases, manufacturing, medical imaging, or virtual reality.

Unlike the New York workshop, the Cambridge workshop was self-supporting. Holding it just before a major vision conference helped us to open participation to a wider international audience: in the end, over eighty people, roughly balanced between Europe and the US, participated in the workshop. Three well-known researchers, Professors Takeo Kanade, Jan Koenderink, and Ram Nevatia

---

[4] "Object Representation in Computer Vision", M. Hebert, J. Ponce, T.E. Boult and A. Gross (eds.), Springer-Verlag, Lecture Notes in Computer Science 994, 1995.

were invited to present their views on object representation. Fifteen additional papers were selected for presentation from forty-five submissions. Each paper was reviewed by three program committee members.

The core of the workshop was a set of three half-day sessions in three broad research areas: geometric and topological representations, appearance-based representations, and 3D representations and applications. Each session consisted of a forty-five minute invited talk, followed by five oral presentations, and an hour-long panel. The following three sections are summaries of the panel discussions. It should be noted that we have deliberately chosen to summarize the ideas that emerged from the panel discussion rather than transcribing the detailed discussions and attributing to each panelist his or her comments.

2 Panel 1: Progress in Geometric Representations

The goal of this panel was to assess the recent advances in geometric object representations within the context of object recognition. This includes object representations based on primitive shapes, part decompositions, weak models based on differential geometry or epipolar parameterizations, and invariants.

Panelists: D. Kriegman, J. Koenderink, R. Nevatia, J. Ponce (moderator), C. Rothwell, L. Van Gool.

Where are we now?

The most successful geometric representation to date is generalized cylinders, particularly Straight Homogeneous Generalized Cylinders (SHGCs). Other representations have made a mark (e.g. superquadrics), but SHGCs are the furthest developed in terms of implementations, understanding and some notion of "parts". In general the representations that have been successful have been volumetric, compact, and with a limited number of parameters.

The current wisdom is that:
- Representation is not "generic" but is task dependent. For example, a different representation would be used for face recognition or recognizing objects at a picnic.
- Representation is only geometric at present — generally a particular geometric class (SHGCs, polyhedra), it is not functional. For example, a glass could currently be described as a Surface of Revolution, and recognized as such. It could not currently be recognized from the functional description "something that can hold a liquid." Nor do we have any idea of how to define a model for a chicken.
- Parts are important. Not, perhaps, the actual primitives themselves, but the manner in which they impose structure. For example, a horse can be described as a set of planes, blobs, or sticks, but although the description starts with these primitives, it is not bound by them — it ends up being a horse.

How to evaluate progress?

A standard way to evaluate progress is to have a test suite. Why isn't this standard practice in the model-based recognition field? In other fields, for example voice recognition, there is an agreed test set, and papers report on the success of their method with this set. A major reason why such a test set has not become standard in vision is that visual recognition has so many variables, some of which are ill-defined (the task of voice recognition is more clearly defined: it might be compared to optical character recognition).

Visual recognition occupies an n dimensional space, where the axes span different measures of difficulty. These include: object type (rigid, articulated..), object shape (polyhedral, curved..), extent of occlusion, extent of background clutter, viewpoint, extent of perspective distortion, etc. These axes are not "orthogonal" and will be influenced by current recognition approaches and successes.

There have been attempts to define test sets with real objects. Inevitably, some researchers have designed strategies which, because they are peculiar to these sets, are not more generally applicable. However, this happens in any benchmarking system; the solution is to refine the test set.

Clearly, as the test sets and competences develop, we will be able to move away from the origin of the n dimensional space, but it is unlikely that this will be achieved by a single recognition strategy. Populating this space will chart progress and also catalogue what we can and cannot do.

Has there been progress?

Even in the absence of such a test suite there is certainly no doubt that there *has* been progress. First in terms of functionality: in the 70's parts could be recognized from range images. Now parts can be recognized from intensity images, where the scenes have textured (i.e., non-trivial) backgrounds. Secondly, there is now a far better understanding of geometry and invariant properties under perspective projection.

This "research" progress has had beneficial spin-offs in applications — good theory leading to good applications. For example, meta-strategies for grouping SHGCs in laboratory images have ported successfully to grouping buildings in satellite images. Even if SHGCs are just another toy world (an extension of blocks world), we have learnt by building recognition systems for SHGCs. This will not be wasted.

3 Panel 2: Alternatives to Geometry

This goal of this panel was to explore the role of non-geometric object representations in object recognition (e.g., appearance models, topological representations, learning methods, hybrid approaches), to evaluate the strengths and weaknesses of these representation schemes, and to compare them to more traditional geometric approaches.

Panelists: D. Forsyth, D. Kriegman, D. Lowe, R. Mohr, S. Shafer, L. Wolff, A. Zisserman (moderator).

General Comments

Since most of the actual discussion focussed on the comparison of geometric and appearance-based methods, it may be useful to define what we mean by these terms in this report.

- Geometric approaches: objects and input images are modelled by a discrete set of geometric features (which may be volume or surface elements, embedded in two- or three-dimensional spaces), and recognition proceeds by matching the model and scene features. Geometric constraints are used to prune inconsistent matches.
- Appearance-based approaches: objects are modelled by (possibly an abstraction of) a set of images, and recognition is performed by matching directly the input image to the model set. The matching process is guided by some measure of similarity between images that may be based on intensity/color, geometry, topology, or a combination of these.

Differences

There are several philosophical and practical differences between geometric and appearance-based approaches to object recognition.

First, appearance-based approaches are often associated with viewer-centered object representations, which capture the appearance of an object by a discrete set of characteristic views. In contrast, most traditional geometric approaches rely on object-centered object descriptions (e.g., some type of 3D model, polyhedron, generalized cylinder, etc..) for modeling and matching.

This may be a rather shallow distinction because a representation of the intensity patterns on a surface could very well be constructed in an object-centered manner, e.g., by explicitly storing the surface reflectance at each vertex of a polyhedral representation. Further, the appearance manifolds of Murase and Nayar's SLAM system are actually continuous, spline-based descriptions of the appearance of an object, very different from discrete representations such as aspect graphs.

A deeper philosophical divergence between the two approaches is in the role that photometric information should play in the object description and recognition processes: conventional geometric systems essentially ignore this information, while some appearance-based techniques, such as the SLAM approach of Murase and Nayar mostly ignore geometric information and focus on matching patterns of intensity and color. It should be noted, however, that appearance-based methods such as the one proposed by Pope and Lowe do not explicitly use intensity or color information, relying instead on topological relationship among simple image features, such as junctions or ellipses. Other approaches, such as the one proposed by Schmid and Mohr, base matching on local photometric descriptors with *geometrically* invariant (or rather quasi-invariant) properties.

Another distinction between the two approaches is that geometric methods tend to analytically model the relationship between the object features and their projections; in contrast, this relationship is traditionally captured in an empirical manner by appearance-based techniques.

However, analytical and structural analyses of appearance are starting to emerge. Beside the invariance properties mentioned earlier, it has been recently recognized by several researchers (Murase and Nayar, Epstein *et al.* in these proceedings, but also Koenderink and others) that the space of all pictures of a Lambertian object with a fixed pose is a three-dimensional vector space. This yields an insight into the structure of the space of photometric appearances. Note that a very similar insight into the structure of the space of geometric appearances can be gained by remarking that the image positions of points on a rigid object observed under affine projection span an eight-dimensional vector space, or that, equivalently, Tomasi and Kanade's observation matrix has rank four.

Limitations and Promises

Interestingly, one of the main reasons for concentrating on geometric information in the past was that the photometric information contained in images is difficult to model and interpret (e.g., the failure of shape from shading). It is interesting to see that intensity- and color-based methods with rudimentary (if any) image formation models are now among the most successful approaches to object recognition. An interesting question is whether their success is due to very controlled imaging conditions. If this were true, the general applicability of appearance-based methods to real-world vision tasks would be severely limited.

A more positive explanation is that, unlike shape-from-shading techniques whose only sources of information besides the input images are (necessarily simplified) image formation models, the appearance-based recognition methods can take advantage of the very rich constraints available in a priori object/scene models. Similarly, line-drawing interpretation, which, like bottom-up shape from shading, relies purely on (simplified) first principles, has had only limited success; while, on the other hand, geometric model-based recognition, which takes advantage of the geometric constraints attached to each model, has been a success story, at least when the number of models is small and their geometry is simple.

One of the limitations of appearance-based techniques noted at the first workshop was that they are global methods and thus rely heavily on powerful bottom-up segmentation algorithms to delineate the objects of interest in an image. This view is challenged by new techniques such as the ones proposed by Pope and Lowe and by Schmid and Mohr at this workshop. Indeed, both methods use local information and are capable of dealing with cluttered backgrounds and imperfect segmentations.

Another interesting question is the representation of object classes: no one has yet proposed an approach for extending appearance-based methods to describe and recognize objects such as chairs or chickens. Of course, very few geometric approaches are capable of addressing the notion of class.[5] In fact, as noted during the panel discussion, state-of-the-art representation schemes, whether geometry- or appearance-based, capture hardly any abstraction from instances.

To the extent that class abstraction implies the representation of symbolic information (which is not a priori obvious) however, it seems at this point that geometric approaches have an edge over their appearance-based counterparts. Nevertheless, it should be recognized that appearance-based methods such as SLAM or Schmid and Mohr's hybrid approaches have demonstrated very impressive results with large object databases. In that context, it may very well be that some notion of object classes can be captured by clustering appearance models in some feature space. More empirical evaluation as proposed by Mundy *et al.* in this workshop is, of course, necessary.

4   Panel 3: Applications

The goal of this panel was to address the representational issues involved in applications areas where object recognition is not the primary task. These issues include digital libraries, human-computer interaction, medical imaging, navigation, and virtual reality. Of particular interest was identifying how specific applications might contribute to the general problem of object representation in computer vision, and, conversely, which approaches have been successfully demonstrated in specific applications.

Panelists: M. Hebert (moderator), D. Huttenlocher, J. Malik, J. Mundy, A. Noble, R. Szeliski.

Issues

Computer vision has been successfully applied to a number of practical problems. Although traditional applications (e.g., inspection) mostly involve 2D image processing, the emergence of new technologies requires complex 3D representations of objects and scenes: advances in medical imaging demand sophisticated 3D registration techniques; the explosion of the virtual reality and multimedia market has generated a need for efficient 3D representation; and the recent interest in digital libraries has demonstrated the importance of generic models of complex objects. Thus, object representation is becoming a central theme in many different application domains besides object recognition, traditionally the favorite area in computer vision research. New applications provide opportunities for new research directions. They also share a number of requirements and constraints that are not addressed directly in traditional computer vision research:

---

[5] With the notable exception of a few systems based on Binford's generalized cylinders, which can handle object classes such as planes, electric motors, or teapots, e.g., Brooks' ACRONYM or more recently Zerroug and Nevatia's program for extracting part-based scene descriptions from photographs. Obviously these systems are still limited to very rigid classes of objects.

- Testing: typically, computer vision systems used in practical applications require operation, and therefore testing, on much larger sets of images than systems developed for proof-of-principle research.

- Domain knowledge: because generality is typically less important than relevance to the task, domain knowledge is used heavily in the design of practical vision systems. Using domain knowledge in a systematic way has not been explored widely in computer vision research. Such usage requires a change of attitude in which choosing the most general or the most elegant approach is not critical; rather choosing the technique with the best performance is the critical factor. Although new research directions, such as "context-based vision" do address this issue, more work is needed in this area.

- Human interaction: once they are transferred to real-world applications, vision systems must be usable by operators who are not familiar with the inner workings of computer vision algorithms. As a result, the extensive "tweaking" of many computer vision techniques (e.g., setting parameters, selecting algorithms, selecting points or edges in images, etc.) is unacceptable and the extent of operator involvement must be minimized. A simple example is image calibration which typically requires a number of manual steps, e.g., selection of features in an image. The number of those steps, and the accuracy with which they need to be performed, must be minimized in order for workers unfamiliar with the underlying technology to use it.

- Integration: higher-level object representations are often studied independently of the early processing required to extract them from raw data. However, in practical applications, the cycle, from raw data to high-level representation, must be addressed as a whole. In particular, issues of segmentation, figure/ground separation, etc. cannot be abstracted out.

It would seem that researchers interested in general representational theories or in generic recognition should have a great deal of interaction with developers of specific applications. For example, instances of general representations should be used in specific applications and the experience gained in using them should be used for refining the general theory.

In reality, however, there seems to be little interaction between the two communities. This panel focuses on identifying successful transfer of representations to applications, on identifying lessons that can be learned from those applications, and on recommendations for future work.

From Theory to Applications

Several examples of successful transitions from computer vision research have been identified. One example is surface registration, a critical component of medical applications. In this area, general representations and matching techniques have proved to be effective in solving a number of data registration problems. Although much work is needed in shape description, in particular for describing deformable shapes and growth models, this is a promising application area for research in 3D object representation.

Another example is virtual reality, in which building representations from visual data is critical. Many recently demonstrated systems directly use results obtained from basic research. This is true both for implicit models (e.g., view interpolation) and for explicit 3D models. Given those initial successes, considerable work remains to be done in comparing those representations in the context of virtual reality applications (e.g., volumetric vs. surface representations). This area is another example of an application in which competing representations may be evaluated and compared.

From Applications to Theory

The use of computer vision in practical applications may provide insight into new approaches, or better understanding of existing approaches. Although this transfer from applications to theory of object representation is more difficult and less common, several instances were identified.

For example, the RADIUS project for the interpretation of aerial imagery used the idea of using context for dealing with limited resolution, and thus poor geometrical description. Context is defined as information that an algorithm requires in order to be successful, but that it cannot itself provide. For example, the number and density of edgels in a context suggests the presence or absence of an object.

Generalizing from specific image interpretation tasks, context-based vision now has implications for more general vision problems and for the object representations used in those problems.

The Future

It is clear from this panel that evaluating the performance of various approaches in the context of specific applications has tremendous value. In particular using well-defined tasks allows for extensive testing and clear evaluation criteria. In order to facilitate transition toward applications, however, researchers need to rethink some of the approaches. In particular, the strongest model applicable to the task must be used, i.e., the model that will give the best performance even if it does not generalize. Similarly, multiple cues or representational levels must be used. For example, it was noted that both appearance and geometry are needed in most applications.

Finally, interaction between researchers and potential users must be developed further. Although, such interaction is fairly strong in some fields, such as in medical applications, it is clear that much more needs to be done for the transfer to other fields to be successful.

Geometric and Topological
Representations

From an Intensity Image to 3-D Segmented Descriptions[1]

Mourad Zerroug and Ramakant Nevatia

Institute for Robotics and Intelligent Systems, University of Southern California, Los Angeles, CA 90089-0273. zerroug@iris.usc.edu

Abstract. We address the inference of 3-D segmented descriptions of complex objects from a single intensity image. Our approach is based on the analysis of the projective properties of a small number of generalized cylinder primitives and their relationships in the image which make up common man-made objects. Past work on this problem has either assumed perfect contours as input or used 2-dimensional shape primitives without relating them to 3-D shape. The method we present explicitly uses the 3-dimensionality of the desired descriptions and directly addresses the segmentation problem in the presence of contour breaks, markings shadows and occlusion. This work has many significant applications including recognition of complex curved objects from a single real intensity image. We demonstrate our method on real images.

1 Introduction

Recovering and representing the shape of a complex object is one of the most fundamental tasks in computer vision. A good shape representation is useful not only for object recognition but also for manipulation, navigation and even learning. We believe that a good way to represent a complex object is by decomposing it into parts and describing the parts and the relationships between them. If the parts are complex, they can be decomposed into simpler parts and described in the same way as the larger object. It is also highly desirable that the parts be described as volumetric primitives. Such a representation is very rich, stable and allows us to handle occlusion and articulation in a natural way.

1.1 Previous Work

The use of simpler parts to describe more complex objects has a long history in computer vision [4,12,18]. Biederman has argued that a similar scheme is used by the human visual system as well [3]. However, in spite of these theories and the obvious advantages of segmented (or part/whole) representations, their use in computer vision systems has been limited. We believe that this is due to the difficulty of actually computing segmented shape description from real data. The part decomposition hierarchy is not given in advance, we must infer it from the observable features in the data. Most of the previous work has used range data [2,8,15,16].

---

1. This research was supported by the Advanced Research Projects Agency, monitored by the Air Force Office of Scientific Research under grant No. F49620-95-1-0457 and grant No. F49620-93-1-0620.

Paper originally published in Proceedings of the 12th International Conference on Pattern Recognition, Jerusalem, pp 108-113, Oct. 1994.

In this paper, we focus on computing segmented volumetric descriptions from a single intensity image. This is a task that humans perform effortlessly. It is also important for computer vision as a single image can be acquired rather easily, without extensive control of illumination or elaborate calibration procedures. Using single intensity images does pose many problems, however. Lack of direct 3-D measurements makes it more difficult to determine discontinuities that may characterize part boundaries. Instead, we must work with intensity boundaries which may correspond to depth boundaries, but also to markings, shadows, specularities and noise. Furthermore, object boundaries are unlikely to be complete due to both poor edge localization and occlusion. These characteristics make the techniques developed for range data and perfect contours (such as based on curvature extrema) [6,1] largely unapplicable to the case of lines extracted from a real intensity image.

Figure 1 shows a sample image. Notice that the boundaries are not all perfect, continuous or even part of the outline of the object (in fact, most are not). Also, notice the partial occlusions. Here, we would like to separate the teapot from the background and describe it as consisting of the arrangement of four parts: the conical pot, the lid, the spout and the handle. Deciding that there is such an object *and* with that composition is a non-trivial problem. Moreover, we would like to recover the 3-D shape of the object.

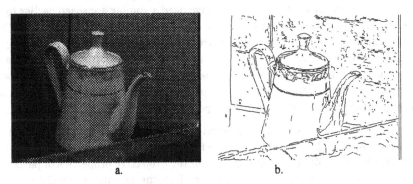

a. b.

Fig.1   Sample real image of a compound object. a. intensity
image. b. edge image

Some previous work has attempted to address part and object segmentation from intensity images. Specifically, the work of [17] and [13] has attempted to solve similar problems to those presented in this paper. However, these efforts relied largely on heuristic properties of observed contours and did not attempt any 3-D recovery (they address a 2-dimensional problem). The method of [10], based on a neural network implementation, addresses geon-based description and recognition from possibly discontinuous boundaries. The boundaries were synthetic and the axial descriptions were assumed given.

1.2 Approach

In this paper, we present an approach that exploits rigorous properties of the contours of 3-D objects to solve the figure/ground (object segmentation) problems, *and* to recover 3-D structure of the objects. The class of objects addressed are those which consist of the arrangement of parts which can be described as generalized cylinders (GCs) [4]. The sub-classes of GCs we allow here are the *straight homogeneous generalized cylinders* (SHGCs) and *planar right generalized cylinders* (PRGCs). SHGCs are obtained by scaling a planar cross-section along a straight axis curve. PRGCs are obtained by scaling a planar cross-section along a curved planar axis curve. More precisely, two sub-classes of PRGCs are addressed: *planar right constant generalized cylinders* (PRCGCs), characterized by a constant sweep, and *circular planar right generalized cylinders* (circular PRGCs), characterized by a circular but varying size cross-section.

The generic joint relationships addressed include those where parts are in contact at their ends (*end-to-end* joints) and those where parts ends are in contact with other parts bodies (*end-to-body* joints). We believe that the above classes of GCs (henceforth *parts*) with such relationships can represent well a large fraction of man-made objects.

Our approach to detecting and describing complex objects is based on the exploitation of the projective properties of the above classes of parts and of their relationships. They consist of geometric *invariant* and *quasi-invariant* and structural properties of the image boundaries of an object (the projection geometry is approximated by orthography in this work). We have used a similar approach earlier to analyze scenes of objects consisting of *single* GC primitives. Our work on SHGCs is described in [24,25,26]. Our analysis of circular PRGCs is presented in [23,24]; the method for recovering them from a single intensity image appears in a separate paper in this conference [27]. Dealing with compound objects introduces many new issues related to the use of joint relationships to effect object-level segmentation and also shape description, including 3-D shape inference.

Our method for object segmentation and description consists of two main levels, the *part level* and the *object level*. The former is concerned with generating part hypotheses using evidence of regularity from the projective properties of parts. The latter analyzes the joint relationships between hypothesized parts to build a geometric context used to hypothesize compound objects, refine hypotheses, and infer (3-D) shape.

Our method and some results are described in the following. First, we provide a brief overview of our part detection and description system and then describe our method for segmentation and description of compound objects.

2 The part level

Most of the methods used in this level have been described elsewhere [23,24,25,26]; due to lack of space, we will summarize them instead of giving details. The classes of parts addressed in this work (SHGCs, PRCGCs and circular PRGCs) are shown in Fig. 2.

Two fundamental aspects characterize our method for detecting parts. First, it uses geometric (orthographic) invariant and quasi-invariant (characterizing the intrinsic geometry of a GC), and structural, properties (characterizing lawful bound-

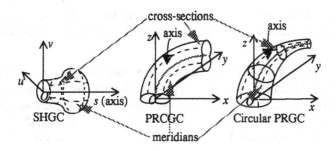

Fig.2   Generalized cylinders used as parts in our method

ary interactions such as junctions) of the above classes of primitives. Second it orga-
nizes the segmentation and description as a multi-level hypothesize-verify process.
The projective properties provide *necessary* conditions that projections of SHGCs
and PRGCs must satisfy in the image. They also give direct relationships between
3-D shapes and computable image descriptions which is useful for recovering volumetric descriptions from a monocular image.

The method for detecting parts consists of searching (aggregates of) boundaries that are likely to project from the same part. This is done by first *detecting* surface patches (parts fragments) that locally satisfy the projective properties of a part.
Then, surface patches with "similar" projective descriptions (with respect to the expected projective properties of a part) are *grouped* into a single part hypothesis. For
example, for an SHGC, the local surface patches must have the same axis projection.
Each part hypothesis is then *verified* for global consistency. Consistency is defined in
terms of both geometric and structural criteria. The geometric criteria consist of enforcing global consistency of the geometry of the part with respect to its geometric
invariants and quasi-invariants. The structural criteria consist of enforcing closure
and associated junctions at the end of a part. They express the fact that the image
of a part may have one of several well defined closure patterns involving specific
junction labeling (that include occlusion junctions) [11,27]. The verification also uses
an *inter-part* filtering whereby weaker part hypotheses (conflicting with "better" hypotheses) are removed.

The hypothesize-verify nature of the part detection method allows us to handle markings, shadows and occlusion. Non-object boundaries (such as surface markings and shadows) are unlikely to survive the successive application of the strong
projective properties. In using the view invariant (and quasi-invariant) properties,
the obtained descriptions do not depend (much) on the particular viewpoint the
scene is viewed from.

Several enhancements, beyond our previously described work, have been
made to the part level in order to handle compound objects. First, cross-sections may
not be visible due to joints between parts. Second, by using a more complete set of
projective properties (those of parts and those of joints), several ambiguities occur
and need to be addressed. The ambiguities are due to the fact that different 3-D

events could produce similar image events (for e.g. junction patterns along a single part or between joined parts). For lack of space, we omit the details of the improvements to the part level.

Figure 3 shows results of the part level on the image of Fig. 1. All four verified parts consist of aggregrates of local surface patches (the pot for e.g. consists of two due to the dividing marking across its surface). Notice that the complete pot has been recovered although its boundary is occluded by both the spout and the front flat object. In this example spurious hypotheses have been rejected at the verification stage.

Fig.3   Results of the part level from the image of Fig. 1.

3 The object level

The object level uses the part hypotheses to form compound objects descriptions. The process is not as direct as simply detecting joints between parts. An inherent issue to monocular analysis of 3-D scenes is the ambiguity of the projective properties. Thus, multiple interpretations are possible from contours alone. Further, since our goal is to produce descriptions in terms of GCs and their relationships, we must also produce descriptions that are as complete as the image allows us to infer. These descriptions could be 3-dimensional if sufficient information is available in the image or otherwise 2-dimensional but corresponding to the *projections of the 3-D descriptions*. This level addressed these issues. It is organized in two main steps: the *detection of compound objects* and the *inference of 3-D shape*. The steps are discussed below.

3.1 Detection of compound objects

To detect compound objects, we have to detect joints between hypothesized parts. Because of the possible ambiguities, an explicit analysis of conflicting interpretations is also carried out. The two step are discussed below.

Detection of joints.

The objective of this step is to identify potential joint relationships between hypothesized parts (whether there is physical contact between parts cannot be firmly concluded from an image). For this, it is useful to analyze the generic image events corresponding to the two types joints addressed, *end-to-end* and *end-to-body* (Fig. 4), and use them to hypothesize those relationships.

end-to-body joints

end-to-end joint

end-to-body joint

Fig.4   Examples of joints between parts

The structural properties of these two types of joints consist of the closure patterns of the joined parts and the observed junction relationships between the joined parts' boundaries. The observed structure depends on the viewing direction, the parts shapes (sweep derivatives and axes curvatures) at the joints and the visibility of their *joint curve* (intersection of their surfaces). In the following we only give the possible image events for each type of joint. The details of the analysis are given in [28].

End-to-end joints: Our model of an end-to-end joint has two possibilities: the two cross-sections have the same size (Fig. 5.a through c) at their contact or have different sizes (Fig. 5.d and e). The different arrangements for both joint closures and events between parts are shown in Fig. 5. The abbreviations for the junctions are as follows: *L-j* stands for *L*-junction, *T-j* for *T*-junction and *3-tgt-j* for three-tangent junction (from the catalog given in [11]; arrow and Y-junctions are currently not used). The *2-tgt-j* corresponds to the *3-tgt-j* case where the upper branch is not observed due to occlusion between the parts.

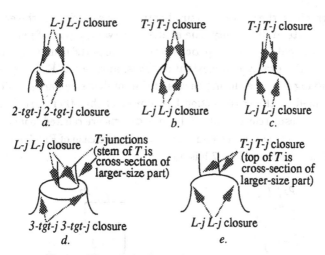

L-j L-j closure   T-j T-j closure   T-j T-j closure

2-tgt-j 2-tgt-j closure   L-j L-j closure   L-j L-j closure
a.   *b.*   *c.*

L-j L-j closure   T-junctions (stem of T is cross-section of larger-size part)   T-j T-j closure (top of T is cross-section of larger-size part)

3-tgt-j 3-tgt-j closure   L-j L-j closure
d.   *e.*

Fig.5   Structural relationships for end-to-end joints; equal-size ends (a through c) and different-size ends (d and e).

In case a., there is no self-occlusion. In case b. there is self-occlusion and the joint curve is visible. In case c. there is self-occlusion and the joint curve is not visible. Cases d. and e. have self-occlusion and differ in the visibility of the joint curve.

End-to-body joints: Our model of end-to-body joint consists of a part's end in contact with another part's body. In Fig. 6.a, the joined part has an L-j L-j closure and T-junctions with the other part's boundaries. In Fig. 6.b, the joined part has T-j T-j closure where the T-junctions are with the other part's boundaries.

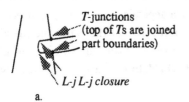

Fig.6   Structural relationships for end-to-body joints;
a. visible joint-curve; b. non-visible joint-curve

In the above, examples with L-j L-j closure could be replaced by any of the image closure patterns that result from the cross-section *facing away* from the viewer and the examples with 3-tgt-j 3-tgt-j closure could be replaced by any of the image closure patterns that result from the cross-section *facing towards* the viewer (see [27]).

To detect joints, the above structural patterns are checked at the end of each part with other parts in its vicinity. The analysis, between a pair of parts, is based on the parts closure patterns, the junctions between the parts' boundaries and on an "extent" analysis. For example, the extent analysis, in case d of Fig. 5 for example, consists of verifying that the closing curves of the smaller size part are all "inside" the region bounded by the cross-section boundaries of the other part's larger size cross-section. The closure constraints of the joined parts are relaxed to include occlusion at most at one side of a part's end. For example, the case of Fig. 7.a is accepted as a joint, whereas the one of Fig. 7.b is not.

Fig.7   Joint detection allows for partial occlusion. a. a joint
is marked between parts $p1$ and $p2$; b. no joint is
marked

Analysis of ambiguities.

Ambiguities include conflicting geometric properties (e.g. the same symmetry relationships for different GC primitives) and conflicting structural properties (e.g. same local generic events for a single part and for joined parts).

This step attempts to identify cases where more than one 3-D interpretation is possible from the given descriptions detected so far. First, since the detected joints provide a global structural context, they are used to filter out certain inconsistent interpretations. For this, joints with visible joint-curve appear to be useful. They can be thought of as non-accidental relationships whose presence in the image suggests that the boundaries of the joint-curve should be interpreted as parts ends, not sides (or limb). For example, while the single part of Fig. 8.a, taken by itself could be interpreted as either an SHGC or a PRCGC, when considered in the joint of Fig. 8.b, can only be interpreted as an SHGC part and when considered in the joint of Fig. 8.c as a PRCGC part.

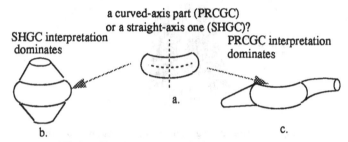

Fig.8   Part perception is linked to joint perception

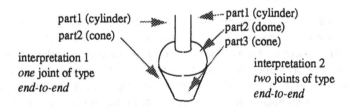

Fig.9   Some ambiguities may persist

Remaining ambiguities are those for which certain image boundaries have different parts and joints interpretations (for e.g boundaries which could be interpreted as either cross-section or side boundaries). Figure 9 gives an example where two interpretations are possible (two parts and one joint or three parts and two joints).

The result at this stage of the method is a set of possible interpretations each of which is represented a set of graphs (one for each compound object) whose nodes are the parts and whose arcs are the joints between the parts. The arcs are labeled partially by the type of joints they represent. This graph is only a representation of the detected objects. Its purpose is not the same as the one in the method of [17] where the graph was used to segment objects made up of ribbons. Although multiple interpretations are a feature of our system, in the examples given in this paper, only one interpretation is found for each image. Figure 10 shows the graph constructed for the parts of Fig. 3.

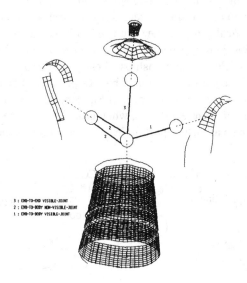

3 : END-TO-END VISIBLE-JOINT
2 : END-TO-BODY NON-VISIBLE-JOINT
1 : END-TO-BODY VISIBLE-JOINT

Fig.10   Resulting graphical representation from the
hypothesized parts of Fig. 3

3.2 Inference of 3-D shape

Recovering 3-D shape of a compound object consists of recovering the intrinsic 3-D description of each of its parts; i.e. its 3-D cross-section, its 3-D axis and the sweep function. This necessitates, for each part, an image description which gives its cross-section(s) and the correspondences between its sides (projections of points on the same cross-section in 3-D), both of which give the projection of the 3-D description. Having these two elements is essential for constraining the 3-D shape of the part [9,19,20,21,23,24,25,26]. Joint relationships between parts should also be used to add further constraints on 3-D shape.

The 3-D shape recovery of SHGCs and curved-axis primitives, as isolated parts, are discussed in [25,26] and [23,27] respectively. Here, we discuss how joints between parts can be used for 3-D shape inference. There are two such uses. The first is to set mutual constraints on the cross-section plane orientation of joined parts. The second is to infer invisible parts cross-sections.

In the former, end-to-end joints with visible joint-curve are used as indicators that the joined parts are cut by the same plane and thus their ends must have the same 3-D orientation. For this, a cut classification of the parts is useful. This classification uses evidence from the observed properties of the hypothesized parts ends, such as linear parallel symmetry [20], line-convergent symmetry [22], local bilateral symmetry and closure by elliptic arcs [14,23], to infer whether a part's cut is likely to be cross-sectional and/or planar. The details of this classification are omitted for lack of space (they are given in [28]). All part ends which must have the same orientation are identified. This is done by traversing the object graph through the arcs labeled as "end-to-end with visible joint curve" and through nodes labeled as having "parallel cuts", such as SHGCs with linearly parallel symmetric ends. The common plane orientation of such parts ends is the average of the individual orientations obtained using the methods described in [20,21,25,26,23].

End-to-end joints with equal-size ends can also be used to infer missing parts cross-sections and consequently make their 3-D recovery possible. A missing part cross-section can be inherited from a joined part if the 3-D shape of this latter has been fully recovered. An example is given in Fig. 11.a. For parts whose cross-sections cannot be so inherited then circular cross-sections are assumed whenever consistent with the observed properties, including local bilateral symmetry and elliptic-arc closures [14,23,28]. Examples are given in Fig. 11.b (for a surface of revolution) and Fig. 11.c (for a circular PRGC).

Each time a part's cross-section(s) is (are) so inferred, the part's 3-D shape is recovered (as described in [25,26 ,23,27]) and the new information is propagated through the end-to-end joints with equal-size ends and visible joint-curve.

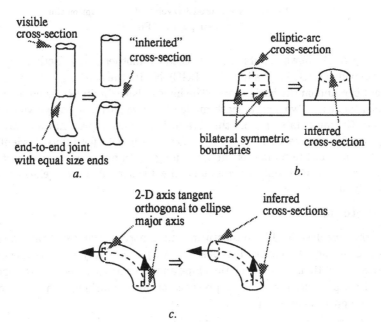

Fig.11   Completion of descriptions: inferring parts cross-sections

A 3-D shape completion step is performed for parts with gaps in their descriptions. This completion consists of filling in the gaps in the 3-D axis by quadratic curves (in its recovered plane) and the gaps in the sweep function by piecewise linear sweeps.

Currently, end-to-body joints are not used for inferring 3-D shape. Their use would require to set explicit differential geometric relationships between the joined parts' surface orientations at their intersection curve [7]. We believe this to be a research topic in its own right.

Figure 12 shows the recovered volumetric descriptions, in terms of cross-sections meridians and axes, of the parts of the object in Fig. 10, shown for different 3-D orientations. The pot and the lid (SHGCs) have been identified as having the same cross-section orientation, the spout (circular PRGC) has been completed in 3-D and the handle (PRCGC) could not be recovered in 3-D because its ends are not visible (its description remains projective)

Fig.12   Recovered 3-D volumetric descriptions for
the descriptions of Fig. 10.

Figure 13 shows results of the method on another image. Both mugs consist of a main body (SHGC) and a handle (PRGC). Notice the markings on the surface of the front cup and on the background. The image also exhibits occlusion between independent objects. Two objects, each made up of two parts joined by two joints, are obtained. The joints labeling and the obtained graphical representations are shown in Fig. 13.c. The recovered 3-D parts are shown in Fig. 13.d for different orientations. The 3-D shape of the handle of the front mug could not be recovered because its cross-sections are not visible. At this stage, the 3-D parts and their relationships are completely identified.

4   Conclusion

The method we described here constitutes an important progress towards realistic monocular 3-D scene analysis. It is based on a rigorous analysis of the properties of GC parts and of their generic relationships and their use to achieve both segmentation and shape description in the presence of noise, markings, shadows, contour breaks and partial occlusion.

We believe that it can be extended to handle other classes of objects and parts such as those having multiple faces (as occurs with polyhedral cross-sections for e.g.) A partial solution is given in [26].

The results of this work have several applications. The descriptions obtained by our system (either the 3-D intrinsic elements of a GC or their projective descrip-

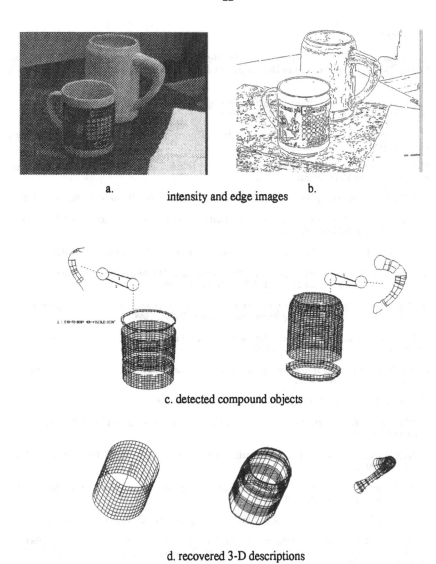

a.   b.
intensity and edge images

c. detected compound objects

d. recovered 3-D descriptions

Fig.13   Additional results of the method

tions) can be used to provide powerful, view-insensitive, indexing keys to large databases of object models for object recognition (such as in [15] for example). In
manipulation, the 3-D descriptions can be used to plan for the grasp and pre-shape
the hand. In navigation, they can be used to select appropriate paths to avoid obsta-
cles, for example, and in learning, the symbolic descriptions can be used to analyze
differences and similarities between newly recovered objects and previously recovered ones.

References

[1] R. Bergevin and M.D. Levine, "Generic Object Recognition: Building and Matching Coarse Descriptions from Line Drawings," in *IEEE Transactions PAMI*, 15, pages 19-36, 1993.

[2] P.J. Besl and R.C. Jain, "Segmentation Through Symbolic Surface Descriptions", In *Proceedings of IEEE CVPR*, pages 77-85, 1986.

[3] I. Biederman, "Recognition by Components: A Theory of Human Image Understanding", *Psychological Review*, 94(2):115-147.

[4] T.O. Binford, "Visual Perception by Computer," *IEEE Conference on Systems and Controls*, December 1971, Miami.

[5] T.O. Binford, "Inferring Surfaces from Images," *Artificial Intelligence*, 17:205-245, 1981.

[6] S. Dickinson, "3-D shape Recovery using Distributed Aspect Matching," *IEEE Transactions PAMI*, 14(2):174-198, 1992.

[7] Do Carmo, "Differential Geometry of Curves and Surfaces," Prentice Hall, 1976.

[8] T.J. Fan, G. Medioni and R. Nevatia, "Recognizing 3-D Objects using Surface Descriptions", *IEEE Transactions PAMI*, 11(11):1140-1157, 1989.

[9] A. Gross and T. Boult, "Recovery of Generalized Cylinders from a Single Intensity View", In *Proc. Image Understanding Workshop*, pages 557-564, 1990.

[10] J.E. Hummel and I. Biederman, "Dynamic Binding in a Neural Network for Shape Recognition" *Psychological Review*, 1992.

[11] J. Malik,"Interpreting line drawings of curved objects," *International Journal of Computer Vision*, 1(1):73-103, 1987.

[12] D. Marr, "Vision", W.H. Freeman and Co. Publishers, 1981

[13] R. Mohan and R. Nevatia, "Perceptual organization for scene segmentation", *IEEE Transactions PAMI*. 1992.

[14] V. Nalwa, "Line drawing interpretation: Bilateral symmetry," *IEEE Transactions PAMI*, 11:1117-1120, 1989.

[15] R. Nevatia and T.O. Binford, "Description and recognition of complex curved objects," *Artificial Intelligence*, 8(1):77-98, 1977.

[16] A. Pentland, "Recognition by Parts," in *Proceedings of the ICCV*, pages 612-620, 1987.

[17] K. Rao and R. Nevatia, "Description of complex objects from incomplete and imperfect data," In *Proceedings of the Image Understanding Workshop*, pages 399-414, Palo Alto, California, May 1989.

[18] L. Roberts, "Machine Perception of Three-Dimensional Solids," MIT Press, 1965.

[19] H. Sato and T.O. Binford, "Finding and recovering SHGC objects in an edge image," *Computer Vision Graphics and Image Processing*, 57(3), pages 346-356, 1993.

[20] F. Ulupinar and R. Nevatia, "Shape from contours: SHGCs," In *Proceedings of ICCV*, pages 582-582, Osaka, Japan, 1990.

[21] F. Ulupinar and R. Nevatia, "Recovering Shape from Contour for Constant Cross Section Generalized Cylinders," In *Proceedings of Computer Vision and Pattern Recognition*, pages 674-676. 1991. Maui, Hawaii.

[22] F. Ulupinar and R. Nevatia, "Perception of 3-D surfaces from 2-D contours," *IEEE Transactions PAMI*, pages 3-18, 15, 1993.

[23] M. Zerroug and R. Nevatia, "Quasi-invariant properties and 3D shape recovery of non-straight, non-constant generalized cylinders", In *Proceedings of IEEE CVPR*, pages 96-103, New York, 1993.

[24] M. Zerroug and R. Nevatia, "Using invariance and quasi-invariance for the segmentation and recovery of curved objects," in *Proceedings of the 2nd ARPA/ESPRIT Workshop on Geometric Invariance in Computer Vision*, The Azores, 1993.

[25] M. Zerroug and R. Nevatia, "Segmentation and 3-D recovery of SHGCs from a single intensity image", *European Conference on Computer Vision*, Stockholm, 1994.

[26] M. Zerroug and R. Nevatia, "Volumetric Descriptions from a Single Intensity Image," to appear in the International Journal of Computer Vision.

[27] M. Zerroug and R. Nevatia, "Segmentation and 3-D recovery of curved axis generalized cylinders from an intensity image", Proceedings of the ICPR, Jerusalem 1994.

[28] Mourad Zerroug. "Segmentation and Inference of 3-D Descriptions from an Intensity Image". Ph.D. dissertation, Institute for Robotics and Intelligent Systems, University of Southern California, 1994.

Recovering Generalized Cylinders by Monocular Vision

Patrick SAYD, Michel DHOME, Jean-Marc LAVEST

LAboratoire des Sciences et Matériaux pour l'Electronique, et d'Automatique,
URA 1793 of the CNRS, Université Blaise-Pascal de Clermont-Ferrand,
F-63177 Aubière Cedex, France

Abstract. In this paper, we propose a theoretical study of the Generalized Cylinders (GC) with constant circular cross-section viewed under perspective projection.

We show that a single view of the object make it possible to infer its 3D shape. Two algorithms are described. The first method is based on the matching between two contour points from a same cross-section and each cross-section is separately modeled. The second algorithm uses a global model, parameterized with a BSpline curve. The model shape is twisted from an initial position until the limb projection fits the image content. With such an approach the matching problem between image points from the same cross-section is avoided and the global shape is estimated through a minimization process. Experiments on real data are provided, which prove the validity of the two approaches.

1 Introduction

In computer vision, many objects classes have been studied in order to recover their shape or their location. Generalized cylinders (GC) have been studied by Shaffer and Kanade [22] Ponce and Chelberg [17], Liu et al. [12] or Sato and Binford [20].

The expression "Generalized Cylinders" was first introduced by Binford [3] in 1971. A generalized cylinder is defined by a space curve, called the axis of the object, and a set of cross-sections. A transformation rule specifies the size and the shape transformation of the cross-section along the axis. The shape of the axis, the way the section changes along the axis means many subsets of generalized cylinders can be defined. The terminology depends on the authors who were interested in this subject; Shaffer [23] or Naeve and Eklundh [14] give exhaustive descriptions of the subclass of generalized cylinders.

Some subclasses of generalized cylinders have been studied for recognition, location or reconstruction problems. Dhome et al. [18] have taken interest in Straight Homogeneous Generalized Cylinders (SHGC), called also solids of revolution. They present some location methods by monocular vision using interpretation of elliptical contours, zero-curvature contour point and limb projection from matched points. Kriegman and Ponce [10] present a method for recognizing and positioning generalized cylinders from contours. Nevatia et al. propose several methods to recognize generalized cylinders in grey level images [13], or to recover their 3D shape by monocular vision [24] or by stereovision [5]. Gross and Boult [8] use the reflectance on the surface to model straight homogeneous generalized cylinders. Saint-Marc and Medioni [19] propose a symmetry detection in grey level images to extract limbs of generalized cylinders.

This paper deals with a particular subset of generalized cylinders which have circular and identical cross-sections, sweeping along any space curve. According to Shaffer's taxonomy, these objects may be called Uniform Circular Generalized Cylinders (UCGC). Nevatia [25] has worked on synthetic images of a near subset (Planar Circular Generalized Cylinders) under scaled orthographic projection. We propose to use a more accurate image formation model : the Perspective Projection. We show that it is possible to model the shape of objects belonging to a subset of generalized cylinders (GC with constant circular cross-section) by analyzing these object contours in a single image. Two approaches are developed :

In the first one, called *local approach* and presented in [21], each object cross-section is reconstructed independently. It is shown that two paired points belonging to the projection of the same object section, enable the orientation and the location in space of this section to be recovered. Finally, all reconstructed sections are connected together to recover the object shape. The main difficulties of this method can be summarized in the following points:

- First of all, matching between two image points corresponding to the limb projection of the same object cross-section is required in order to constrain the equations involved in the modeling process. It is not a tricky problem and false matching leads to inconsistent 3D object sections.

- As each section is estimated independently of the others, a chaining process is required to connect all the object cross-sections and also remove inconsistent reconstruction.

- This method cannot be applied to generalized cylinders, which present contour discontinuities.

In order to address these previous failings, the second approach, called *global approach* is presented. A cubic BSpline curve is used to model the object axis. During an initialization step, the model is first of all supposed to be flat and located in a plane somewhere in front of the camera. Then, the BSpline curve is twisted until the generated model becomes coherent with the contours extracted from the grey level image. The questions to address are :

- how to initialize the control vertices of the 3D BSpline curve.

- how to match an image contour point to the 3D axis.

- how to move the control vertices in order to obtain the correct model.

The paper is organized as follows :

First of all, we briefly describe the geometrical properties induced by a UCGC viewed under perspective projection. It is shown that a given number of constraint planes can be extracted from image data. Included in this first section, a short background of BSpline curve is realized. These functions will be used in both modeling approaches.

In a second part, the local approach is presented. The modeling algorithm is detailled and resultst on grey level image are presented.

Drawbacks of this first method lead us to introduce the second one : the global approach based on a deformable model parametrized with a BSpline curve.

Finally, advantages given by this second approach are discussed in the last part.

2 Geometrical Properties of UCGC's Limbs and Modeling Tools

Definition : *The limbs (also called contour generators) are particular points of the object surface depending on the view point. The rays of light going from the limbs points to the optical center of the camera are tangent to the object surface.*

In this article, 2-D information is denoted by lower case and 3-D information by upper case. Let us denote:

- P a limb point of the UCGC's surface and p its projection in the image.
- $\overrightarrow{N}$ the normal to the surface at point P and $\overrightarrow{t}$ the tangent to the contour at p.
- C the center of the section and R the radius of the section.

2.1 Translation of Interpretation Plane

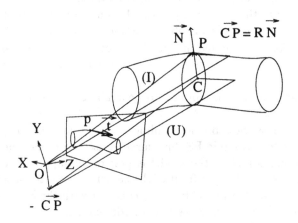

Fig. 1. Translation of interpretation plane

An interpretation plane (I) can be associated to each contour point p. This plane contains the optical center O and the contour tangent $\overrightarrow{t}$ at point p. Its normal is computed from image data:

$$\overrightarrow{N} = \frac{\overrightarrow{Op} \wedge \overrightarrow{t}}{\|\overrightarrow{Op} \wedge \overrightarrow{t}\|} = (N_x, N_y, N_z)^t \qquad (1)$$

This plane (I) contains the limb point P. Using definition of limbs, (I) is tangent to the surface at P. So, the normal to (I) is the normal to the surface at P. The section being circular and the surface normal pointing outside the object, $\overrightarrow{CP} = R.\overrightarrow{N}$

So, a translation of (I) by $-\overrightarrow{CP}$ leads to a new plane (U) parallel to (I) which contains C. The equation connected to the plane (U) is given by:

$$N_x.X + N_y.Y + N_z.Z + R = 0 \qquad (2)$$

2.2 Normal Plane

A normal plane (Π) can be defined for each contour point p. This plane contains the optical center O, the contour point p and is orthogonal to the interpretation plane associated to p. This plane contains the limb point P and the normal to the surface $\overrightarrow{N}$ at P (the normal to the interpretation plane). The normal to (Π) is given by:

$$\overrightarrow{N_\pi} = \frac{\overrightarrow{Op} \wedge \overrightarrow{N}}{\|\overrightarrow{Op} \wedge \overrightarrow{N}\|} = (N_{\pi_x}, N_{\pi_y}, N_{\pi_z})^t \qquad (3)$$

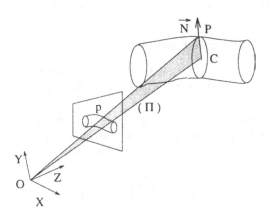

Fig. 2. Normal plane

2.3 Symmetry Plane

Let S be the plane containing the optical center O, the cross-section center C, and the cross-section normal $\overrightarrow{N_S}$. S is a plane of symmetry for the points

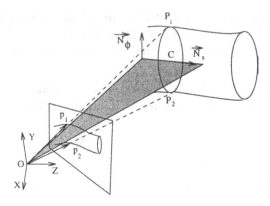

Fig. 3. Plane of symmetry for two matched points

P_1 and P_2 belonging to the same cross-section. S is easily computed from the location of the limb points projection (p_1, p_2). Its normal is given by :

$$\overrightarrow{N_\phi} = \frac{N_{\phi'}}{\|N_{\phi'}\|} \quad with \quad \overrightarrow{N_{\phi'}} = \frac{\overrightarrow{Op_1}}{\|\overrightarrow{Op_1}\|} - \frac{\overrightarrow{Op_2}}{\|\overrightarrow{Op_2}\|} \tag{4}$$

By definition, this plane contains the cross-section center C.

2.4 Background: BSpline Curve

The BSpline Curves are used in both presented methods. The continuity and derivability properties of these functions are useful for describing the UCGC axis.

Bartels et al.[2] give relevant information about BSpline curves. The BSpline curves are defined by the combination of a sequence of blending polynomials functions $B_{i,k}(u)$ with a set of $M + 1$ control vertices V_i :

$$p(u) = \sum_{i=0}^{M} V_i . B_{i,k}(u) \tag{5}$$

The parameter k controls the degree of the $B_{i,k}$ polynomials which are defined recursively by the following expressions:

$$\begin{cases} B_{1,k}(u) = 1 \ if \ t_i \leq u \leq t_{i+1} \\ \qquad\quad = 0 \quad otherwise \\ B_{i,k}(u) = \dfrac{u - t_i}{t_{i+k-1} - t_i} B_{i,k-1}(u) + \dfrac{t_{i+k} - u}{t_{i+k} - t_{i+1}} B_{i+1,k-1}(u) \end{cases} \tag{6}$$

The degree of resulting polynomial in u is controlled by $(k - 1)$. The knot values (t_i) relate the parameter u to the control vertices. If $(M+1)$ is the number of control vertices and $(k-1)$ the degree of the BSpline, the number of knots is given by $(M+k+1)$. Piegl [15] shows that the knots distribution influences the shape of the BSpline curves. In this paper, we use uniform BSpline, i.e with a constant knot spacing.

The BSpline curve tangent is given by the derivate of the polynomials $B_{i,k}$, with respect to u :

$$p'(u) = \sum_{i=0}^{M} V_i \cdot \frac{\partial B_{i,k}(u)}{\partial u} \tag{7}$$

with :

$$\frac{\partial B_{i,k}(u)}{\partial u} = (k-1)\left[\frac{B_{i,k-1}(u)}{t_{i+k-1} - t_i} + \frac{B_{i+1,k-1}(u)}{t_{i+k} - t_{i+1}}\right] \tag{8}$$

We use cubic BSpline ($k=4$), which gives $\mathcal{C}^2$ curves.

3 UCGC Modelling : a Local Approach

3.1 Principle

In this algorithm, the location of cross-sections is recovered independently to the others. Let us consider two image paired points, which correspond to the projection of two limb points belonging to the same object cross-section. To reconstruct this cross-section, which means finding its orientation and location, five distinct planes can be defined. The center C of the cross-section is located at the intersection of these different planes (see section 2).

– Two planes defined from Interpretation Planes
For each contour point, the translation of the interpretation plane gives a plane which contains the section center. So, for a couple of contour points, from the same object cross-section, the cross-section center belongs to the intersection of the two translated planes (U_1) et (U_2) (see section 2.1).

– Two Normal Planes
For each contour point, a normal plane can be defined. It contains the cross-section center. So, for a couple of contour points, from the same object cross-section, the cross-section center belongs to the intersection of the two normal planes (Π_1) et (Π_2) (see section 2.2).

– A Symmetry Plane
This plane S, computed from the two points from the projection of a cross-section, contains the cross-section center (see section 2.3).

Given two image paired points, corresponding to the projection of two object limb points, which belong to the same UCGC cross-section:

- To recover the cross-section orientation $\overrightarrow{N_s}$, the normal $\overrightarrow{N_s}$ is computed. The normal vectors to the two interpretation planes and to the symmetry plane belong to the object section plane. The $\overrightarrow{N_s}$ coordinates (U, V, W) are given by the solution of the following linear system :

$$\begin{cases} N_{1_x}.U + N_{1_y}.V + N_{1_z}.W = 0 \\ N_{2_x}.U + N_{2_y}.V + N_{2_z}.W = 0 \\ N_{\phi_x}.U + N_{\phi_y}.V + N_{\phi_z}.W = 0 \end{cases} \tag{9}$$

- The location of cross-section center is obtained by solving the linear system constructed from the five constraint planes:

$$\begin{cases} N_{1_x}.X + N_{1_y}.Y + N_{1_z}.Z = -R \\ N_{2_x}.X + N_{2_y}.Y + N_{2_z}.Z = -R \\ N_{\pi_{1_x}}.X + N_{\pi_{1_y}}.Y + N_{\pi_{1_z}}.Z = 0 \\ N_{\pi_{2_x}}.X + N_{\pi_{2_y}}.Y + N_{\pi_{2_z}}.Z = 0 \\ N_{\phi_x}.X + N_{\phi_y}.Y + N_{\phi_z}.Z = 0 \end{cases} \tag{10}$$

This system is solved by a least squared method. The over-determination of the system (five planes rather than three) make it possible to obtain a more robust solution. As we can notice from the last system, if the radius R is unknown the reconstruction is done up to a scale factor.

3.2 Matching Process

The problem is now to select the different couples of paired points from all image contour points .

The external contour is sampled and for each contour point p_k, we search the point p_s from the same cross-section projection. In order to find p_s, all the possible couples (p_k, p_s) are formed and the five associated planes are computed. The 3D point, C, solution of the system (10) allows us to compute the following criterion, the sum of distances of C to the five planes :

$$E(k, j) = |\overrightarrow{N_j}.\overrightarrow{OC} + R| + |\overrightarrow{N_k}.\overrightarrow{OC} + R| + |\overrightarrow{N_{\pi_j}}.\overrightarrow{OC}|$$
$$+ |\overrightarrow{N_{\pi_k}}.\overrightarrow{OC}| + |\overrightarrow{N_\phi}.\overrightarrow{OC}| \tag{11}$$

If the paired points are the correct ones (they come for the same object cross-section), the criterion $E(k, j)$ must be close to 0.

In practice, paired points which present a local minimum value of the criterion $E(k, j)$ (below a given threshold) are saved as candidate pairs. Note that bad pairs lead to the reconstruction of sections disconnected to the right axis.

To reduce the computational time and to avoid false matching, the search zone of the matching point is limited by the following constraints :

– The segment line $[p_k, p_s]$ cannot cross an external contour.

– p_k and p_s have to respect symmetry properties described by Brady and Asada [4]. The vectorial sum of the contour tangent to p_k and to p_s give with $\overrightarrow{p_k p_s}$ an angle close to $90°$:

$$(\overrightarrow{t_k} + \overrightarrow{t_s}).\overrightarrow{p_k p_s} < \varepsilon$$

The following figures present an example of matching from a grey level image. The grey level image, the search zone associated to p_k, the matching criterion and the reconstructed section are presented.

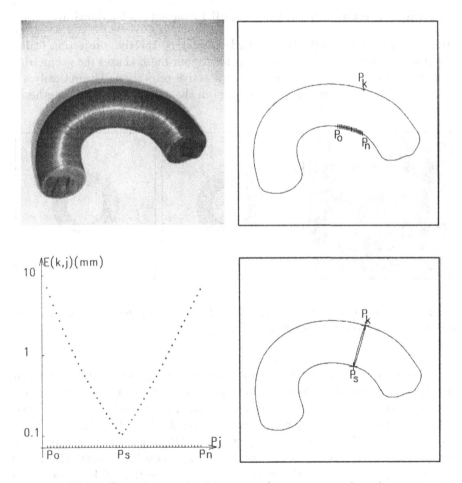

Fig. 4. From the grey level image to the reconstructed section

3.3 Smoothing Process

The 3D chaining process has two goals. First of all, it allows us to eliminate the disconnected sections, which are provided by bad pairs of limb points. It also means that we can order the 3D reconstructed data to be smoothed. This algorithm computes the 3D distance between two points and verifies the continuity of the consecutive normal cross-sections $\overrightarrow{N_s}$.

After this process, the longest chain is kept. In this way, most false matchings which usually give small 3D chains are eliminated.

In order to avoid noise in reconstructed data, the longest chain is smoothed using a BSpline curve [16] [9]. The BSpline is sampled to obtain the cross-sections (the orientation is given by derivating the BSpline).

3.4 Improvements Given by the Full Perspective Projection

Many reconstruction methods are based on weak perspective projection (called also scaled orthographic projection). As for us, our method uses the geometrical properties of UCGC viewed under full perspective projection. Theoretically, the weak perspective approximation is justified in the case of long focal lengthes.

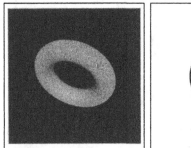

Fig. 5. *Synthetic image (left), Model obtained under perspective projection (center), under scaled orthographic projection (right).*

Using synthetical data, it is possible to compare these two ways of reconstruction. In Figure 5, we have drawn a synthetical image of a torus (the radius of cross-sections is 0.03 m, the radius of the circular axis is 0,10 m, The object is located at 0,8 m in front of the camera and the focal lengths used in this experiment are 7,5 mm (500 pixels), 18,75 mm (1250 pixels) and 30 mm (2000 pixels). Figure 5 represents the reconstruction results using respectively the full perspective projection and the weak perspective projection with a focal length of 18.75. The general shapes of these two UCGC are visually closed.

In order to verify the coplanarity of the set of reconstructed cross-section centres, we have computed (using a least square approximation) the best plane that fits these data. The standard variation σ_D of the distances between data and these planes are given in Table 1.

Secondly, we have computed the best circle that fits the reconstructed cross-sections centres. The estimated radius (to compare with 10 cm) and σ_C, the standard deviation of the distances between circles and data, are presented.

Whatever the focal length, the results obtained under full perspective projection are better than weak perspective.

Table 1. Comparison between weak perspective projection and full perspective projection

Focal lengths	Estimated parameters	Weak perspective projection	Full perspective projection
7,5 mm	σ_D (mm)	1,40	0,05
	R_t (cm)	9,957	9,999
	σ_C (mm)	0,92	0.02
18,75 mm	σ_D (mm)	0,52	0,12
	R_t (cm)	9,992	9.999
	σ_C (mm)	0,26	0,05
30 mm	σ_D (mm)	0,23	0,19
	R_t (cm)	9,996	9.999
	σ_C (mm)	0,11	0.08

3.5 Experimental Results

In each experiment, real grey level images are used. The two objects are supple pipes located at 1.5 m in front of the camera and their diameter is 8 cm.

The different steps of the process applied to these images are presented. From the grey level images, an edge detector is applied to extract the external contours of the viewed objects. After the contour fitting, the object cross-sections are reconstructed and chained. Finally, the noisy reconstructed axis is smoothed using a BSpline curve. In both cases, the reconstructed models are close to the corresponding objects.

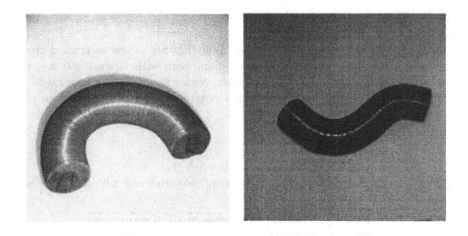

Fig. 6. Grey level images

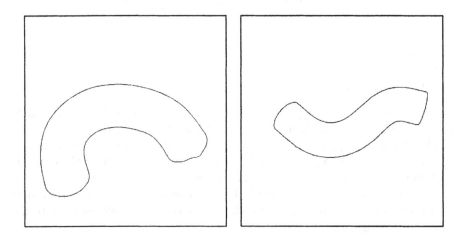

Fig. 7. External contours

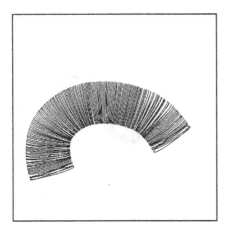 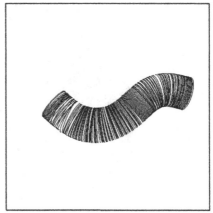

Fig. 8. Reconstructed cross-sections

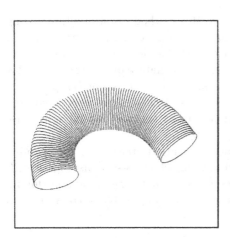 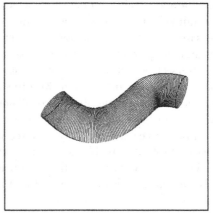

Fig. 9. Models obtained after smoothing process

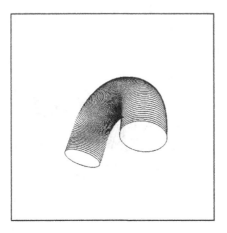

Fig. 10. Different attitudes for the two models

3.6 Limits of the Local Approach

There are a certain number of drawbacks to this method. First of all, matching is required between two image points corresponding to the limb projection of the same object cross-section in order to constrain the equations involved in the modeling process. The matching criterion estimate requires the search for the intersection between the five constraint planes for each couple (p_k, p_j). False matchings lead to inconsistent 3D object sections. A chaining process is required to remove these inconsistent reconstructions. It is a tricky problem : some false sections can be relatively close to the right ones.

Finally, as each section is estimated independently to the others, the chaining process make it possible to connect all the object cross-sections together. If the UCGC view presents discontinuities (cusps), matching cannot be found. Only partial models are recovered and it is not always possible to connect them together.

4 UCGC Modelling : a Global Approach

The purpose here is to find the 3D object axis coherent with the contours. After the initialization of the BSpline curve, an iterative process is applied. The matching between the contour points and the axis, followed by an estimate of the control vertex coordinates, are repeated until the corrections to apply to the control vertices become insignificant.

4.1 Initialization of the 3D Axis

The axis being modeled by a cubic BSpline, the first step consists in the initialization of the BSpline. This means that we have to find the starting location of the control vertices.

First of all, we suppose that the contours, resulting from the projection of the limbs, are extracted from grey level image and chained.

Then, a Duda and Hart [6] recursive splitting process is applied on the contours. We try to match each summit of the resulting polygon with another contour point. The condition that needs to be respected to match two points p_1, p_2 (see Figure 11) is: *the angles Θ_1 and Θ_2 defined respectively between the tangent $\vec{t_1}$ at p_1 and the vector $\overrightarrow{p_1 p_2}$ and between the tangent $\vec{t_2}$ at p_2 and the vector $\overrightarrow{p_1 p_2}$ are close to $90°$.*

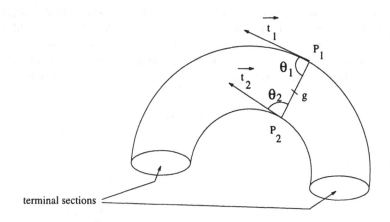

Fig. 11. Initialization of control vertices location

For each pair of points, the middle point g of the segment $[p_i, p_j]$ is retained as the initial projection of a control vertex of the BSpline.

If a summit s_i cannot be matched, the contour points located between s_{i-1} and s_{i+1} are eliminated. Thus, the contour points from the projection of the terminal cross-section of the viewed object can be eliminated.

The initial set of $(M+1)$ 3D control vertices is finally obtained by an inverse perspective projection; all points are located in a plane Q parallel to the image plane in front of the camera (see Figure 12).

Even with this summary initialization step, the BSpline curve is pertinent enough. The vertices are naturally concentrated by the Duda and Hart recursive splitting process around the maxima of curvature, which correspond to sensitive zones for the modeling process.

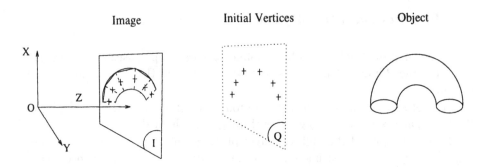

Fig. 12. Inverse projection of control vertices

4.2 Matching Between Contour Points and 3D Axis.

In order to constrain the location of BSpline control vertices, a matching needs to be performed between image and model points. For each contour point p, we search for the intersection point P_s between the normal plane associated to p and the model axis found at a step k of the iterative process (see Figure 13). According to the notation given in section 2.2, P_s verifies the equation of the normal planes :

$$N_{\pi_x} x_s + N_{\pi_y} y_s + N_{\pi_z} z_s = 0 \qquad (12)$$

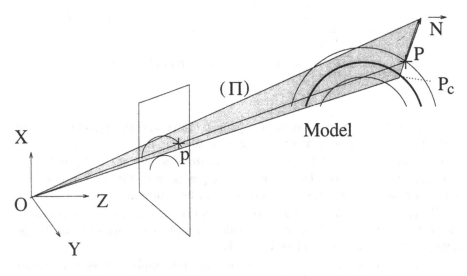

Fig. 13. Matching between a contour point and the 3D axis

The P_s coordinates can be expressed as a function of parameter u:

$$x_s(u) = \sum_{i=0}^{M} X_i.B_{i,4}(u)$$

$$y_s(u) = \sum_{i=0}^{M} Y_i.B_{i,4}(u)$$

$$z_s(u) = \sum_{i=0}^{M} Z_i.B_{i,4}(u) \tag{13}$$

where (X_i, Y_i, Z_i) are the coordinates of the control vertex V_i.

For each contour point, we are looking for the parameter u that gives a point P_s on the BSpline, which belongs to the normal plane. The signed expression E, which represents the distance of P_s to the normal plane, must be equal to zero :

$$E = N_{\pi_x} x_s(u) + N_{\pi_y} y_s(u) + N_{\pi_z} z_s(u) = 0 \tag{14}$$

The expressions $x_s(u), y_s(u), z_s(u)$ being cubic polynomials, E is not linear with respect to u. So, a Newton-Raphson iterative process is used.

At each step k, we have:

$$E = (E)_k + \left(\frac{dE}{du}\right)_k \Delta u \tag{15}$$

$\left(\dfrac{dE}{du}\right)_k$ corresponds to the value of the first derivative of the expression E estimated at the current value of u. At the solution, $E = 0$, the parameter u is updated at each iteration by the computed quantity Δu given by:

$$\Delta u = -\frac{(E)_k}{\left(\dfrac{dE}{du}\right)_k} \tag{16}$$

In practice, less than 10 iterations are required to find the solution of a given point p. The result of this matching is used to point the normal surface outside of the object.

4.3 Axis Deformation by Optimization of the Control Vertices Location

We search for the optimal location of the control vertices with respect to the constraints given by the contour points. At the previous step, each contour point p is matched with a point P_s on the axis, corresponding to the center of a given section. This 3D point must verify the two constraints described in section 2. First of all, it has to belong to the plane (U), from the translation of the interpretation plane (see section 2.1). Then, it has to belong to the normal plane Π (see Section 2.2).

According to the notations used in section 2, these conditions can be written for an axis point $P_s(x_s(u), y_s(u), z_s(u))$:

$$N_x.x_s(u) + N_y.y_s(u) + N_z.z_s(u) = -R$$
$$N_{\pi_x}.x_s(u) + N_{\pi_y}.y_s(u) + N_{\pi_z}.z_s(u) = 0 \qquad (17)$$

It can also be written in function of the unknown control vertices coordinates of the BSpline curve :

$$N_x.\sum_{i=0}^{M} X_i.B_{i,4}(u) + N_y.\sum_{i=0}^{M} Y_i.B_{i,4}(u) + N_z.\sum_{i=0}^{M} Z_i.B_{i,4}(u) = -R$$

$$N_{\pi_x}.\sum_{i=0}^{M} X_i.B_{i,4}(u) + N_{\pi_y}.\sum_{i=0}^{M} Y_i.B_{i,4}(u) + N_{\pi_z}.\sum_{i=0}^{M} Z_i.B_{i,4}(u) = 0 \qquad (18)$$

We add a term of regularization to smooth the BSpline curve. As proposed by Arbogast [1], this minimizes the third derivative of the BSpline curve (see section 2.4) :

$$\frac{\partial^3 OP}{\partial u^3} = \sum_{i=0}^{M} X_i.\frac{\partial^3 B_{i,4}(u)}{\partial u^3} \qquad (19)$$

The criterion, which is to be minimized according to a least square approach, is given by:

$$C = \sum_{l=1}^{L}\left[\left(\overrightarrow{N}.\overrightarrow{OP} + R\right)^2 + \left(\overrightarrow{N_\pi}.\overrightarrow{OP}\right)^2 + \lambda\left(\frac{\partial^3 OP}{\partial u^3}\right)^2\right] \qquad (20)$$

where P, $\overrightarrow{N}$, $\overrightarrow{N}_\pi$ and u are calculated from the lth contour point.

To minimize C, we take the partial derivatives of C with respect to the control vertices coordinates $(X_0, ..., X_M, Y_0, ..., Y_M, Z_0, ..., Z_M)$ at zero.

Using $B_i = B_{i,4}(u)$ and $\dfrac{\partial^3 B_{i,4}(u)}{\partial u^3} = B_i^{(3)}$ it can be written as follows:

$$\frac{\partial C}{\partial X_k} = \sum_{l=1}^{L} 2B_k[(N_x^2 + N_{\pi_x}^2)x_s + (N_x.N_y + N_{\pi_x}.N_{\pi_y})y_s$$

$$+ (N_x.N_z + N_{\pi_x}.N_{\pi_z})z_s + R.N_x] + \lambda B_k^{(3)}\sum_{i=0}^{M} X_i.B_i^{(3)}$$

$$\frac{\partial C}{\partial Y_k} = \sum_{l=1}^{L} 2B_k[(N_x.N_y + N_{\pi_x}.N_{\pi_y})x_s + (N_y^2 + N_{\pi_y}^2)y_s$$

$$+ (N_y.N_z + N_{\pi_y}.N_{\pi_z})z_s + R.N_y] + \lambda B_k^{(3)} \sum_{i=0}^{M} Y_i.B_i^{(3)}$$

$$\frac{\partial C}{\partial Z_k} = \sum_{l=1}^{L} 2B_k[(N_x.N_z + N_{\pi_x}.N_{\pi_z})x_s + (N_z.N_y + N_{\pi_z}.N_{\pi_y})y_s$$

$$+ (N_z^2 + N_{\pi_z}^2)z_s + R.N_z] + \lambda B_k^{(3)} \sum_{i=0}^{M} Z_i.B_i^{(3)} \tag{21}$$

x_s can be written in function of $(X_0, ..., X_M)$, y_s in function of $(Y_0, ..., Y_M)$ and z_s in function of $(Z_0, ..., Z_M)$. It leads to :

$$\frac{\partial C}{\partial X_k} = 2.\sum_{l=1}^{L} B_k[(N_x^2 + N_{\pi_x}^2)(B_0.X_0 + ... + B_M.X_M)$$

$$+ (N_x.N_y + N_{\pi_x}.N_{\pi_y})(B_0.Y_0 + ... + B_M.Y_M)$$

$$+ (N_x.N_z + N_{\pi_x}.N_{\pi_z})(B_0.Z_0 + ... + B_M.Z_M)$$

$$+ R.N_x]$$

$$+ \lambda B_k^{(3)}[B_0^{(3)}.X_0 + ... + B_M^{(3)}.X_M]$$

$$\frac{\partial C}{\partial Y_k} = 2.\sum_{l=1}^{L} B_k[(N_x.N_y + N_{\pi_x}.N_{\pi_y})(B_0.X_0 + ... + B_M.X_M)$$

$$+ (N_y^2 + N_{\pi_y}^2)(B_0.Y_0 + ... + B_M.Y_M)$$

$$+ (N_y.N_z + N_{\pi_y}.N_{\pi_z})(B_0.Z_0 + ... + B_M.Z_M)$$

$$+ R.N_y]$$

$$+ \lambda B_k^{(3)}[B_0^{(3)}.Y_0 + ... + B_M^{(3)}.Y_M]$$

$$\frac{\partial C}{\partial Z_k} = 2.\sum_{l=1}^{L} B_k[(N_x.N_z + N_{\pi_x}.N_{\pi_z})(B_0.X_0 + ... + B_M.X_M)$$

$$+ (N_z.N_y + N_{\pi_z}.N_{\pi_y})(B_0.Y_0 + ... + B_M.Y_M)$$

$$+ (N_z^2 + N_{\pi_z}^2)(B_0.Z_0 + ... + B_M.Z_M)$$

$$+ R.N_z]$$

$$+ \lambda B_k^{(3)}[B_0^{(3)}.Z_0 + ... + B_M^{(3)}.Z_M] \tag{22}$$

Finally, we have to solve $A.S = B$, where A is a square symmetric matrix of size $3(M+1) \times 3(M+1)$, where partial derivatives are expressed with respect to the vertices coordinates. S is the unknown vector that gives the new coordinates of the control vertices and B the second member where the value R of the object radius is introduced. As can be seen, if R is unknown a priori, the location is performed up to a scale factor.

The vectors S and B are written as:

$$S = \begin{pmatrix} X_0 \\ \cdot \\ \cdot \\ X_M \\ Y_0 \\ \cdot \\ Y_M \\ Z_0 \\ \cdot \\ \cdot \\ Z_M \end{pmatrix} \quad B = \begin{pmatrix} \sum_{l=1}^{L} B_0.R.N_x \\ \cdot \\ \cdot \\ \sum_{l=1}^{L} B_M.R.N_x \\ \sum_{l=1}^{L} B_0.R.N_y \\ \cdot \\ \cdot \\ \sum_{l=1}^{L} B_M.R.N_y \\ \sum_{l=1}^{L} B_0.R.N_z \\ \cdot \\ \cdot \\ \sum_{l=1}^{L} B_M.R.N_z \end{pmatrix} \quad (23)$$

After initialization of the BSpline curve, an iterative process is applied. The matching between the contour points and the axis, followed by an estimate of the control vertices coordinates, are repeated until the corrections to apply to the control vertices become insignificant. The object sections are obtained by sampling of the BSpline. The sections orientation is given by the derivation of the BSpline polynomials.

4.4 Experimental Results

The following results are obtained on the grey level images presented in section 3.5. The main steps of the process applied to these objects are presented. From the grey level images, an edge detector is applied to extract the external contours of the viewed objects.

Figure 14 shows the initial choice of the control vertices obtained after the contour splitting. Figure 15 shows the initial model projection. The flat model is put in a plane ($Z = 3m$) in front of the camera.

Then, the Figures 16 and 17 present the model obtained after respectively 1 and 10 iteration. We can note that the Z coordinates of the control vertices are almost corrected after the first iteration. But errors are still visible along the model shape. Then the model is twisted to obtain a good fit between the model and the image data. After 100 iterations, the projected model fits the contour given by the object (Figure 18 and Figure 19).

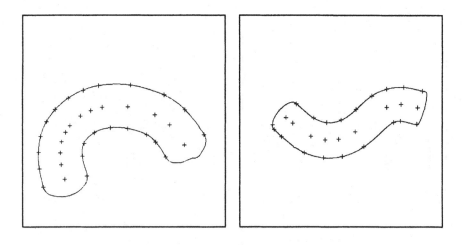

Fig. 14. Results of the contour splitting and projection of the initial vertices

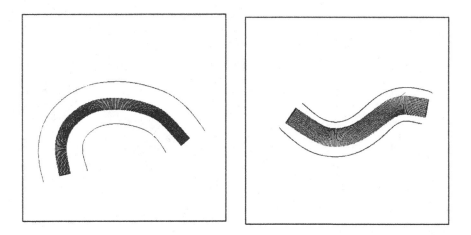

Fig. 15. Retained contour points and initial models location

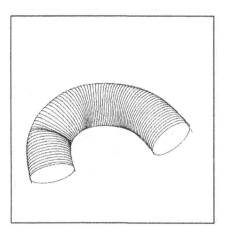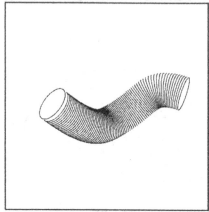

Fig. 16. Obtained models after 1 iteration

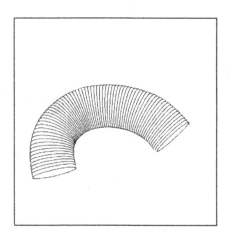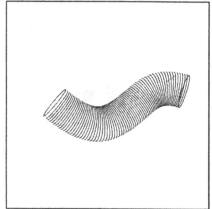

Fig. 17. Obtained models after 10 iterations

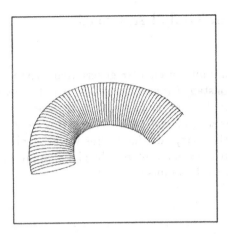
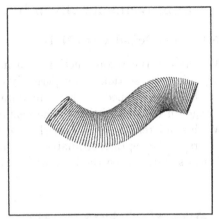

Fig. 18. Obtained models after 100 iterations

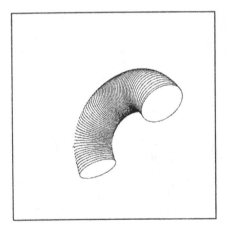
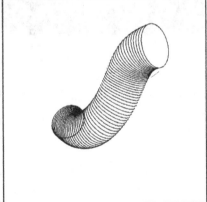

Fig. 19. Reconstructed models observed from a different point of view

5 Improvements Given by the Global Approach

5.1 Less Sensitive to Noise

We present the reconstruction of a non-uniform circular generalized cylinder. The diameter section is not perfectly constant (from 3 cm to 3.5 cm). The two methods are applied on the same data : the contour points from the grey level image. In the view attitude, the results are coherent with contour (Figure 20). With a different point of view (Figure 21), it appears that radius variation gives errors on the depth coordinates of the cross-section centers. The global approach is less sensitive than the local approach to the radius variation.

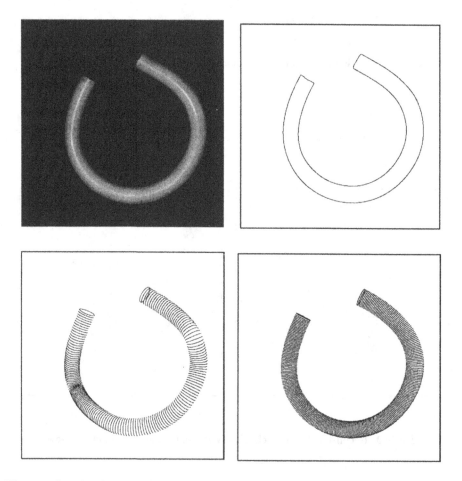

Fig. 20. Grey-level image and external contour. Results of the local algorithm (left) and of the global algorithm (right)

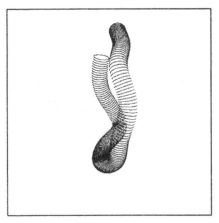
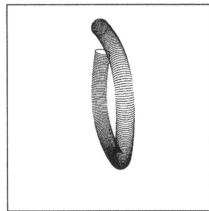

Fig. 21. The two models in a different attitude : local(left) and global(right) approach

5.2 Reconstruction of UCGC with Cusps

In this second experiment, the UCGC projection gives a contour discontinuity (cusps). With the first method, matchings cannot be found around this discontinuity, so only a partial model would be recovered. As shown in the following images, second method make it possible to recover an accurate model.

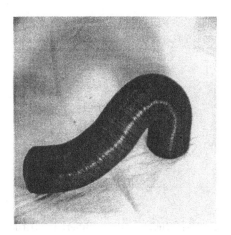
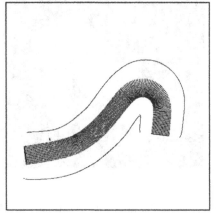

Fig. 22. Grey level image and initial model projection

49

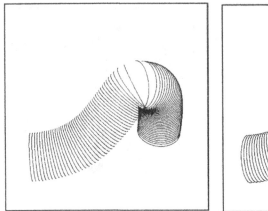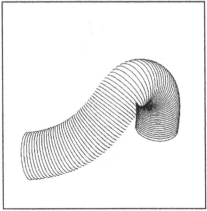

Fig. 23. Model after 1 iteration (left) and after 10 iterations (right)

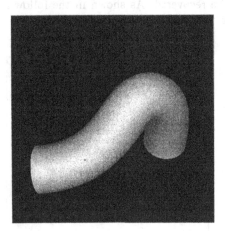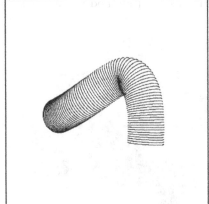

Fig. 24. Model after 100 iterations : synthetic image (left) and in a different attitude (right)

6 Conclusion

In this article we have presented two algorithms to model Uniform Circular Generalized Cylinders (UCGC). Using the properties of UCGC, viewed under perspective projection, geometrical constraints make it possible to recover the object shape.

The originality of the global method is mainly to use a deformable model, built with a BSpline curve. It gives a more robust algorithm to recover UCGC. The choice of cubic BSpline functions means that any axis can be defined. The results computed from real grey level images show the improvements given by this approach.

The modeling algorithm could be adapted to a multi-image system. All the constraint planes given by each image could be expressed in a global frame in order to compute the location of the control vertices. This approach would allow us to recover the radius of the sections, which could change along the axis.

References

1. Arbogast E. : Representation of Contours and their Segmentation, Technical Report RR 115, LIFIA, Grenoble, France (1990).
2. Bartels, R., Beatty, J., Barsky, B. : An Introduction to Splines for use in Computer Graphics and Geometric Modeling, Morgan Kaufmann Publishers (1987).
3. Binford T.O., Visual Perception by Computer, Proceeding of IEEE Conference On Systems and Control, Miami (1971).
4. Brady, M., Asada, H. : Smoothed Local symmetries and Their Implementation, International Journal of Robotics Research, **Vol. 3(3)** (1984) 36–61.
5. Chung, R.C.K., Nevatia, R. : Recovering LSHGCs and SHGCs from Stereo, Proceeding of Computer Vision and Pattern Recognition, New-York (1993) 42–48
6. Duda, R.O., Hart, P.E. : Pattern Classification and Scene Analysis, Wiley, New-York (1973).
7. Glachet R. : Modélisation géométrique par apprentissage de cylindre droits généralisés Homogènes en vision monoculaire, thèse de doctorat de l'Université Blaise Pascal de Clermont-Ferrand, (1992).
8. Gross, A.D., Boult, T.E. : Recovery of SHGC From a Single Intensity View, IEEE Trans. On Pattern Analysis Machine Intelligence, **Vol. 18(2)** (1996) 161–180 and Erratum in **Vol. 18(4)** (1996) 471–479.
9. Gueziec A., Ayache, N. : Smoothing and Matching of 3D Space Curves, International Journal of Computer Vision, **Vol. 12(1)** (1994) 79–104.
10. Kriegman, D.J., Ponce, J. : On Recognizing and Positioning Curved 3D Objects from Image Contour, IEEE Trans. On Pattern Analysis Machine Intelligence, **Vol. 12(12)** (1990) 1127–1137.
11. Lavest, J.M., Glachet, R., Dhome, M., Lapreste, J.T. : Modelling Solids of Revolution by Monocular Vision, Proceeding of Computer Vision and Pattern Recognition, Hawaii (1991) 690–691.
12. Liu, J., Mundy; J., Forsyth, D., Zisserman, A. et Rothwell, C. : Efficient Recognition of Rotationally Symmetric Surfaces and Straight Homogeneous Cylinders, Proceeding of Computer Vision and Pattern Recognition, New-York (1993) 123–128, .

13. Mohan, R., Nevatia, R. : Perceptual Organisation for Scene Segmentation and Description, IEEE Trans. On Pattern Analysis Machine Intelligence **Vol.12** (1992) 616–635

14. Naeve A., Eklundh J.O. : Representing Generalized Cylinders, Proceeding of Europe-China Workshop on Geometrical Modelling and Invariants For Computer Vision, Xi'An, China, (1995) 63–70.

15. Piegl, L., Tiller, W. : Curve and surface Construction Using Rational BSpline, Computer-Aided Design **Vol. 19(9)** (1987) 485–498.

16. Plass, M., Stone, M. : Curve-Fitting with Piecewise Parametric Cubics, Siggraph'83, Computer Graphics **Vol. 17(3)** (1983) 229–239.

17. Ponce, J., et Chelberg, D. : Finding the Limbs and the Cusps of Generalized Cylinders, International Journal of Computer Vision, **Vol. 1(3)** (1987) 195–209.

18. Richetin, M., Dhome, M., Lapresté, J.T., Rives, G. : Inverse Perspective Transform Using Zero-Curvature Contour Points : Application to the Localization of Some Generalized Cylinders from a Single View, IEEE Trans. On Pattern Analysis Machine Intelligence, **Vol.13(2)** (1991) 185–191.

19. Saint-Marc, P., Medioni, G. : Spline Contour Representation and Symmetry Detection, Proceeding of First European Conference on Computer Vision, Antibes, France (1990) 604–606

20. Sato, H., Binford, T.O. : Finding and Recovering SHGC Objects in an Edge Image, Computer Vision Graphics and Image Processing, **Vol.57(3)** (1993) 346–356

21. Sayd, P., Dhome, M., Lavest, J.M : Reconstruction of Uniform Circular Generalized Cylinders under Full Perspective Projection, Theory and Applications of Images Analysis II, World Scientific Publishing (1996) 169–182

22. Shaffer, S.A., Kanade, T.O. : The Theory of Straight Homogeneous Cylinders, Technical Report CS-083-105, Carnegie Mellon University (1983).

23. Shaffer, S.A. : Shadows and Silhouette in Computer Vision, Kluwer, Boston (1985).

24. Ulupinar, F., Nevatia, R. : Shape from Contour : Straight Homogeneous Generalized Cones, Proceeding of International Conference on Computer Vision, Osaka, Japan (1990) 582–586.

25. Zeroug, M., Nevatia, R.: Three-Dimensional Descriptions Based on the Analysis of the Invariant and Quasi-Invariant Properties of Some Curved-Axis Generalized Cylinders, IEEE Trans. On Pattern Analysis Machine Intelligence, **Vol.18(3)** (1996) 237–253.

Combinatorial Geometry for Shape Representation and Indexing

Stefan Carlsson

Dept. of Numerical Analysis and Computing Science, Royal Institute of Technology,
S-100 44 Stockholm, Sweden, stefanc@bion.kth.se

Abstract. Combinatorial geometry is the study of order and incidence
properties of groups of geometric features. Ordering properties for point
sets in 2-D and 3-D can be seen as a generalization of ordering properties
in 1-D and incidences are configurations of features that are non-generic
such as collinearity of points. By defining qualitative shape properties using combinatorial geometry we get a common framework for metric and
qualitative representations. Order and incidence form a natural hierarchy
together with metric representations in terms of increasing abstraction

　　　　Metric　　　==>　　　Order　　　==>　　　Incidence

The problem of recognition can be structured in a similar hierarchy ranging from the recognition of specific objects from specific viewpoints ,using
calibrated cameras to that of calibration free, view independent recognition of generic objects. Order and incidence relations have invariance
properties that make them especially interesting for general recognition
problems.

We present an algorithm for 3-D object hypothesis generation from single
images. The combinatorial properties of triplets of line segments are used
to define an index to a model library. This library consists of line segment
triplets for object model views. Every indexing of a model triplet by an
image triplet is accumulated into a matching matrix between image and
model line segments. From this matrix we can evaluate the strength of
an hypothesis that a specific object is present.

1 Invariance Problems of Object Recognition

Visual object recognition in its most general form is the problem of deciding the
presence in the scene of an object of a certain class using information from a
single or multiple images. By using geometric shape as a descriptor in 2-D and
3-D, recognition can be made independent of illumination to a large degree. This
is not the only cause of variation of the appearance of an object however. The
main problem of shape based recognition is due to the fact that an object of a
certain class can project to different image shapes mainly due to:

- class variation
- view variation
- camera parameter variation

The problem of generic object recognition can be said to be that of designing descriptors that are un affected by all these variations. Object recognition can of course also be formulated as a less general problem, e.g. recognizing a specific instance of an object independent of viewpoint. Partial information about viewpoint can be assumed as can information about camera calibration. Recognition problems can be ordered depending on the degree of invariance that is required from it, ranging from the recognition of a specific object in a specific view with a calibrated camera to that of generic viewpoint and calibration independent object recognition. We will argue that this hierarchy has a natural counterpart at the representational side, i.e. that there is a natural ordering of transformation groups and structural shape descriptions ranging from euclidean geometry and metric structure to that of projective geometry and qualitative structure.

1.1 Class Variation

The recognition of an object of a specific category as opposed to the recognition of a specific instance of that category imposes very different constraints on object representation and modeling. A specific instance of an object category can be described by its metric properties. If camera viewpoint and calibration is known, or has been computed, the metric properties of its image can be determined. Recognition can then be based on the evaluation of metric differences between an image and views of specific object instances stored in an object library. [13, 19, 24] The metric description of an object of a certain category will vary from instance to instance within the category. If we want to recognize object categories, we would like to use descriptors that are invariant to this within class variation. For this purpose it has been suggested to use qualitative object descriptors based on relational structure of an objects subparts. [3, 5, 9, 26, 29, 42] Relational structures are represented by graphs and the problem of recognition becomes a problem of matching a subgraph extracted from image data with those in the object library. A critical problem in these approaches is the definition and extraction of the subparts of the object. This is in general a highly non-trivial problem in computer vision.

1.2 View Variation

Viewpoint variation of an image of an object can be approached by a variety of methods:

- Computing view invariant object descriptors from a single image
- Explicit 3-D reconstruction of the object from single or multiple image data
- Using a set of representative views of the object and matching of the image to this set.

For non planar objects, view independent 3-D descriptors can be extracted from a single image, provided the object is sufficiently constrained. Typically this has been applied to polyhedral objects and generalized cylinders. [22, 27, 36, 39]

These methods generally requires elaborate grouping of image data. This also applies to so called quasi invariants, i.e. image descriptors that show non zero but small variation due to change of viewpoint [2][41] By limiting the object instances to be planar in 3-D, the choice of projectively invariant descriptors actually ensures that view variation is eliminated. [34, 35].

A more general way to overcome view variation is to use multiple cameras and reconstruct the object in 3-D. [32] If this reconstruction is based on projective metrics, the effect of internal camera parameters will be cancelled [10, 14, 28]. Introducing multiple cameras means that images have to be matched in order to achieve reconstruction. Besides being extra time consuming, errors in the matching will introduce errors in the 3-D reconstruction which will affect recognition performance.

If 3-D object recognition is to be achieved using a single image, either the relative pose of the object has to be determined before matching [19], or the object has to be represented by multiple views. Representing the object by a set of characteristic views, transforms the 3-D recognition problem to a problem in 2-D [6]. Qualitatively different views of an object can be represented by the so called aspect graph [20, 21]. The nodes in the aspect graph represents views of an object whose relational or qualitative structure remains invariant to small changes in viewpoint. It has been argued that all qualitatively different views of an object have to be stored for efficient recognition. The main drawback of multiple view representations has therefore been the large memory requirements and the extra complexity of retrieval due to the extensive search that has to be performed.

1.3 Camera Parameter Variation

The variation due to changing camera calibration is probably the one that has been most successfully attacked leading to so called calibration free methods for recognition and reconstruction. [10, 14, 28, 34] By choosing a sufficiently general metric for the image descriptors, the effect of internal camera parameter variation can be eliminated. For a linear pinhole camera this requires projective metric but affine metric is sufficient in many cases. The transformation groups euclidean, affine and projective form a natural hierarchy in terms of subgroups to the general linear group.

1.4 Image Representation Hierarchies for Transformations and Structure

The invariance problem of object recognition is the problem of maintaining recognition performance while going from specific objects, views and camera parameters to generic ones. Depending on the recognition problem at hand, ranging from a specific object seen from a specific viewpoint with a camera with known calibration to that of recognition of a generic object from arbitrary viewpoint with an un calibrated camera, different representations of image data should be chosen. Image representations can be organized in a hierarchy corresponding to the recognition problem , ranging from euclidean transformation group and metric structure to the projective group and relational structure.

Specific object instance	Generic object
Specific View	General view
Specific Camera Parameters	General camera
Metric structure	Relational structure
Euclidean geometry	Projective geometry

A certain choice of geometry and structure for description defines an equivalence class of objects and images, where each member of the equivalence class have the same description parameters. E.g. by choosing projective transformations and metric structure we get the equivalence class of images related by a general linear transformation. It is the purpose of this paper to show that relational structure can be defined for each transformation group by considering qualitative properties of feature configurations that are invariant to the transformations in question. Metric and relational descriptors will be seen to be very closely linked in a hierarchy, similar to that of the transformation groups.

Just as the generalization of the transform group admits larger equivalence classes of camera parameters, and to some degree objects and viewpoints, so will the generalization of structural descriptors from metric to qualitative.

As a practical application, qualitative structure will be defined for pairs of line segments. The effect of image degradation on qualitative geometry will taken into account in a recognition algorithm based on indexing qualitative models directly from image segment data.

2 Combinatorial Geometry for Qualitative Shape Representation

Relational image or object descriptors are based on relations between the subparts of the image or the object. The extraction of relational structure therefore in general requires substantial image analysis and grouping of low level features in order to identify the subparts. This requirement is often a major drawback in these systems since it in general requires very good image data. Since the process is in general sequential, proceeding from low level to high level analysis, a failure at some step will propagate to the later steps.

In order to define qualitative structure, it is in principle not necessary to study high level object descriptions. Qualitative properties of geometric features can be defined at a very low level, for representational use. Early work in computer vision [7], [18] was concerned with junction labelling in order to infer 3-D properties of primarily polyhedral but also [25] more general kinds of objects.

The work in [16] and [24] uses qualitative relational properties of image features such as parallelism, for perceptual grouping, and relational graphs based on qualitative image feature relations have been considered in [8], [17].

Qualitative and quantitative relations between features can be treated in a common framework as algebraic constraint relations [30, 31]. Similar to the work in [30, 31] our representation will be formulated in terms of algebraic constraints among groups of features. We will make a clear distinction between quantitative (metric) constraints and qualitative (combinatorial) constraints , and see how they relate to transformation properties of various linear subgroups. The formal definition of combinatorial structure given will permit us to analyze relations between 3-D and 2-D representations which will be of importance in order to understand viewpoint invariance properties and 2-D to 3-D inference problems.

The simplest example of combinatorial geometric structure is probably the left right ordering of a set of points on a line. We will see how the concept of ordering can be generalised to higher dimensions and more complex features. This will permit us to define qualitative properties of algebraically defined collections of features in a formal way.

2.1 Point sets

The image properties of general 3-D objects that in general remain invariant over change of viewpoint and general perspective camera projection, are typically incidence relations such as intersections and tangencies of lines and curves or collinearities of points. For points, lines and algebraic curves such incidence relations can be expressed algebraically as a polynomial in homogeneous coordinates being zero.

Three points in the plane can line up so that they are collinear. This is a qualitative property of the three points which can be expressed algebraically as:

$$[p_1, p_2, p_3] = 0 \tag{1}$$

where $[\ldots]$ denotes the determinant of the 3×3 matrix formed by the homogeneous coordinates p_1, p_2 and p_3 of the points.

By using the fact that the determinant of a linear transformation can be factored out in the expression above we see that the incidence relation is invariant over linear transformations. Typically all incidence relations such as collinearity, tangency and common intersection can be written in terms of polynomials in determinants being zero, and they are therefore invariant over general linear transformations. For point sets we can talk about the *incidence structure* of the set, by listing all subsets of points that are collinear or alternatively by evaluating all determinants of triplets and note whether they are zero or not.

Linear invariant *metric structure* for point sets on the other hand, is based on invariants which can be computed as cross ratios. For five points in an image, a linear invariant can be computed as:

$$I = \frac{[p_1\ p_2\ p_5]\ [p_3\ p_4\ p_5]}{[p_1\ p_3\ p_5]\ [p_2\ p_4\ p_5]} \tag{2}$$

We see that both metric and incidence linear invariant properties are expressed in the same language, as polynomials in brackets. Using the bracket formalism we can define yet another qualitative property of point sets that can be seen as a generalization of the concept of order in 1-D. Consider the two cases of configurations of four points in general position in 2 D shown in fig. 1.

Fig. 1. *Two qualitatively different configurations of four points*

These two configurations can be said to be qualitatively different, but there are no incidence relations involved. Since any four points in generic position the plane can be mapped to any other four points, the two configurations are actually linearly equivalent. However, if we introduce the *oriented* homogeneous coordinates:

$$\bar{p} = w \begin{pmatrix} x \\ y \\ 1 \end{pmatrix} \qquad w > 0 \qquad (3)$$

where x and y are the euclidean image coordinates, we can consider the sign of determinants formed by three points:

$$sign[\bar{p}_1, \bar{p}_2, \bar{p}_3] \qquad (4)$$

This sign will be positive or negative depending on whether point $\bar{p}_3$ is on the left or right side of the directed line $\bar{p}_1 \, \bar{p}_2$

Fig. 2. *The sign of the determinant $[\bar{p}_1, \bar{p}_2, \bar{p}_3]$ is positive or negative depending on whether the point $\bar{p}_3$ is to the left or to the right of the line $\bar{p}_1 \, \bar{p}_2$*

It will be invariant for the class of linear transformations that preserve the orientation of the homogeneous coordinates. This class is sufficiently general to be of practical interest since it contains e.g. all view transformations associated with viewing a planar surface while keeping the camera on the same side as the surface. By considering this subclass of linear transformation we can talk about *oriented projective geometry* [38].

The signs of all determinants of triplets of oriented homogeneous coordinates can be said to represent the *order structure* of the point set. [12] The *incidence* and *order* structures together define the *combinatorial structure* of the point set represented by the mappings:

$$sign[\bar{p}_i, \bar{p}_j, \bar{p}_k] \rightarrow \{-1, 0, 1\} \tag{5}$$

for every combination i, j, k of points in the set.

As an example, the configurations in fig 1 will have the representations shown in fig. 2.1. Point sets with order and incidence structure defined in this way are examples of *oriented matroids* [4] that have received attention mainly in mathematics and computational geometry

The order structure of the point set can be said to be intermediate between the metric and the incidence representation, and the metric structure includes both the order and the incidence structures. This induces a natural hierarchy of the representations ranging from metric to qualitative:

$$metric \quad \Rightarrow \quad order \quad \Rightarrow \quad incidence$$

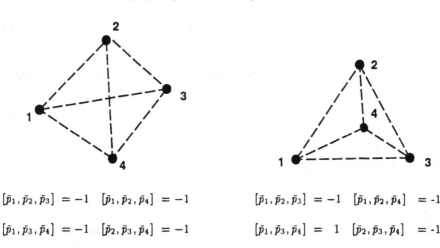

$$[\bar{p}_1, \bar{p}_2, \bar{p}_3] = -1 \quad [\bar{p}_1, \bar{p}_2, \bar{p}_4] = -1$$

$$[\bar{p}_1, \bar{p}_3, \bar{p}_4] = -1 \quad [\bar{p}_2, \bar{p}_3, \bar{p}_4] = -1$$

$$[\bar{p}_1, \bar{p}_2, \bar{p}_3] = -1 \quad [\bar{p}_1, \bar{p}_2, \bar{p}_4] = -1$$

$$[\bar{p}_1, \bar{p}_3, \bar{p}_4] = 1 \quad [\bar{p}_2, \bar{p}_3, \bar{p}_4] = -1$$

Fig. 3. *Representations of combinatorial structure of two point sets using brackets*

Different geometric shapes with the same combinatorial structure form equivalence classes. It is interesting to note that these equivalence classes correlate quite well with "perceptual" equivalence classes. Shapes with the same combinatorial structure tend to be perceived as similar. (Fig. 4)

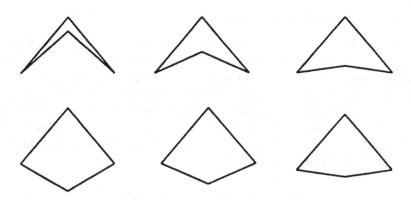

Fig. 4. *The equivalence classes formed by considering combinatorial geometric proper-ties correlate well with perceptual equivalence classes*

2.2 General algebraic features

Combinatorial structure can be defined not only for sets of points but for any set of algebraically defined features. Combinatorial geometric properties for point sets was derived by considering the sign of the 3 point collinearity incidence relation. In the same way we can consider signs of incidence relations involving combinations of points. lines, conics and general algebraic curves. Let $a^T x = 0$ and $x^T A x = 0$ be the equations for a line and a conic described by the homogeneous 3− vector a and 3×3 matrix A respectively.

Points and Lines The fact that a point p is coincident with the line can be expressed as:

$$a^T p = 0$$

The sign of $a^T p$ determines on which side of the (oriented) line the point p is.

Points and Conics A point p is coincident with a conic A if:

$$p^T A p = 0$$

The sign of $p^T A p$ represents the qualitative relation between the points and the conic. E.g. if the conic is an ellipse, the sign says whether the point is inside or outside the ellipse.

Lines and Conics A line a is tangent to a conic A if:

$$a^T A^{-1} a = 0$$

The sign of $a^T A^{-1} a$ gives information about whether the line intersects the conic or not.

2.3 Euclidean and Affine Structure

Combinatorial structure, i.e. incidence and order relations, can be defined also for the case of euclidean and affine transformation groups by considering their invariance properties. Since these groups are subgroups of the general linear (projective) group, any incidence and order structure that is invariant in the projective case will also be invariant in the affine and euclidean cases. In the same way, since the euclidean group is a subgroup of the affine, any affine property will be invariant in the euclidean case. There is therefore a natural hierarchy among the transformation groups [11]:

$$euclidean \quad \Rightarrow \quad affine \quad \Rightarrow \quad projective$$

Starting with incidence relations, we can define order properties of point configurations in the euclidean and affine cases. Parallelism of lines is an invariant property under the affine group. Four points, two each on parallel lines, will therefore be an affine incidence structure, fig 5. The points on the lines induce a direction of the lines. General four point configurations can therefore be divided into two classes depending on the position of the crossing of the lines, fig. 5.

Since angle is invariant under euclidean transformations, Three points can therefore be classified depending on whether they form an angle larger or less than 90 degrees. fig. 5.

2.4 Combinatorial Structure from 3-D to 2-D

In the same way as combinatorial structure of three points in 2-D can be computed from the determinant of their oriented homogeneous coordinates, we can compute combinatorial structure in 3-D for four points by considering the mapping:

$$sign[\bar{P}_i, \bar{P}_j, \bar{P}_k, \bar{P}_l] \quad \rightarrow \quad \{-1, 0, 1\} \tag{6}$$

for every combination i, j, k, l of four points in 3-D.

The first three points i, j and k defines a plane in 3-D with an orientation. The sign of the bracket then depends on which side of this plane the point l is, and 0 means that the four points are coplanar in space.

For image point sets that are perspective projections of point sets in 3-D, the combinatorial structure is easily computed from the structure of the set in 3-D and knowledge of the camera projection point. If we take the first point as the projection point of the camera, we actually have:

$$sign[\bar{p}_i, \bar{p}_j, \bar{p}_k] = sign[\bar{P}_i, \bar{P}_j, \bar{P}_k, \bar{P}_0] \tag{7}$$

where $\bar{p}_i, \bar{p}_j, \bar{p}_k$ are the oriented image coordinates of the 3-D points $\bar{P}_i, \bar{P}_j, \bar{P}_k$. The projected combinatorial structure of three points in space depends on whether the projection point is above or below the oriented plane defined by the three points in space. This can be seen from fig. 2.4. Eq. 7 can be considered as a

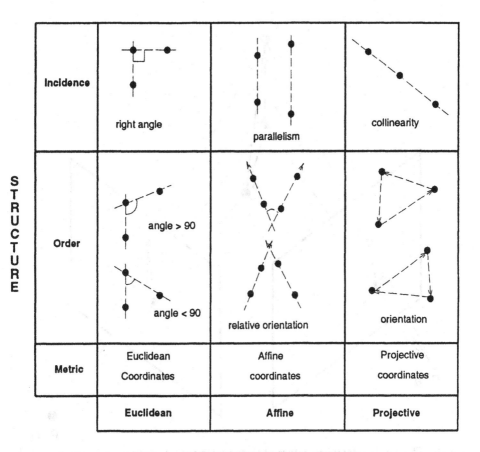

	Euclidean	Affine	Projective

TRANSFORMATION GROUP

Fig. 5. *Hierarchies of transformation groups and structural descriptions for planar point sets (Note that for the case of the projective group and order structure we actually consider the restriction to oriented projective transformations which strictly do not form a group*

"projection" equation for combinatorial structure. The importance of this will be obvious in the last chapter when we study the possibilities of 3-D inference of combinatorial structure from observed image data.

This is of course very much in analogy with the aspect graph concept, where the viewspace is partitioned into regions where the image is stable with respect to small perturbations of the viewpoint. In the case of combinatorial structure, the viewspace is partitioned into cells bounded by planes formed by taking three points in space. Within each such cell combinatorial structure is strictly view-

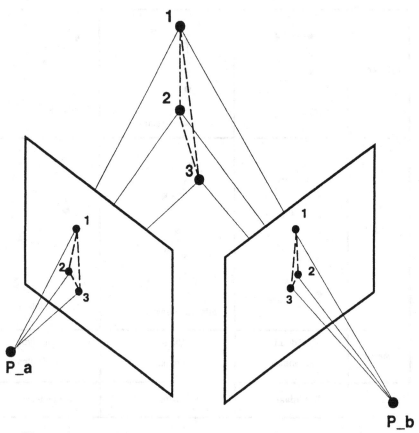

Fig. 6. *The combinatorial structure of three image points depends on which side of the plane spanned by the three points in space, the viewpoint is.*

point invariant. Combinatorial geometry can be seen as giving a strict definition of the concept of topological image structure that is used to distinguish views in the definition of the aspect graph.

2.5 Line segments

An interesting set of features for which combinatorial structure can be defined is that of oriented line segments. For structured scenes, line segments can in general be extracted in a robust manner and are very rich scene descriptors for typical man made objects and structured scenes in general. A pair of line segments is just a special case of four points pairwise grouped together. For two directed line segments with endpoints expressed by the oriented homogeneous coordinates $\bar{a}_1, \bar{a}_2$ and $\bar{b}_1, \bar{b}_2$ respectively, the combinatorial structure is represented by the signs of the four determinants:

$$[\bar{a}_1, \bar{a}_2, \bar{b}_1] \quad [\bar{a}_1, \bar{a}_2, \bar{b}_2] \quad [\bar{a}_1, \bar{b}_1, \bar{b}_2] \quad [\bar{a}_2, \bar{b}_1, \bar{b}_2]$$

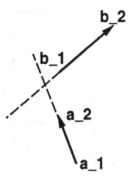

Fig. 7. *Two line segments with endpoints*

By noting the association of the value to the set $\{-1, 0, +1\}$ we would get a total of $3^4 = 81$ combinatorially different combinations. Due to dependencies among the determinants, this number is reduced to 51 Figure 8 shows half of these. The other half is obtained by permutation of the two segments.

In most practical image cases where the extent of the object is small compared to the distance to the camera, parallelism of lines in 3-D will be preserved in the image. Including parallelism, which is invariant under the affine group, into the combinatorial structure is therefore in general of great value.

For the case of a rectangular block fig. 9 which projects 9 line segments we get the combinatorial structure as shown in the table. The entries of the table are the relations valid between pairs of segments and they are coded using the numbers of fig. 8. The prime ' denotes that the segments are reversed in relative orientation compared to that of fig. 8. (p p) and (p -p) denotes parallel and anti parallel segments respectively.

2.6 View variation of combinatorial structure

For two line segments in general position in 3-D, the projective combinatorial image structure is determined by the position of the viewpoint relative to the planes formed by the sides of the tetrahedron that is constructed by joining the endpoints of the two segments to each other. A change in combinatorial structure is associated with the viewpoint moving from one side of a plane to another. If the viewpoint is coincident with the plane the combinatorial image structure is of incident type, i.e. there will be at least one determinant formed from the endpoints that is zero. This can be seen as a formalisation of the aspect graph concept for line segment pairs where structurally stable viewpoints are separated by accidental ones.

For segments that are coplanar in 3-D, the tetrahedron degenerates to a planar figure. For those pairs, combinatorial image structure is invariant over 180 degrees change of viewpoint, i.e. as long as the viewpoint stays at the same side of the plane formed by the line segment pair in 3-D. The content of 3-D coincident line segments of a certain shape therefore determines the invariance of projected combinatorial structure. An extreme case is of course when all seg-

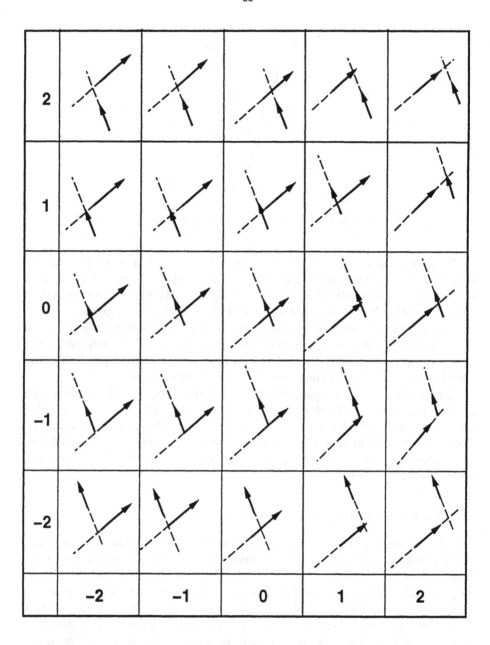

Fig. 8. *Combinatorially different structures of two directed line segments in 2-D*

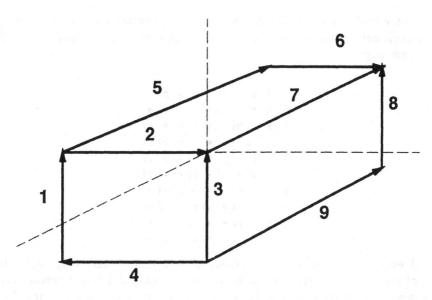

.	1	2	3	4	5	6	7	8	9
1	-	(1 -1)	(p p)	(-1 1)'	(1 -1)	(2 -2)	(0 -2)	(p p)	(-2 -2)
2	(-1 1)'	-	(1 1)'	(p -p)	(-1 1)'	(p p)	(1 -1)'	(2 0)'	(2 2)
3	(p p)	(1 1)	-	(-1 -1)'	(2 0)	(2 -2)	(1 -1)	(p p)	(-1 -1)
4	(1 -1)	(p -p)	(-1 -1)	-	(2 -2)	(p -p)	(2 -2)	(-2 -2)	(-1 -1)
5	(-1 1)'	(1 -1)	(0 2)'	(-2 2)'	-	(1 -1)	(p p)	(2 2)'	(p p)
6	(-2 2)'	(p p)	(-2 2)'	(p -p)	(-1 1)'	-	(1 1)'	(1 1)'	(2 2)'
7	(-2 0)'	(-1 1)	(-1 1)'	(-2 2)'	(p p)	(1 1)	-	(1 1)'	(p p)
8	(p p)	(0 2)'	(p p)	(-2 -2)'	(2 2)	(1 1)	(1 1)	-	(-1 1)
9	(-2 -2)'	(2 2)'	(-1 -1)'	(-1 -1)'	(p p)	(2 2)	(p p)	(1 -1)'	-

Fig. 9. *Combinatorial relations for line segment pairs of an image of a rectangular block*

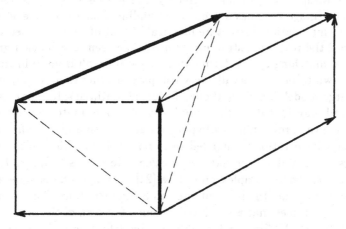

Fig. 10. *Combinatorial image structure for a 3-D line segment pair is determined by the position of the viewpoint relative to the planes formed by joining the segment endpoints.*

ments are coplanar giving 180 degrees viewpoint invariance for the whole shape. The table below indicates with *vi* that two segments in the rectangular box has this property.

```
      1  2  3  4  5  6  7  8  9
1  −  vi vi vi vi  *  *  *  *
2  vi −  vi vi vi vi  *  *  vi
3  vi vi −  vi  *  *  vi vi vi
4  vi vi vi −  *  vi  *  *  vi
5  vi vi  *  vi −  vi vi  *  vi
6  *  vi  *  vi vi −  vi vi  *
7  *  vi vi  *  vi vi −  vi vi
8  vi  *  vi  *  *  vi vi −  vi
9  *  *  vi vi vi  *  vi vi −
```

If we restrict ourselves to the range of viewpoints where only segments 1 - 9 of the block are visible we se that there are only 3 different transitions of combinatorial image structure that can occur due to change of viewpoint. Those transitions are associated with the elongation of segments 2 3 and 7 intersecting the intersection points of segments 8-9, 5-6 and 1-4 respectively.

The definition of combinatorial structure for line segment pairs can be extended to any number of line segments. In the algorithm presented below we will use groups of three segments

3 3-D Object Recognition by Indexing View Models

3.1 Hypothesis Generation from Model Library

Given that a set of objects and views are represented by their combinatorial relations between groups of line segments, the problem of object recognition using a single image becomes a problem of finding a subset of line segments in the image with a relational structure identical to that of a specific member of the library. Since the relational structure table can be seen as a labeled graph, the image model matching can be formulated as a sub graph isomorphism problem which is known to be intrinsically very complex, leading to very long execution times for large model libraries although faster sub optimal solutions exist [1, 15]. The standard way to reduce the complexity of model matching has therefore been to divide the problem into a hypothesis generation and a verification part. The hypothesis generation is intended to be relatively fast but imprecise in that several possible candidates for matching are generated. These can then be verified using a relatively more complex procedure [24]. The hypothesis generation can be made very efficient if it is formulated as an indexing problem where model data is stored in tables that are indexed by some function of the observed image data. [8, 17, 23, 34, 37] Since index tables are by definition discrete, the discrete nature of the combinatorial structure representation is ideally suited for this technique.

In this work we will be primarily concerned with the hypothesis generation part using the combinatorial structure of groups of line segments for indexing into a table containing object identity and view of specific objects. Specifically we will consider triplets of line segments and form an index from the combinatorial structure of the three segments. The information about the view models is stored in an index table. An entry in the index table contains all model view segment triplets that have the same index, i.e. combinatorial structure.

Indexing the index table starts with extraction of line segments i_1, i_2, i_3 from an image. Just as every pair of line segments can have a finite number of combinatorial relations so can every triplet of segments. Based on the combinatorial structure of the line segment triplet an index is computed. This index is used as an address to the index table and for each model view triplet m_1, m_2, m_3 found at that address, an association matrix Q is increased at positions $(i_1, m_1), (i_2, m_2)$, and (i_3, m_3). The final result of the indexing is therefore an association matrix where the value of the element $Q(i, m)$ indicates the strength of the hypothesis of associating image segment i with model view segment m.

The indexing scheme, leading to a table of associations between image line segments and model line segments is presented in fig. 11

In order to find a unique number measuring the strength of the match between a certain model view and the image, we first find the maximum over i of the image - model view association matrix $Q(i, m)$ for each model segment m and then sum over all model segments:

$$S = \sum_m \max_i Q(i, m) \tag{8}$$

In order to rank the scores for different view models, they have to be normalized w.r.t. the number of segments in the model.

3.2 Finite Precision and Fragmentary data

The evaluation of the combinatorial relations for line segments using finite precision can in practise not be done exact. Incidence between segments can only be evaluated with a certain tolerance. E.g. we will consider two segment endpoints coincident if they are within a certain euclidean distance of each other.

An important practical problem is the fact that parts of line segments are sometimes missing, giving rise to open ended segments. This means that sometimes a certain incidence relation is transformed into another. Fig 12 illustrates this for a case when a part of a segment in (A) has been lost (B). In this case we note the length d_2 of the missing segment and compare it with length d_1 of the original segment. If d_2 is very small compared to d_1 there is a strong chance that the segment is a result of a deletion of a part of a longer segment. For the situation (B) in fig 12 we therefore assign multiple segment relations with different weights. Incidence type (A) is given weight $1 - d2/d1$ and incidence type (B) is given weight $d2/d1$

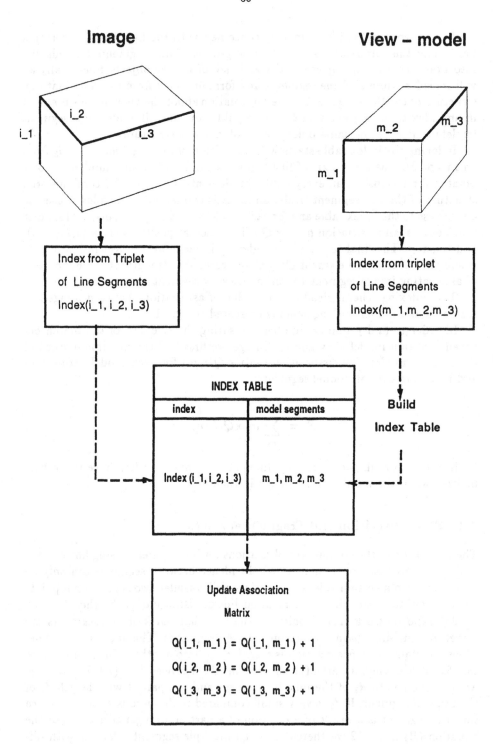

Fig. 11. *Scheme for computation of associations between image and model line segments*

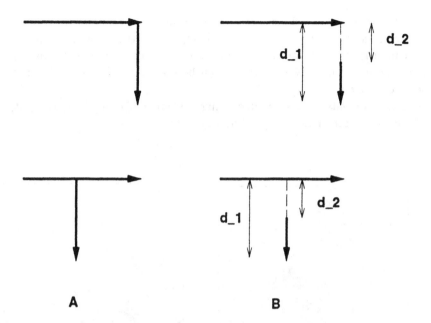

Fig. 12. *Deletions of parts of line segments changes the combinatorial relation between segment pairs*

3.3 Experimental Results

The crucial questions w.r.t. performance of the representation and indexing scheme are:

- How many view-models are hypothesized by a certain image ?
- How many views of a certain object have to be stored in order to achieve view invariance ?
- How does generality vs. specificity of the model affect the performance ?

We have made some preliminary experiments using pictures of objects taken from a camera with view angle varying from 0 to 180 degrees. The objects are 3 different chairs, denoted chair1, chair2, and chair3, and a bin. The pictures of the objects used are shown in figs. 13

For imaging situations where there is a well defined vertical direction in the image such as the case when the optical axis of the camera is parallel to a planar ground, it is of interest to include verticality or non-verticality as a property of a line segment. This means that we note for each line segment if it is parallel to a vertical reference line or not. This has been done in the following experiments. This means that triplet matchings between image and model, as described in the previous section, are rejected unless each segment is correctly matched w.r.t. verticality.

We have evaluated the matching scores for each image against two different models: A "generic" chair model and a more specific bin model. These are shown together with the matching scores for each image in the polar diagrams of figs. 14, 15, 16 and 17 where the polar angle represents the camera viewpoint. Note that in order to compare the matching score between the different models they must be normalized. Normalization can be done e.g. relative to the maximum score over the images.

Line segments were extracted using a Canny-edge detector, the output of which was screened for straight line segments.

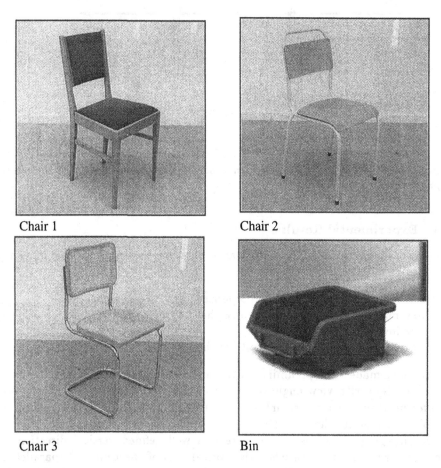

Chair 1   Chair 2

Chair 3   Bin

Fig. 13. *Examples of grey value images*

The experimental results seem to verify the assertions about view and class invariance w.r.t. models. Looking at fig.14 15 and 16 we see that all three chairs have matching scores for all viewpoints that are larger than those from the bin pictures. The scores have well define minima at the "degenerate" viewpoints 0, 90 and 180 degrees. For other viewpoints performance is rather invariant.

This invariance reflects the view invariance of the combinatorial relations of the chair. Note that the model of the chair admits essentially two distinct 3-D interpretations where the shape of the 3-D chair is similar but the viewpoint is shifted. These two interpretations are related by a "necker-reversal". The two viewpoints corresponds essentially to the two local maxima of the scores for the chairs. The variation of the scores is of course due to view variation of projected combinatorial structure but also due to variation in performance of the line extractor.

From 17 we see that a more specific model and an object with less symmetry, results in a matching score versus view angle with a definite peak at the "correct" angle. These results illuminate the inherent conflict between invariance and discriminabilty that exists for representations. In order to increase the discrimination power relative to other objects, models can be made more specific but with a loss of viewpoint invariance for the object under consideration.

The experiments certainly verify that a limited set of view models, say 3-4, can be useful at a hypothesis generation stage of a recognition system. The most essential of these would correspond to the most characteristic and stable views, [40] as in the model examples. These can be identified as those views where combinatorial image structure shows least variation. A certain degree of instance invariance within a category also seems possible at the price of reduced discriminability w.r.t. other categories.

4 Summary and Conclusions

We have argued that in order to improve the invariance properties of shape based 3-D object recognition systems we have to consider qualitative image representations as opposed to quantitative (metric). In order for such recognition systems to be robust, qualitative properties should be defined at feature level, thus avoiding complex and error prone grouping processes. We have presented a scheme for object representation based on *combinatorial geometry* , which can be described as qualitative properties of groups of geometric features that are invariant over certain transformations. In the case of projective transformations, combinatorial geometry consists of incidence and order properties of the feature groups. For the case that the features consists of line segments, we have implemented a 3-D recognition scheme based on indexing view models from triplets of line segments. The view variation properties of this representation has been analyzed and experiments have verified that a high degree of view invariance and a certain degree of invariance over object instances in a certain object class, can be achieved for the purpose of generating hypothesis of 3-D objects. This scheme should be seen as a hypothesis generation part in a more complete recognition system which would include a verification part. The system could also be complemented with metric models in order to increase the specificity of recognition.

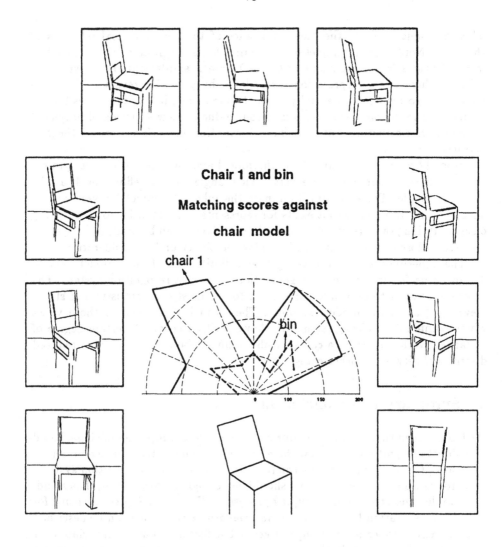

Chair 1 and bin

Matching scores against

chair model

Fig. 14.

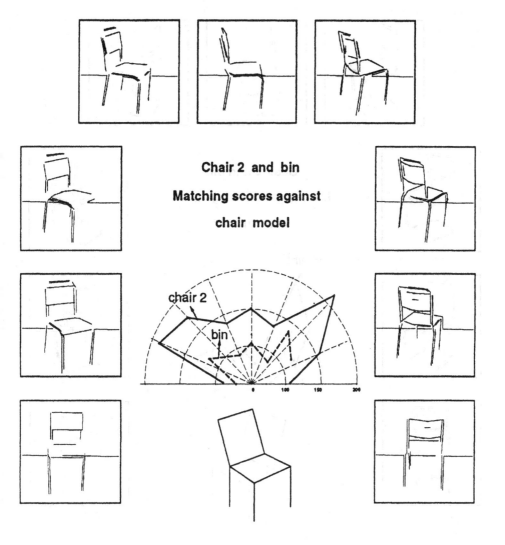

Chair 2 and bin

Matching scores against

chair model

Fig. 15.

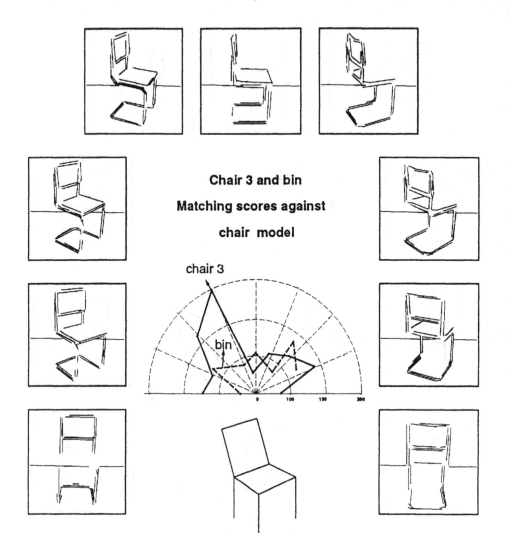

Chair 3 and bin

Matching scores against

chair model

Fig. 16.

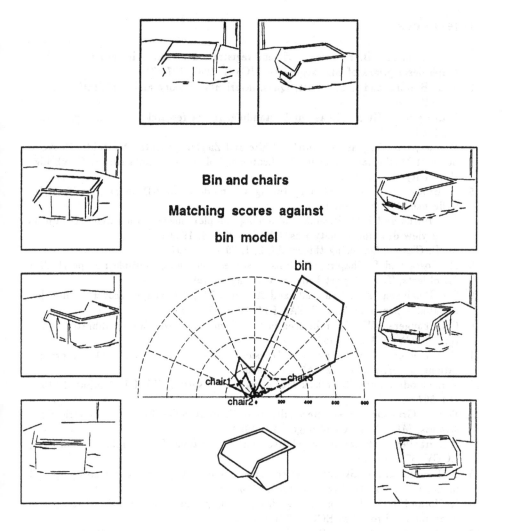

Bin and chairs

Matching scores against

bin model

Fig. 17.

References

1. N. Ayache and B. Faverjon, Efficient registration of stereo images by matching graph descriptions of edge segments, IJCV 1, 1987, 107-131.
2. T.O. Binford and T.S. Levitt, Quasi-invariants: Theory and exploitation, IUW, 819-829. 1993
3. I. Biederman, Human image understanding: recent research and a theory, CVGIP 32, 1985, 29-73.
4. Bjorner, Las Vergnas, Sturmfels, White and Ziegler, Oriented Matroids, Encyclopedia of Mathematics and its Applications, Vol. 46, C.G.Rota editor, Cambridge University Press, 1993
5. R. Brooks, Symbolic reasoning among 3-D models and 2-D images, Artificial Intelligence, 17: 285- 348, 1981
6. J.B. Burns and E.M. Riseman, Matching complex images to multiple 3D objects using view description networks, CVPR, 328-334. 1992
7. M. B. Clowes, On seeing things, AI, 2, 1, 79 - 116, 1971
8. M. Costa and L. Shapiro, Analysis of scenes containing multiple non-polyhedral 3D objects, Tech. Report Univ. of Washington 1995
9. S.J. Dickinson, A.P. Pentland, and A. Rosenfeld, 3-D shape recovery using distributed aspect matching, T-PAMI 14, 1992, 174-198.
10. O.D. Faugeras, What can be seen in three dimensions with an uncalibrated stereo rig?, ECCV, 563-578. 1992
11. O.D. Faugeras Stratification of 3-D vision: projective, affine, and metric representations, JOSA-A mar, 12, 465–484, 1995,
12. J. E. Goodman and R. Pollack Multidimensional sorting SIAM J. Comput. 12 1983 484–507
13. W.E.L. Grimson, Object Recognition by Computer-The Role of Geometric Constraints, MIT Press, Cambridge, MA, 1990.
14. R.I. Hartley, Estimation of relative camera positions for uncalibrated cameras, ECCV, 579-587. 1992
15. R. Horaud and T. Skordas, Stereo correspondence through feature grouping and maximal cliques, T- PAMI 11, 1989, 1168-1180.
16. R. Horaud, F. Veillon, and T. Skordas, Finding geometric and relational structures in an image, Proc. 1:st ECCV, 374-384. 1990
17. R. Horaud and H. Sossa, Polyhedral object recognition by indexing. Pattern recognition, Vol. 28, No. 12 pp. 1855 - 1870, 1995
18. D. Huffman, Impossible objects as nonsense sentences, Machine Intelligence, 6, 1971
19. D.P. Huttenlocher and S. Ullman, Recognizing solid objects by alignment with an image, IJCV 5, 1990, 195-212.
20. J. J. Koenderink and A. von Doorn, The singularities of the visual mapping, Biological Cybernetics, 24, 51 - 59, 1976
21. J. J. Koenderink and A. von Doorn, The internal representation of solid shape with respect to vision Biological Cybernetics, 32, 211 - 216, 1979
22. D.J. Kriegman and J. Ponce, On recognizing and positioning curved 3-D objects from image contours, T-PAMI 12, 1990, 1127-1137.
23. Lamdan, Y., Schwartz, J.T. and Wolfson, H.J. Object recognition by affine invariant matching. In: Proc. CVPR-88, pp. 335-344. (1988)
24. D.G. Lowe, Perceptual Organization and Visual Recognition, Kluwer, Boston, 1984.

25. J. Malik, Interpreting line drawings of curved objects, IJCV 1, 1987, 73-103.
26. D. Marr and K. Nishihara, Representation and recognition of the spatial organization of three dimensional shapes, Proc. Roy. Soc. B 200: 269 - 294 1987
27. R. Mohan and R. Nevatia, Perceptual organization for scene segmentation and description, T-PAMI 14, 1992, 616-635.
28. R. Mohr, Projective Geometry and Computer Vision, Handbook of Pattern Recognition and Computer Vision, (1992)
29. P.G. Mulgaonkar, L.G. Shapiro, and R.M. Haralick, Matching "sticks, plates, and blobs" objects using geometric and relational constraints, IVC 2, 1984, 85-98.
30. J.L. Mundy, P. Vrobel, and R. Joynson, Constraint- based modeling, IUW, 425-442. 1989
31. V.D. Nguyen, J.L. Mundy, and D. Kapur, Modeling generic polyhedral objects with constraints, CVPR, 479-485. 1991
32. S.B. Pollard, J. Porrill, and J.E.W. Mayhew, Recovering partial 3D wire frames descriptions from stereo dates, IVC 9, 1991, 58-65.
33. K. Pulli, TRIBORS: A triplet based object recognition system, Tech. Report Univ. of Washington 1995
34. Rothwell, C.A., Zisserman, A.P., Mundy, J.L., Forsyth, D.A., Efficient Model Library Access by Projectively Invariant Indexing Functions, In: Proc. CVPR-92, pp. 109-114. (1992)
35. Rothwell, C.A., Zisserman, A.P., Forsyth, D.A., Mundy, J.L. Canonical frames for planar object recognition. Proc. of 2nd ECCV, pp. 757 - 772. (1992)
36. Rothwell, C.A., Forsyth, D.A, Zisserman, A.P., Mundy, J.L. Extracting projective structure from single perspective views of 3D point sets, ICCV, 573-582. 1993
37. Stein F., and Medioni G., Structural Indexing: Efficient 2-D object Recognition, IEEE Trans. on Pattern Analysis and Machine Intelligence, 14, pp. 1198 - 1204, (1992)
38. J. Stolfi Oriented Projective Geometry: A Framework for Geometric Computations Academic Press New York, NY 1991
39. K. Sugihara, Machine Interpretation of Line Drawings, MIT Press, Cambridge, MA, 1986.
40. D. Weinshall, M. Werman, and Y. Gdalyahu, Canonical views, or the stability and likelihood of images of 3D objects, IUW, 967-971. 1994
41. M. Zerroug and R. Nevatia, Quasi-invariant properties and 3-D shape recovery of non-straight, non-constant generalized cylinders, CVPR, 96-103. 1993
42. S. Zhang, G.D. Sullivan, and K.D. Baker, The automatic construction of a view-independent relational model for 3-D object recognition, T-PAMI 15, 1993, 531-544.

Representing Objects Using Topology

Charlie Rothwell,[1] Joe Mundy[2] and Bill Hoffman[2]

[1] INRIA, 2004, Route des Lucioles, BP-93, Sophia Antipolis, 06902 CEDEX, France
[2] General Electric CRD, 1, River Road, Schenectady, New York 12301, USA

Abstract. In this paper we call for the revival of the study of topological representations in computer vision. Topology allows us to express the connectivity relationships which exist between different primitives in images and in scenes. Although topology was once of significant interest in vision, it has recently become over-shadowed by geometric considerations. We believe that it has a very important role to play in visual processing.

First we introduce an object-oriented class hierarchy which records the topological descriptions which exist in images and scenes. Once we have shown how the image topology relates to that of the scene, we demonstrate how it can be extracted from raw images. Subsequent to this, we describe how the newly found topological descriptions can be employed to facilitate feature grouping, the recognition of polyhedra, and the evaluation of recognition hypothesis which result from a mature object recognition system.

1 Introduction

The two main goals of this paper are to demonstrate that the local topology of object contours in an image can often be recovered, and that even incomplete feature topology can greatly improve the effectiveness of grouping and recognition processes. Throughout this work the term *topology* refers to the relationship between image features implied by *incidence* or *connection*. For example, the relationship between two line segments joined by a common vertex is a topological relation in that the segments are incident at the vertex. These local topological relations can be extended to form global topological notions of *boundary* and *closure*. It is emphasized that we are not suggesting that an object be represented solely in terms of topology, but rather that a combination of topological and geometric structure is required to achieve a successful representation.

The exploitation of topological concepts in object representation extends back to the earliest stages of computer vision research. Guzman used the combination of topological and geometric rules to decode the projections of complex three dimensional scenes [10]. An example of his work is shown in Fig. 1 where his program, SEE, used rules about junction alignment to group together matching portions of 3D polyhedra. One example of a rule used in SEE is also illustrated in Fig. 1 where T-junctions should form collinear matching pairs when the edge of an object is occluded by another object.

A few years later, a series of important discoveries about the combined topological and geometric properties of polyhedral scenes generated a formal basis for the interpretation of line drawings (Huffman [14], Clowes [5], Waltz [32] and Kanade [16]). These works demonstrated that a three-dimensional interpretation of a polyhedral scene could be derived completely from the relationship between image vertices and edges the of the

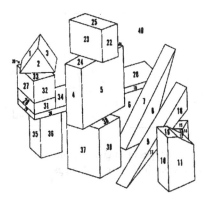

Fig. 1. A scene of a complexity typical of that which could be tackled by Guzman's SEE program (taken from [10]).

projected 3D object boundaries. The edges of a projected object boundary are labeled as {concave, convex, occluding}. A junction (vertex) is labeled as {"T", "Y", Arrow, "L"}, depending on the angle between edges incident at the junction, and on its order (the number of concurrent edges).

By propagating constraint rules on the labelings around the topological graph of the projected object, it is usually the case that only a few consistent labelings are possible. The resulting edge labels provide a 3D interpretation of the object projection which can support effective object-library indexing. Waltz also showed that junction and edge information from shadow boundaries can be used to restrict 3D object interpretations.

These results have been extended to curved surfaces by Shapira and Freeman [28], and Malik [18]. In the case of arbitrary curved surfaces, the junction rules are less constraining since curved surfaces have arbitrarily many degrees of freedom. Still, there are only two classes of discontinuities of mappings between surfaces, the *cusp* and *fold*, as elaborated in the important paper by Whitney [33]. Examples of folds and cusps are given in Fig. 2. Whitney's theoretical results were used by Malik in deriving his junction catalog. Recent work by Zerroug and Nevatia [34] has demonstrated that the junction properties at intersections of generalized cylinders can be used effectively for the interpretation of composite curved objects. The evidence for a join of two cylinders if often simply the existence of appropriate junctions in the neighborhood of the intersection. Some examples from Zerroug and Nevatia's paper are shown in Fig. 3.

In general however, this extensive body of topological machinery has not had much impact on the representations used in object recognition. The key reason is that standard boundary segmentation algorithms, for instance the Canny [4] edgel-detector, are not robust at recovering correct junction topologies from real images with complex illuminations, dense surface textures, and significant object occlusions. Under these circumstances, the topology of the extracted image features bears little relation to the surface topology of the underlying objects. This fragmentary representation is to be expected, since the signal properties of boundary intensity discontinuities are not well-represented

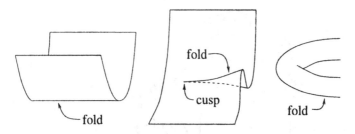

Fig. 2. Curved surfaces generate two types of singular feature under projection to the image. These are either **folds**, or their terminations which are called **cusps**.

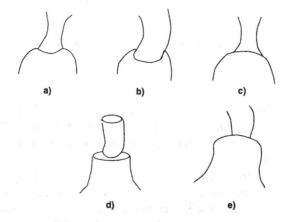

Fig. 3. The junction configurations for end-to-end joints used by Zerroug and Nevatia. The configurations are classified in terms of the type of junctions which close the projected surface regions of the occluding and occluded part. The classification also is distinguished by the larger or smaller part in the foreground or background. (a) Larger foreground closed by two tangent junctions, background by L-junctions. (b) Smaller foreground closed by L-junctions, background by T-junctions. (c) Larger foreground closed by L-junctions, background by T-junctions. (d) Both foreground and background closures visible, larger part closed by three-tangent junctions, smaller part by L-junctions, occlusion manifested by T-junctions. (e) L-junction closure of the foreground part, T-junction closure of background.

by a step-edgel model near junctions and in regions of high-texture density. Though, as we shall see later in this article, there are other important factors which contribute to the fragmentation of the edgel description.

Some attempts have been made to construct a consistent junction labeling by searching combinations of topological repairs [17, 27]. This approach can overcome a modest number of topological errors, but the case for actual images is far worse and the combinatorial explosion of labelings quickly becomes unmanageable. As a result, object representations developed over the last decade or so have emphasized the geometric aspects of object boundaries (Grimson [9], Ayache and Faugeras [1], and Lowe [17]). From geometric relations alone, it is only possible to group small numbers of features in each step of perceptual organization. For example, in the vertex-pair recognition system of

Mundy and Heller [20], a set of junction pair features, taken from a 3D object model, are exhaustively grouped to find feasible object poses using pose-consistency voting. While this approach can successfully group objects in a complex textured background, the combinatorial cost is high (the algorithm complexity is proportional to MN^2, where M is the number of model vertex-pairs and N is the number of vertices (junctions) in the segmented image).

As interest shifts to the recognition of general curved object classes, (for instance Zisserman, *et al.* [36]), such brute-force model-based grouping algorithms are not practical. The constraints imposed by a class property, such as symmetry, require a larger number of grouped features to support the evaluation of the constraint relation. For instance, in the case of a rotationally symmetric object, as shown in Fig. 4, it is necessary to isolate at least one pair of matched concavities (delineated by bitangencies) on the surface outline in order to test the hypothesis of rotational symmetry on the projected object boundary. Even this modest grouping goal can be challenging in the presence of complex background texture as is present in the figure.

Fig. 4. Images such as this pose tremendous problems for feature extractors. As well as having low contrast features, we are challenged by the presence of regular textures and miscellaneous lighting effects.

Binford and Levitt have suggested that the feature combination problem can be reduced by the use of quasi-invariants [2]. Fewer features are required to test a quasi-invariant geometric relation than a full invariant geometric relation. For example, in the rotational symmetry case just described, the lines joining symmetrical points on each side of the object are approximately parallel for most viewpoints. Thus, bitangent points which potentially match can be grouped according to this parallelism constraint, which eliminates a large fraction of incorrect correspondences. This use of quasi-invariants reduces the complexity of grouping curved object features, but does not avoid the inevitable intractability of grouping even four or five boundary features simultaneously against a cluttered background.

Nevertheless, the LEWIS object recognition system, which is based on invariant object representation [26], has successfully grouped a combinatorially large number of features to support planar invariant computations. Plane projective geometric relationships require the grouping of a minimum number of features, such as two conics, four collinear points, a conic and three lines (or points), or five lines (or points). In order to avoid trying all combinations of such features in an image, which is combinatorially impractical, LEWIS exploits the topological relations between the segments along the object boundary. This strategy results in a combinatorial cost in feature grouping which is nearly linear rather than $O(N^m)$, where N is the number of features in an image and m is the number of features required in a group.

These promising results suggest that it is time to reassess the use of topology in object representation. Object classes impose both topological and geometric constraints on their perspective projections. It is now clear that the perceptual organization of features of such objects from the scene background is not tractable without exploiting both forms of constraint. At the same time, the use of topology in the representation of composite structures is essential to define part–whole junctions and to define occlusion events which are a major key to discovering the three dimensional structure of a scene.

Therefore, it is the central thesis of this paper that topology cannot be ignored in object representation, and efforts must be redoubled to recover an accurate topological description from real images, with all the inherent complexity. It is argued that much better topology recovery is possible with relatively minor modifications to existing segmentation algorithms. Furthermore, it is demonstrated that it is feasible to follow boundaries through very complex topological structures at much lower combinatorial cost than required for geometric grouping alone.

Thus, in Sect. 2 we introduce the topological structures which are of use to us, and then briefly review how they are implemented within an object-oriented programming environment. In Sect. 3 we draw together the topological relationships between structures in the three-dimensional world and those in the image, showing above all, that image topology provides a key to the *shapes* of objects in a scene. We then show in Sect. 4 how the topological descriptions can be extracted from images prior to discussing in Sect. 5 through to Sect. 7 how to use them within vision systems.

2   The Topological Hierarchy

The topology objects of interest in computer vision can be expressed within a vertical hierarchy which develops at the lowest level from the simplest object called a *vertex* (merely a single point) up towards complex structures called *blocks* (which represent complete three-dimensional shapes). Throughout the rest of this section we review the hierarchy.

All of the objects are in principle embedded within three-dimensional space. However, the data structures we use have in themselves no notion of the dimensionality of the space in which they reside. It is worth emphasizing that concepts of shape, such as those which portray whether an object is planar, relate strictly to the *geometry* of an object, rather than to the *topology*. Thus, we can in both in practice and in theory decouple ourselves from concerns about the actual forms of objects whilst we analyse the purely topological properties. However, towards the end of this section, we that see topology and geometry are actually very closely related; our data structures permit the transition from one representation to another.

The topological hierarchy is described by an *inferior-superior* stratification composed of the following objects:

- **Vertex:** this is simply a point in space. It has neither extent, nor understands its relationship to other objects. The *order* of a vertex is the number of *edges* (see below) to which it is attached.
- **0-Chain:** is an ordered set of vertices. Through a 0-chain a vertex can determine which other vertices are its neighbours. A vertex can lie on an arbitrary number of 0-chains, and so can have any number of immediate neighbours.
- **Edge:** is the simplest non-trivial 0-chain. As such it is a pair of vertices, and describes a one dimensional manifold (which geometrically may or may not be a linear space). All that an edge does is to record the connectivity between a pair of vertices. It cannot immediately determine which other edges are its neighbours, though of course, through a deductive chain of queries, we can easily determine which edges are neighbouring.
- **1-Chain:** represents an ordered list of edges. An obvious inferior of a 1-chain is the 0-chain which records the list of vertices lying at the ends of each edge. A special case of a 1-chain is a **1-cycle**; this is when the first and last vertex of any 0-chain inferior to the 1-chain are equal (any such 0-chain will be a **0-cycle**). All vertices lie on two edges within a 1-cycle.
- **Face:** is composed of 1-cycles. It always lies within a two dimensional manifold. A plane polygon is an example of a face, but so are each of the two strips which make up a tennis ball (these are curved surfaces, each one bounded by a closed curve). Another example is that of the area lying between one plane polygon and another lying entirely within the first (such a face is composed of two 1-cycles).
- **2-Chain:** is a an ordered chain of faces (the ordering is no longer necessarily expressible as a linear sequence). Each pair of adjacent faces in the chain share at least one edge. A typical 2-chain is the net of six squares used to make a cube out of a piece cardboard. A special case of a 2-chain is a **2-cycle** within which each edge must lie on two faces. Such an object separates a pair of closed topological spaces in three-dimensions (note that up to this point in the hierarchy the objects do not divide three-dimensional space into different sectors).
- **Block:** is composed of a set of 2-cycles. It encloses a volume in three-dimensional space, though it might cause space to be broken up into a number of different parts.

This hierarchy is implemented using the object-oriented tools available in C++. Each of the seven different topological structures are represented by a C++ class. The superior objects are deemed responsible for the management of the lower order objects, and so maintenance of any topological description is governed by the highest level. This manifests itself, for instance, by a vertex being unaware of its neighbours; such adjacency information is carried by the superior 0-chain.

The data structures allow ready access to a number of different properties of the objects, but more importantly allows the formation of responses to a number of different topological queries. One such typical question would be: on which face does this vertex lie? Conversely we might want to know which vertices lie on a given face. As a further example, we might from the vision point of view be very interested in knowing which edges (representing edgel-chains) intersect at a given vertex (an image junction).

It is worth emphasizing the fact that *1-cycles are perhaps the most important object available to us as vision system builders. They provide a complete and accurate representation of all of the boundary information present in an image. Such boundary data provides an efficient way to access region information within the image, and hence leads to the grouping and segmentation of object structures.*

Finally, the class structures provide a method of easy maintenance of the topological description. It is for instance very easy to connect together a pair of previously unconnected vertices by an edge, or in fact just to merge two proximal vertices. Similarly the complexity of the description can be reduced, such as a vertex being connected to three edges becoming disconnected to one of them through the creation of a new vertex.

2.1 Adding Geometry

No member of the topology hierarchy stores any *geometrical* information. A vertex is merely a point in three-dimensional space, and is without coordinates. Likewise, an edge is simply a one-dimensional space which links a pair of vertices. From the topological point of view we do not care what shape an edge takes. We could for convenience assume that it is linear, though this might not always be suitable, nor even work in practice. For instance should the two vertices bounding the edge be *the same* vertex, we would be unable to recording the presence of the edge using only a straight line.

All of our topological classes therefore satisfy the demands which are traditionally made of topological objects. However, from the vision point of view we must also include notions of geometry such as coordinate frames, and just as importantly *shape*. Consequently, we have adapted our C++ topology object hierarchy to account readily for geometry by allowing each topological class to point (quite literally) towards a geometric object.

The geometric objects (again these are C++ classes) contain all of the shape information normally required within a vision application. However, above this we are now able to ask topologically oriented questions of the geometric objects, and so have a superior processing capability than purely geometric systems. What is more, the topological queries are posed at a distinct level which saves us from cluttering our geometric intuitions (and vice versa). Thus, we have the flexibility to analyse only the topology, or only the geometry, or both.

The Geometry Classes The most primitive geometric feature is a point. The information it carries is little more than its (x, y, z) coordinates in three-dimensional space.

Linking a vertex to a geometric point means that the vertex may be restricted to lie in a particular place in space (the image plane), or in may be marked clearly as being a scene feature (in 3D space).

There is a range of geometric primitives used to represent the geometry associated with edges. The three main categories are *parametric curves*, *algebraic curves*, and *digital curves*. All of these curves are one-dimensional objects. We return to representing edges in the next paragraph. In the mean time, continuing up the geometric hierarchy, we include a range of different surface primitives (again parametric or implicit) which are bound topologically to faces. Finally, there is the geometric representation allied with blocks: these are volumes bounded by surfaces.

The Geometry of Edges Throughout the rest of this paper we make significant use of the geometry associated with edges. Above we said that these fall into three principal classes of one-dimensional geometric object: the parametric curve; the implicit algebraic curve; and the digital curve. The first of these covers objects such as splines, and in fact any curve which can be parametrized by a single variable. Common instances are lines and conics, though in practice we normally use an implicit representation of such primitives.

The most common example of an implicit algebraic curve is a line. This is represented by the three dependent parameters $(a, b, c)^\top$ which derive from the line constraint equation $ax + by + c = 0$. Additionally associated with a line are its endpoints (note that we are thus actually dealing with a *line-segment* rather than the pure geometric object called a *line*). In fact, all of our one-dimensional geometric objects have start and end points. Similarly, there exist implicit conics represented by the six dependent variables $(a, b, c, d, e, f)^\top$. These parameters derive from the conic constraint equation $ax^2 + bxy + cy^2 + dx + ey + f = 0$. Again there are endpoints for the conic curve as we normally represent only segments of conics, and not only the entire sections.

The final class of geometric object related to edges is based on what we call *digital geometry*. At the highest level, digital curves are ordered lists of samples of a given curve (hence the term digital). Such classes are ideal for representing the curve descriptions recovered from feature detectors such as the Canny filter [4], as each sample point is simply a sub-pixel edge location.[3] A *digital curve* also has a start and endpoint. Additionally, it may have an associated implicit or parametric representation. This representation might, for instance, be the result of fitting a line or a conic to the underlying edgel data. Consequently, due to the versatility and the quantity of visual information which can be contained within a digital curve, we realise that they form one of the fundamental objects within our vision applications.

---

[3] Note that throughout this paper we refer to the image feature corresponding to a high contrast point of the intensity surface, and often called an *edge*, as an *edgel*. The usage of this term is widespread and is an extension of the terminology *pixel* and *voxel* to encompass objects called 'edge-elements'. For our purposes, it is very important to ensure that the nomenclature for such an image primitive is well understood. This will then save confusion with the completely different feature called a *topological edge*. Consequently, we shall also use terminology such as edgel-detector (for instance the Canny), or edgel-chain (as list of edgels).

3   3D Topology and Its Projection to the Image

In the case of polyhedral objects [5, 14, 32], the measured image boundaries correspond to the projections of the unoccluded 3D object boundaries. Therefore, under the assumptions of generic viewpoint, the ideal topology network of the image projection is closely related to that of the original object with the following alterations:

- T-junction vertices are generated by occlusions;
- vertex order remains constant or can be reduced by self-occlusion of one or more edges incident at a vertex;
- edges can be split by other occluding boundaries;
- faces can be eliminated or split by occluding surfaces.

This mapping of topology from the scene to the image under projection is highly constrained under ideal conditions and leads to a compact set of vertex labeling rules which are based on the occupancy of volume and exhaustive enumeration of possible viewpoints [32].

In the case of curved objects, there is the additional complexity that a single surface region (or *face*) can be self-occluding. That is, the occluding boundary does not necessarily correspond to the projection of an *edge* or *1-chain* of the 3D object topology. Malik [18] has developed a rigorous analysis of the projection of general curved objects which is based on the earlier polyhedral analysis with the addition of the key results of Whitney [33] who showed that there are only two types of self-occlusion under image projection, the cusp and the fold (see Fig. 2). Thus, for curved surfaces, the projected topology network can include both edges and vertices which are not present in the original object topology. For example, the drawing of the projection of the torus in Fig. 5 contains a number of edges which are not present on the surface in 3D; they have been generated by the self-occluding surface. These edges form a number of contours in the image description which are not represented at all in the object.

These edges, which are called *folds* and are generated by the projection of *limbs* of the surface, cause the image topology to differ considerably from the scene topology. However, once we have eliminated them from the image representation, and also those edges which are partially occluded, we can state the following theorem:

Theorem 1. *The topology of the image is a subset of the topology of the scene.*

Proof. The proof of this theorem is straightforward. First we appeal to [5, 14, 32] and thus know that the theorem is true for all edges generated by the intersection of pairs of surfaces except for edges caused by partial occlusion. Such occluded edges have however been suppressed (a pre-requisite of the theorem), and so need no further consideration. Finally, we have to consider the folds generated by the projection of curved surfaces. However, these have also been eliminated and so do not affect the image topology.

Identification of Folds   Building on the work of Whitney [33] and Malik [18], we can derive an algorithm which identifies folds in the image description. As in [18, 33], we make the assumption that we work with bounded surfaces which are C^3 continuous, though we include the additional assumption of working with objects having trihedral vertices (these are the only objects which are in fact stable, and so the assumption is reasonable). This last requirement limits the order of the vertices in the image.

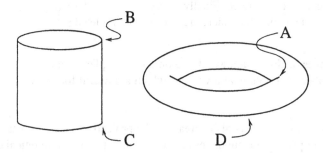

Fig. 5. There are four ways in which folds form junctions with other edges. The descriptions of these types are given in the text.

There are in fact only four ways in which folds form junctions with other edges; these are shown in Fig. 5. The first type are terminations (in fact they are cusps), as shown by the order one vertex (A). If we assume that all three-dimensional object features are contained within 1-cycles (and so *there are no order one vertices in the scene*, which is true for all finite sized objects), then the presence of a termination must have resulted from the termination of a fold. The second type is the tri-tangent junction (B): all three edges entering the vertex have the same slope, and two of the edges have the same curvature. The fold is the edge which has the differing curvature. This type of structure is unique to the presence of a fold. There are also L-type junctions (C) which connect a pair of edges of equal tangency; these result from the occlusion of one of the edges of a tri-tangent junction. The final class of junctions are the *virtual vertices* of (D): this is where a single edge with the same start and end vertex is generated by the projection of a limb to the image.

Modification of the Image Topology Now that we can identify limbs (at least theoretically), we propose an algorithm for modifying the image topology so that it becomes exactly a subset of that of the scene. First, we identify all of the T-junctions, and split them into two parts. The first part is a vertex of order two which links the two edges forming the cross of the T, and the second part involves the creation of a vertex of order one which terminates the stem of the T.

We then identify all of the limbs as suggested by Fig. 5. Type (A) junctions require no attention. At type (B) junctions we break off the fold and attach it to a new order one vertex; the original vertex thus becomes order two. At type (C) corners we separate the edges and terminate each of them by order one vertices. Likewise, for the virtual vertices of type (D) we split the vertex into two, each of which are order one.[4]

---

[4] This last category can cause the disruption of image topology which results from the projection

Once we have identified all of the possible limbs, we suppress the order one vertices which are derived from T-junctions or the terminations of folds; we also remove the edges to which they are attached. This eliminates a number of folds and the edges which are partially occluded. We then repeat the segmentation procedure until no more order one vertices are generated. Finally, we remove all order two vertices by merging their bounding edges (of which there are two) into a single edge.

The Invariances of the Topology Based on the principles stated above, it is possible to state a number of topological invariants which are useful for the analysis of object projections:

1. The order of a vertex can only decrease under projection. New vertices can be generated under projection, but under restricted geometric and topological conditions:
   - a T-junction generated by an occluding boundary;
   - the terminal vertex of a self occluding fold or cusp;
   - a three-tangent junction where a limb is tangent to an occluding edge;
   - a tangent L-junction which is the occlusion of a three-tangent junction;
   - a virtual junction resulting from the projection of a closed limb.
2. An object edge is decomposed into a 1-chain by occlusion, but edges in the 1-chain are projections of subsets of the object edge. Otherwise an edge projects in its entirety to an edge.
3. The convex hull of the image projection is the projection of the convex hull of the object.

The last point requires some development. In fact, we make extensive use of the convex hull property in all of our considerations, though often we do so only implicitly. First, it is worth emphasizing that notions of topology are meaningless within any pure projective space, such as in the projective plane. However, once we place limitations on which part of the projective plane can be used (by watching carefully what happens to the ideal line), we can impose a topology and similarly exploit properties such as the convex hull of a set of data.

The reason for the invariances are in fact principally due to the linearity of projections. Overall, *so long as we place certain restrictions on the type of projection we permit*, we can ensure that all points within a given ball of a point in the pre-image map projectively to a ball in the image. The limitation we employ is that the points within the bounds of the object do not interfere with the ideal points in the image frame. This restriction is always satisfied with real cameras, and so yields the topology. In fact, these notions have already surfaced in the use of projective geometry in navigation and recognition tasks such as those described by Robert and Faugeras [21], Morin [19], and Rothwell [26].

---

of actual scene edges (such as isolated faces consisting of only a single edge). However, we do not feel that damaging these structures is immediately a concern as they do not provide as rich an image structure as the rest of the topological description. Anyway, we do not actually discard any of the original topological information as it will be useful during later processing phases, we only 'suppress' it for the time-being.

4 Deriving Topological Structures from Images

Our discussion up to this point has been focussed on the theoretical relationships between the topological structures of three-dimensional objects and their projections into images. In this section we proceed to show that striving for a topological description is not such a foolish occupation: this is said in light of the incredible wealth of literature claiming that extracting scene topology is too difficult, and that in the few instances where interesting topology exists, that it is usually erroneous.

This last statement needs to be justified in the light of edgel-detection. For many years it has been claimed by a significant proportion of the vision community that edgel-detectors record junction topology unfaithfully, and that there is never much hope of producing an edgel-detector which could recover even a significant number of junctions. One great sufferer of this line of thinking has been the Canny filter [4] which provides extremely useful information in certain parts of the image, but breaks down where the step-edgel model is inapplicable (though the breakdown is not so much caused by the incorrectness of the step model). Of course many other edgel-filters have similar faults, whether they are based on the direct estimation of image derivatives or should they exploit a morphological approach.

There is thus ultimately a belief that edgel-detectors can produce only curve segments (that is *geometry*) without the possession of any real topology. Subsequent processes are therefore required to derive a topological representation; these processes in general come under the term *grouping*. The lowest level of grouping is the linking of edgels together into extended contours. Even some recent papers which examine linking [6] would rather destroy all topological information (even 8-way neighbourhood relations) in the fundamental edgel representation than use any information which might be erroneous. It is only after the removal of the topology that attempts are made to reconnect the pieces together using traditional grouping strategies such as those discussed by Binford [3], Lowe [17], Sarkar and Boyer [27], or Dickson [8]. Unfortunately, the methods used to reconstitute the curves are usually based on heuristics such as measures of intensity differences, or curvatures, and more often than not with the global intention of preventing the formation of anything more than low order connectivity. We believe that this line of bottom-up reasoning is detrimental to the success of low-level vision.

We claim rather that even the edgel-level description provides an opportunity for top-down processing for the verification our low-level topology descriptions extracted from simple intensity gradient measures. This of course means that a significant measure of topology must be available in the edgel-chains, though the ability to deduce such a representation will be demonstrated in this section.

4.1 An Algorithm for Edgel Detection

An evaluation of the processing steps of the Canny [4] edgel-detector has led us to produce a modification which preserves the Canny's basic behaviour on extended image curves, but provides far superior connectivity relationships near object junctions. The main difference found in our algorithm is that non-maximal suppression is not used as a primary filter for the edgels. We reject the use of classical non-maximal suppression because the orientations used as estimates of the normals to the directions of maximal gradient are often so poorly defined that their use damages the resultant edgel description. The details we review here are presented more thoroughly in the [23]; only a brief summary of the algorithm is given in the following paragraphs.

Initial Processing The intensity image I is smoothed with a discretely sampled, separable, two-dimensional isotropic Gaussian to yield a smoothed image S. The directional derivatives S_x and S_y are measured using central finite difference operators of the form $[-1, 0, 1]$, and then $|\nabla S|$ and θ (the normal direction to the intensity gradient) are computed for every point. For convenience we denote $|\nabla S|$ by the discrete function N.

Sub-Pixel Localization For each pixel having $N > t_l$ (t_l is a pre-set threshold which is fixed cautiously low and represents only 'sensor noise'), we estimate where an actual edgel would lie using the sub-pixel interpolation technique suggested by Canny [4]. These locations are found by estimating the zero-crossing of the second derivative in the direction normal to the contour tangent. Traditionally non-maximal suppression would be done at this stage: a pixel would be marked as an edgel if it has a value of N greater than the (interpolated) gradient values perpendicular to the edgel direction. We assign all edgels which would have passed non-maximal suppression to a base set Σ_0 on which to build up the topological description of the image. However, we do not use non-maximal suppression as a means to classification (as would be done conventionally).

Although we are still fixed to a quantized grid for the image, every potential edgel point now has a geometrically accurate location which is on the whole identical to that which would be recovered by a Canny.

Thresholding of N Once N and θ are known, we are principally faced with a classification problem between edgels and other points. Commonly, the partition would be made by thresholding against a single threshold. We use a novel approach to select edgels given that *a priori* we have no knowledge of what their strengths should be. This is done through dynamic thresholding, that is we threshold by a value which varies across the image. A threshold surface (called $T_{x,y}$ on the discrete image domain) is computed and used to classify pixels as edgels whenever $N_{x,y} > \alpha T_{x,y}$ (the use of the constant $0 < \alpha \leq 1$ is described below).

$T_{x,y}$ is defined using the members of the set Σ_0. These edgels provide a good indication of the strengths of edgels locally in the image as all of them are ridge points of the gradient surface (as they would have passed non-maximal suppression). We therefore define $T_{x,y}$ to take the value of $N_{x,y}$ for each $(x, y) \in \Sigma_0$, and then form a piecewise planar interpolated surface for all other (x, y). The parameter α is introduced to account for the fact that edgel strengths can become slightly diminished close to junctions (we use $\alpha = 0.9$). All pixels that pass the threshold test are included in the set Σ; obviously $\Sigma_0 \subset \Sigma$.

Edgel Thinning Σ represents connected subsets which need to be thinned. Thinning is based on the Tsao-Fu algorithm [31] which eliminates elements from a set based on change of *genus*. It works in a way that does not shorten edgel-chains which have a free end (i.e. edgels which are connected only to one other edgel). However, Tsao-Fu thinning treats all members of the set equally, and so a strong edgel may be deleted in preference to a weaker one. So that we preserve the localisation of ridges, we order the members of Σ and remove the weaker elements first. The thinned set is called Σ_t.

Retrieving the Topological Description The topological description is extracted from Σ_t and placed into our topology-geometry class structures. Vertices are placed at edgels which have either only a single neighbour (in which case they represent the end of a dangling edgel-chain), or are edgels which have more than two edgels connected to them (junctions). Topologically, a corner point defined by the meeting of two edgel-chains is contained within a single edge. At this stage we do not attempt to *segment* the edges at such corner points, though this can be done at a later stage using various fitting techniques. Once the vertices have been extracted, we follow the edgel-chains between them using an 8-way following algorithm. As each edgel is extracted, its sub-pixel location is written to a list managed by the topological edge of which it is part.

4.2 Examples

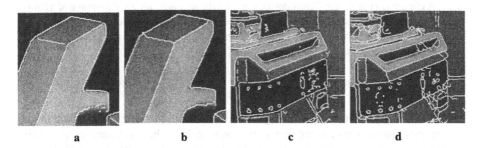

a   b   c   d

Fig. 6. In (a) and (c) are two simple scenes in which a Canny edgel-detector fails to extract the image topology correctly. The outputs of the new algorithm are shown in (b) and (d); note that the connectivity of the edgel-chains is much better, and yet the geometry of the extended edges is unchanged.

Figure 6 shows the improvement in topology around contour junctions when we contrast our edgel-detector to the Canny. The Canny output was tuned to give the best topological results possible, and so we believe that the comparison is fair (in fact the thresholds for the Canny were lower than that used for our edgel-detector). In (b), the topology of all of the junctions is recovered correctly except in the bottom left corner where a shadow has obscured the edge. Note that the geometry around the junctions is far from perfect, though at least we have recovered the topology correctly. Importantly, the topological edgel-description indicates the presence (and complexity) of junctions.

Figure 6d shows the quality of the topological description extracted from the image of a robot. Although the representation is complex around the fine-detailed parts of the scene, it is generally correct. We also see how reliably the method returns edgels in the original intensity surface: a set of edgels has been recovered towards the right of the vent just above the centre of the image. These contours are present in the intensity surface and are actually stronger than some of those of the right-hand boundary of the object (they result from mutual illumination effects). The Canny operator (Fig. 6c) was able to pick out some of the points on the vent, but was unable to recover the complete structure.

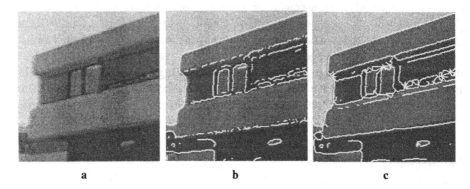

a   b   c

Fig. 7. A comparison of (b) the Canny detector and (c) the new algorithm for an image of a building. The main difficulty for the Canny operator is posed by the weak boundary contrast on interior edges which produces poor definition of edgel orientations. The new algorithm recovers a much better topological description.

Another comparison is given in Fig. 7. This is a small part of a scene which might be used for a navigation or scene reconstruction task. Note that the Canny output in (b) has many gaps along the edges of the building (again we choose the thresholds to give the maximum possible contour continuity). Although the edgels corresponding to the building edges are generally stronger than both the Canny thresholds, the estimate of the orientation which is crucial for non-maximal suppression is poor. Subsequently, many edgels are filtered out. However, Fig. 7c shows the result for the new algorithm which uses conventional non-maximal suppression only to assign the classification thresholds, and then uses topologically correct thinning to derive the final description. As shown, we recover a significant proportion of the image topology.

Given the quality of the topological representation recovered by the edgel-detector we can immediately invoke different types of top-down processing to evaluate and enhance the description. The first of these is the analysis of each junction in the topology network (each vertex). Due to the limitations of space, we simply mention in passing that we can test whether each vertex in the description actually corresponds reasonably to a junction; this is done using a range of different junction detectors such as those of [7, 13, 22]. Another process which we employ is the removal of very small closed regions (faces) in the edge description which are artifacts of the image surface rather than being representative of real three-dimensional structures.

5 Developing the Topological Representation

The edgel-detector we described in Sect. 4 provides a markedly improved representation of the image topology. Consequently, it substantially eases the processing stages that we do after edgel-detection. For instance, a good knowledge of image topology reduces the amount of work which has to be done during edge grouping. Grouping is the process used to associate image features together that are themselves too weak to be useful recognition or navigation cues, but once combined provide strong visual constraints. Binford [3] and Lowe [17] demonstrated early examples of grouping mechanisms; a more recent review is contained in the paper by Sarkar and Boyer [27]. Typically, grouping uses geometric measures such as proximity or orientation to associate features rather than employing topology. As such it must be a bottom-up process. As shown by Dickson in [8] and his implementation of geometric grouping strategies via Bayesian nets, grouping without connectivity relationships leads to algorithms having exponential execution times. Such algorithms run too slowly to be really useful for computer vision.

We ourselves have experienced great anxiety in trying to use different geometric grouping methods on aerial surveillance images when we had no more connectivity than that provided by the Canny filter. However, we now have the benefit of having a much richer low-level edgel-description based around a topological framework. Such input to a recognition algorithm provides improved recognition capabilities and at a smaller cost. In the following sections we discuss two simple approaches to grouping which can only exist in the presence of topology. Incidentally, these grouping strategies are intended to be used within a invariant-indexing recognition system, using any number of the measures such as those discussed by [35].

5.1 Grouping by Regions

The consistency of three dimensional topology and its projection in the image (discussed in Sect. 3) suggests a very powerful grouping cue. First, we must realise that many of the most prevalent structures in the world are built on topological faces (whether planar faces, or curved surfaces). Then, we see that most of these faces will project to faces in the image, and so a direct route to three-dimensional structure is the recovery of the faces in the the image. Such faces are quite simply closed regions.

Given the set of closed regions in the image we have an immediate grouping of scene primitives: the edges and vertices that make up the faces. Should the individual features from a single face be insufficient to provide a suitable shape cue we can group together closed regions into higher order structures (an example of this type of processing is done for polyhedra in Sect. 6). The regions we extract can be filtered prior to the formation of feature groups for indexing using a number of different measures: typical examples would be removal of small regions in the segmentation which have dimensions of only a few pixels, or the elimination of those regions which are not based principally on useful algebraic structures in the images (for the recognition of buildings these structures might be straight lines).

In Fig. 8 we show how the extraction of closed regions can be used to make the fundamental object primitives 'pop-out' of the scene and hence drive an application such

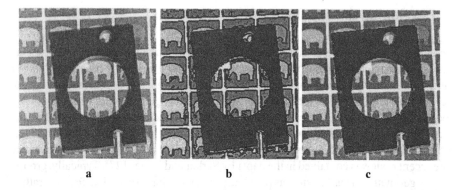

Fig. 8. (a) shows a small part of an image which contains dense background texture and hence has a complicated segmentation structure. From the large number of edgel-chains in (b), it would appear to be a difficult task to recover the boundary of the part lying on top of the texture. However, the regions analyzer picks out the two main boundaries of the object both quickly and correctly as shown in (c).

as recognition. The figure demonstrates the power of the method on a very difficult example where there is a complex background texture and thus a large number of feature groups. However, the two principal regions are extracted at a very low computation cost. Quite simply, without the use of connectivity relationships we would have been unable to disambiguate the useful features from the clutter without resorting to an exhaustive search.

5.2 Grouping Using Connectivity

Another simple approach to grouping is to extract 1-chains from the image and use these as a basis for forming suitable feature groups for indexing. This type of grouping strategy is reminiscent of the linear grouping algorithm in [26].

However, if we are able to extract a complex representation of the scene topology we might well find ourselves overwhelmed by the number of 1-chains in the image. Nevertheless, due the redundancy and local aspects of many of a shape descriptions (again the reader is referred to [26] for a discussion of these terms), we can ignore many of the candidate chains and still have confidence that enough measures will be extracted to leave us capable of identifying objects. We thus invoke a greedy process for the extraction of 1-chains. First we pre-filter the topology network using the regions analyzer and prevent the extraction of any 1-chains based on closed regions. We then search the image for vertices of order one (these are vertices connected to single edges). Subsequently, using depth first search we extract 1-chains from each of the order one vertices in such a way as to visit each vertex only once. In fact, each vertex is marked as being visited once it has been included in a 1-chain. Filtering in this way yield a significant number of linear feature sets which can then be used as a basis for the formation of feature groups from sets of consecutive primitives within each 1-chain. An example of feature group extraction using this approach is given in Fig. 9.

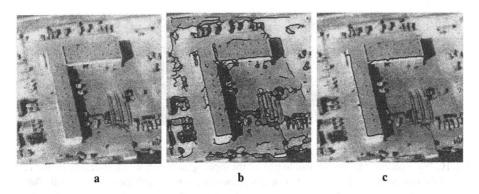

a   b   c

Fig. 9. This image segment has a relatively complex edgel-structure (b), though because of similarities in contrast many edgel-detectors have difficulty in extracting the boundary of the building. However, the edgel-detector we discussed in Sect. 4 finds all but one side of the outer boundary of the building without any difficulty. Then, using the 1-chain extraction procedure we can identify the single long chain shown in (c).

6 Using Topology for the Extraction of Polyhedra

Section 4 demonstrated how a topological description can be extracted from the image at the edgel-chain level. We then showed in Sect. 5 how such a description can ease the task of grouping features together to form cliques which are useful as recognition cues. In this section we demonstrate how the topology extracted from the edgel-detector simplifies a task which could be considered to be one of the early processing phases in the recognition of simple polyhedral objects (such as described by Sugihara [29]). Although we do not go into the details of the design of an object recognition system (nor in fact demonstrate that we are actually recognizing objects), we show how we can derive templates of model instances, which are topologically correct line drawings, in scenes which can later be used for recognition. More complete details about the recognition process for polyhedra (both the extraction of features and the measurement of invariants) are given in [25].

In Figs 10 and 11 we show how knowledge of the image topology makes the extraction of polyhedral models from images a relatively simple task. No searching for image associations is required for either of the examples; the edgel-detector provides all of the connectivity and junction information. This means that higher-level processes do not have to expend valuable time organizing the input data. Briefly, the order of processing is: first the detection of edgels using the topological algorithm; the representation of straight edgel-chains by lines; the extraction of topological structures commensurate with the polyhedral that we wish to recognize (topological faces); and finally a process which ensures the geometric consistency of the edgel and line descriptions (using a polyhedral snake composed of straight lines and junctions). In practice we would then proceed to match the geometric description to models in a library, though this process is beyond the scope of this paper.

It is worth remarking that the key step in the whole of the process is the extraction of the faces of the polyhedra. If this fails we are unable to recover the structure of the

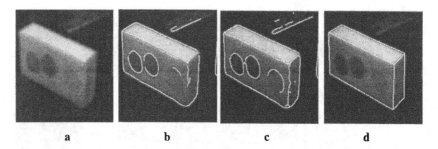

Fig. 10. For a polyhedron we are able to extract the topologically correct edgel-chains shown in (c) and then use a snake-like process to derive a geometrically reasonable description in (d). This process succeeds even though the image features are very small and so are easily confounded (the smaller faces have widths of about eight pixels). For comparison, Canny output is shown in (b).

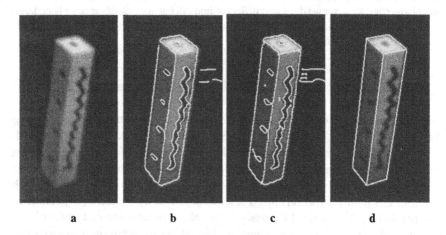

Fig. 11. A second example showing the extraction of a polyhedral description from the intensity image via a topologically reliable edgel description. Again (b) shows Canny output, (c) our edgel description, and (d) the recovered polyhedron.

polyhedra. However, finding the polyhedral faces is not so difficult because they already exist within the topological description of the image: they are simply the closed regions in the image. We can therefore resort to the grouping algorithm discussed in Sect. 5, and simply extract all of the regions from the image. In general there are not many of these (and they are certainly far less numerous than the number of structures we would have had to consider if we had done traditional grouping using geometry), and so the feature extraction process is efficient. Once we have found the regions in the image we need only determine whether they are globally polygonal, and if not we discard them from further consideration.

It is our belief that a grouping strategy for polyhedral features could not be put in place efficiently without the use of topological measures. A reliance on geometric grouping would lead in general to an exponential growth in the number of feature clusters, and hence a remarkable degradation in performance (cf. Lowe [17]).

7   Hypothesis Verification Using Topology

The verification of object hypotheses in object recognition is the final and crucial stage
of processing within a recognition system. When verification performs insufficiently,
we become burdened with false positives and false negatives which detract significantly
from system performance. Thus, it is evident that we should make every effort to employ
the best verification strategies available to us.

In the past verification has been based solely on geometric measures. Typically, when
an object instance has been hypothesized in a scene, its evaluation is performed by projecting the object model into the image, and testing whether features on the projected
model can be explained suitably by observed image features. Necessary constraints are
usually based on ensuring that the model features lie sufficiently close to image features,
and that the feature orientations are similar. Typical examples of this approach for verifying hypotheses can be found in the system of Thompson and Mundy [30], in the detailed
book on recognition strategies by Grimson [9], or in the more recent work on feature
matching by Huttenlocher, *et al.* [15].

We now believe that geometric model descriptions are insufficient for recognition
tasks, and that *topology* should also be exploited. Given that we are able to extract and
store such topological descriptions, we can evaluate recognition hypotheses at both the
geometric and topological levels. In this section we briefly show how consistency of
model-image topology leads to dramatically enhanced recognition performances. The
work presented is a summary of that found in [24].

7.1   Recognizing Objects Using Invariants

In order to evaluate the effects of using topology within recognition we have had to work
with an implemented recognition system. The recognition system we use is based on
LEWIS [26] which recognizes planar objects using an indexing strategy base on a set
of planar projective invariants. LEWIS performs the following processing sequence:

1. **Edgel detection:** edgels are detected in the image using the edgel-detector reviewed
   earlier in this article.
2. **Feature fitting:** lines and ellipses are fitted to the edgel-chains.
3. **Grouping:** lines are grouped by connectivity into feature groups, and conics by
   proximity. The goal of the grouping process is to derive sets of invariant feature
   groups of the following three forms: groups of five lines; three lines and a conic;
   and pairs of conics. Each of these feature groups possesses a number of projective
   invariants which are suitable invariant descriptions for the sets of plane algebraic
   features.
4. **Indexing:** the invariants for each of the feature groups are used to *index* into index-
   spaces. The index-spaces (one for each type of invariant) are represented as hash
   tables. Should the invariants for an image feature group match those for a model,
   then a hypothesis is constructed which expresses the model-image correspondence
   formally.
5. **Formation of extended hypotheses:** the invariant feature groups provide only local descriptions of objects, and do not encompass all of the features of a model. The

distribution of such feature groups around the boundary of an object frequently leads to the formation of a number of hypotheses for a single object in a scene. Compatible hypotheses from different feature groups should be merged together prior to verification to form *extended hypotheses*.

6. **Verification:** the model-to-image transformation is computed using the corresponding model and image features from the extended hypotheses. A projective transformation must be found which maps all of the model algebraic features sufficiently closely to the image features (lines and ellipses), otherwise the hypothesis is rejected out-of-hand. The entire set of model features is then projected into the image and compared to the image data. The model features are represented by edgel-sets recovered from an acquisition image of the model object and include the complete topological description. An extended hypothesis is accepted if more than fifty percent of the model edgels are found to project to within five pixels of image edgel data of the right orientation (the orientations must differ by no more than fifteen degrees).

The original implementation of LEWIS also included additional stages of verification based on pose computation. We do not include them here because we believe that a more complete analysis of the quality of the hypotheses with regard to topology can be achieved prior to resorting to pose-based computations. The results we have found vindicate such an assumption.

Results demonstrating the effectiveness of LEWIS recognizing planar objects are given in [26].

7.2   That which Geometry Cannot Provide

Placing all of our confidence in geometry makes it, in the first instance, very difficult to reason about *negative evidence*. Suppose that we have failed to observe a match for a model feature in the image given a specific hypothesis resulting from indexing. Using geometry alone we cannot determine why this should be the case. It may be due to occlusion, bad segmentation, or the wrong choice of model in the hypothesis. For instance, consider the example of the recognition failure shown in Fig. 12: half of the circular feature of the model has not been matched, but through geometric reasoning this failure does not detract from the overall acceptance of the recognition hypothesis. What we should have realized is that some of the circular feature has been matched, but not all of it. This being the case, we should update our hypothesis to account for the inconsistency. This is, in short, an implicit resort to the use of *topological reasoning*.

Geometry can also let us down with reference to *positive evidence*. The incorrect hypothesis in Fig. 13 is accepted due to a large number of texture elements present in the image. However, most of the elements are independent at the topological level, and so should not be grouped together (they could not be explicable by a single object hypothesis).

Here we have considered only two possible failure mechanisms which have their faults rooted in the limitations of geometric reasoning. In summary, geometry does not allow us to make inferences spread over even local neighbourhoods (yet alone globally). Every single feature point, which for LEWIS means every single model edgel, has to be

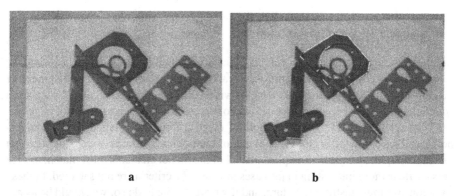

Fig. 12. A model from the model base, which has a five-line configuration similar to that of the bracket seen in the image, is hypothesized as corresponding to features of the bracket (on the strength of the single five-line invariants matching). Back projection of the model recovers a 55.0% level of image support, which is sufficient under the verification criteria to be marked as accepted (more than 50%). Projected model features marked in white found support in the image, those in back or grey did not.

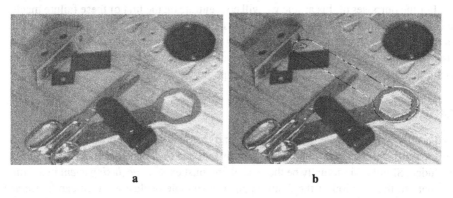

Fig. 13. In this case an invariant configuration causes the estimation of the incorrect pose for an object. This is due to the symmetry of the model (the correct pose is also recovered by another hypothesis). Under verification, and due to the presence of significant linear texture, the poor-pose hypothesis is accepted with 55.2% image support.

treated independently. What is required is that the acceptance or rejection of a match for a particular model edgel has significant bearing on the treatment of its neighbours.

7.3 Making use of Topology

Assume that we have a recognition hypothesis. Let the hypothesis propose a correspondence between a single projected model edgel and an edgel in the image (the correspondence is derived from the satisfaction of the distance and orientation constraints). From the match, we can make the following suppositions:

1. Every model edgel topologically adjacent to the matched model feature should also find image support;

2. Every image edgel topologically adjacent to the matched image feature should also find model support.

Similarly, we can infer that adjacent model features should fail to find matches when the original model edgel of the hypothesis has failed to find a correspondence. (Remember that the adjacent features may be elements within a single topological object such as an edgel-chain, or may be within distinct but connected objects such as a pair of edges.)

We have little need to worry if both of these criteria are satisfied for a given hypothesis. Due to the local-global behaviour of topological relations, we can be sure that one matching model element implies that all of the model elements must match, or vice versa. Life is more interesting in the cases in which the criteria are not satisfied. In these situations we must justify the failure, and if we are unable to do so, we should be aware that there is a problem. Briefly, the main reasons for failure are:

– Occlusions have prevented an image feature from being measurable;
– Poor segmentation has broken the image topology;
– The hypothesis is false.

For the purposes of this article we will concentrate on the first of these failure mechanisms. It is admitted that poor segmentations are common, as are false hypotheses. The causes of poor segmentation are not entirely avoided by our improved edgel detection algorithm, but it is still the case that whatever topology can be recovered is a positive factor in verifying and rejecting incorrect hypotheses.

7.4 Analysing Occlusion Events

The primary inference we make when we find that neighbouring projected model features have different levels of image support is that one of the correspondences has been occluded. Should this actually be the case, there must exist an *occlusion event* in the image near to the location of the feature pair. If a suitable occlusion event can be found in the image, then we can then enhance our confidence in the hypothesis and turn the negative evidence of the lack of a match into a form of positive evidence for an occlusion event in the right place. Consequently, we can continue to measure the quality of a hypothesis by totalling the degree of positive image support.

The occlusion events of interest in LEWIS are principally *T-junctions*. We have so far examined two ways of measuring T-junctions:

1. Examine the edgel description contained in the original gradient image; this is simply the raw output of the edgel-detector. Mark an occlusion event as existing wherever a vertex takes on the form of a T-junction. We assume that such a feature exists whenever any pair of edges entering a vertex of order of greater than or equal to three have orientations differing by 180 ± 20 degrees.

2. Fit a parametric model to the intensity surface around each hypothesized occlusion event. Appropriate model fitting techniques could be based on the algorithms of Hueckel [13], Rohr [22], or Deriche and Blaszka [7]. In fact we use the algorithm of [7] in our tests. Other suitable possibilities for modelling the structure of the junctions would be the methods of Heitger and von der Heydt [12] or Forstner [11]. If

the parametric model has a form similar to a T-junction, then one is marked as being present.

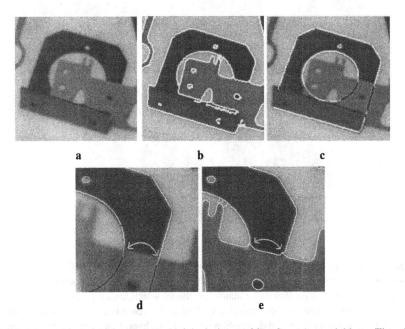

Fig. 14. In (a) we show part of an image which includes an object from the model base. The edgels found by the edgel-detector are shown in (b), and the (correct) projection of the matched model in (c). The recognition hypothesis suggests the presence of occlusions at the transition zones from white to grey of the projected model; this is shown in more detail by the arrows in (d) for two of the occlusion regions. Near to these occlusion points are triple junctions in the edgel description which have the form of a T. This is shown in (f) which is a close-up of (b), and so the hypothesis appears to be correct.

As a first example of how to exploit topology during verification, Fig. 14 shows how a set of edgel data from the edgel-detector hypothesizes the presence of T-junctions near to where occlusion events should be found on the strength of a specific recognition hypothesis. This is an example of positive support for an object hypothesis, with the measurement of the low-level junction description enhancing the acceptance of a hypothesis. Similarly, in Fig. 15 we demonstrate an example of negative evidence which renders a hypothesis unlikely (or perhaps may even cause it to be rejected outright). The projected model curves are shown initially following a contour in the image, but at one stage (due to an incorrect model being hypothesized), the projected model curve is far from the corresponding image curve. The edgel-detector fails to find any junctions near the point of departure and so one can start to doubt whether the original hypothesis is correct. Given the knowledge that a junction is, or is not present, we can update the score attributed to the hypothesis and then provide a more precise evaluation of the image.

An alternative approach evaluates the presence of T-junctions using parametric junc-

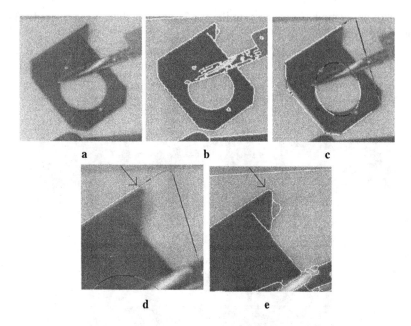

a   b   c

d   e

Fig. 15. For the image section in (a) we have computed the edgels shown in (b). From these and a subsequent fitting process we compute invariants to form the incorrect hypothesis shown in (c). Indexing has incorrectly hypothesized a match to a model which finds significant image support. Near the white-grey transition of the projected outline in (d), which is a close-up of (c), we would hope to find a junction in the edgel description (that is if the hypothesis were correct). However, as shown in (f), the failure of the edgel-detector to record a junction in the right place suggests that the hypothesis is false.

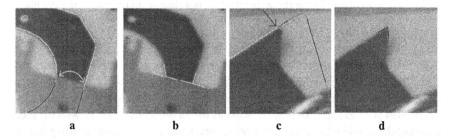

a   b   c   d

Fig. 16. In (a) the close-up of the correct hypothesis shown in Fig. 14d is depicted, and the output of the Deriche-Blaszka junction detector is shown in (b). The left-most T-junction possesses an angle of 179.3 degrees between the straight part of the T, and the right-hand junction 175.8 degrees. Both of these are sufficiently close to 180 degrees, and so can be accepted as T-junctions. However, the Deriche-Blaszka filter fails to find a suitable T-junction near the potential occlusion event already discussed in Fig. 15. Instead it finds an L-junction which is shown in (d). Such a feature has no relation to an occlusion event, and so the original hypothesis is likely to be false.

tion model fitting methods applied to the intensity surface. In Fig. 16 we show similar examples to those for the edgel-junction reasoning, but this time the presence of T-junctions is evaluated using the Deriche-Blaszka operator [7]. In the first case a pair of suitable T-junctions are found and shown superimposed in the figure, and in the second case no such fit could be found and so the hypothesis is marked as being unlikely. Thus we see that parametric model fitting can enhance the understanding of recognition hypotheses when directed by prior topological reasoning.

These examples provide a first level of attack on the frailties of hypothesis verification strategies based solely on geometric methods. There exist also a number of other topological relationships which can be used during verification, for instance:

- Any pair of image topology objects which have been matched to the same projected model object should be connected. Exception to this ruling is only allowed on account of poor segmentation. Otherwise, the hypothesis must be incorrect.
- Every model feature should recover significant support from the image. Similarly, every matched image feature should also achieve good support from a single model feature. Thus, if only a small percentage of an image feature is matched, we can suppose that the correspondence results from chance rather than by way of a correct hypothesis.
- There should be a uniqueness in the mapping from image correspondences to model features. A single image feature cannot reasonably explain two model primitives simultaneously. In LEWIS, uniqueness was not enforced because of the fragility of the verification strategy. Now we believe that sufficiently high confidences in recognition hypotheses can be recovered that multiple interpretations should be suppressed.

7.5 Re-Evaluating Hypotheses

Given that we can use topological information to turn negative evidence (principally from occlusions) into position evidence, we can update the verification scores of hypotheses and subsequently reorder our confidences in each hypothesis. In fact, we find that re-evaluated good hypotheses tend to recover scores in excess of the 90% confidence level in the projections of their outlines, whereas before LEWIS had to be content with anything above 50%. Raising our acceptance thresholds means that we can dramatically improve our chances of the correct classification of images. One example of the change in confidence levels is shown in Fig. 17. The original hypothesis is shown in (a) with the model edgels recovering image support marked in white, and those having unsuitable support in grey or black. Originally, LEWIS only marked 70.5% of the projected model was marked as being white, though this being more than 50% meant that the hypothesis was accepted. After occlusion analysis we can explain far more of the outline: Fig. 17b shows in white all of the model elements which either found direct image support, or were suitably explained by occlusion events. The new verification score (percentage of white pixels) is 93.6%, representing a very high confidence in the validity of the hypothesis.

A final example (more examples are given in [24]) is shown in Fig. 18. Amongst the hypotheses which LEWIS recovers are two accounting for the same set of image features: the first one shown in (b) is the correct one with 70.7% image support; and the

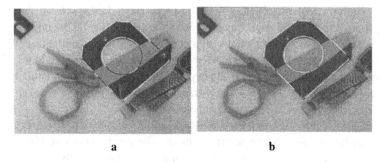

a   b

Fig. 17. The original verification algorithm when applied to the scene produced only a 70.5% score for the hypothesis shown in (a). However, after the prediction and verification of the various T-junctions bounding the invisible part of the hypothesized object, we increase the verification score to 93.6%. Edgels which were previously unmatched, but are now marked as positively occluded, are depicted in white in (b). The overall verification score is incremented by the number of edgels in the positively identified occluded regions.

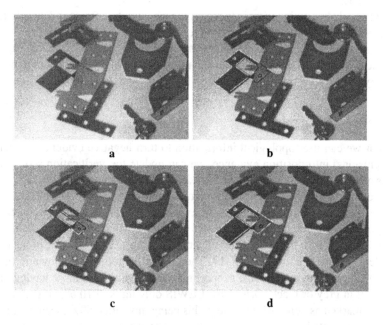

Fig. 18. LEWIS recognizes the lock-striker plate correctly in (b) by finding 70.7% image support. However, it also supposes an incorrect hypothesis with a score of 68.2% in (c). After including topological analysis the hypothesis in (b) is transformed to that in (d) with 91.3% support (the hypothesis in (c) remains unchanged). Such a high level of support shows a strong preference for the correct hypothesis and thus vindicates the use of topological reasoning.

second one is depicted in (c) as having 68.2% support. As these hypotheses have such close confidence levels one cannot realistically hope to distinguish between them. However, once we have employed the various levels of topological processing, the hypothesis of Fig. 18b has its support level raised to 91.3%, and yet that of (c) is unchanged. Consequently we can be sure which hypothesis is really correct and use the requirement for a unique partition of image features to eliminate the weaker hypothesis.

8   Conclusions

Our main goal in presenting this paper is to re-awaken a dormant interest in the use of topological representations in vision. Some of the earliest work in the field concentrated on the understanding of junction topology, however in the last decade there has been a predominant use of only geometrical representations for tasks such as object recognition. Of course, both the topological and geometrical descriptions have their foundations in the theory of projections, but we believe that their properties should be married together in order to gain the richest object representation scheme possible.

One reason for the original rejection of a topological representation was due to its fragility under early visual processing. However, we have demonstrated in this paper that much ground can be gained by realising why standard techniques of feature extraction fail to recover correct topological descriptions. Of course, these are early days, however we believe that the techniques we have employed to produce a topologically superior edgel-detector will most probably extend to other levels of processing.

Given that we can extract topological representations from images we have demonstrated how three specific tasks can be simplified when driven by topological considerations. These three are only a small sample of the many problems in vision which could be tackled via topological reasoning. The first task we considered was grouping, a process which runs very slowly in complex scenes when we rely on only geometric constraints. However, given the connectivity of features as an input, grouping can become more of a top-down process than bottom-up. The second process we considered was the extraction of polyhedra from images. Using only the topological description of polyhedra were are able to derive a representation suitable for recognition which provides a ready framework for the computation of shape cues such as geometric invariants. Finally, we have shown how recognition hypothesis verification can be made more robust when we consider topological measures in addition to the traditional geometric ones. In brief, the use of connectivity relationships allow us to broaden the scope of any local geometric consideration to the encompassing of object descriptions as a whole.

Acknowledgments

We would like to thank Thierry Blaszka for allowing us to use his implementation of the Deriche-Blaszka junction filter. CAR is funded by a Human Capital and Mobility grant from the European Community, whilst GE receives funding under DARPA contract MDA972-91-C-0053.

References

1. N. Ayache and O. Faugeras. HYPER: A New Approach for the Recognition and Positioning of Two-Dimensional Objects. *IEEE Transactions on Pattern Analysis and Machine Intelligence*, 8(1):44–54, January 1986.

2. T. Binford and T. Levitt. Quasi-invariants: Theory and explanation. In *Proceedings of the ARPA Image Understanding Workshop*, pages 819–829, Washington, DC, April 1993.

3. T. Binford. Inferring surfaces from images. *Artificial Intelligence Journal*, (Special Edition on Computer Vision) 17:205–244, 1981.

4. J. Canny. Finding edges and lines in images. Technical Report AI-TR-720, Massachusets Institute of Technology, Artificial Intelligence Laboratory, June 1983.

5. M. Clowes. On seeing things. *Artificial Intelligence Journal*, 2:79–116, 1971.

6. I. Cox, J. Rehg, and S. Hingorani. A bayesian multiple hypothesis approach to contour grouping. In G. Sandini, editor, *Proceedings of the 2nd European Conference on Computer Vision*, volume 588 of *Lecture Notes in Computer Science*, pages 72–77, Santa Margherita Ligure, Italy, May 1992. Springer-Verlag.

7. R. Deriche and T. Blaszka. Recovering and characterizing image features using an efficient model based approach. In *Proceedings of the International Conference on Computer Vision and Pattern Recognition*, pages 530–535, New-York, June 1993.

8. J. Dickson. *Image Structure and Model-Based Vision*. PhD thesis, Department of Engineering Science, Oxford University, Oxford, UK, 1991.

9. W. E. L. Grimson. *Object Recognition by Computer: The Role of Geometric Constraints*. The MIT Press, Cambridge, Massachusetts, London, England, 1990.

10. A. Guzman. Decomposition of a visual scene into three-dimensional bodies. In *FJCC*, volume 33, pages 291–304, 1968.

11. R. Haralick and L. Shapiro. *Computer and Robot Vision*, volume 1. Addison-Wesley, 1992.

12. F. Heitger and R. von der Heydt. A computational model of neural contour processing: Figure-ground segregation and illusory contours. In *Proceedings of the 4th Proc. International Conference on Computer Vision*, pages 32–40, Berlin, Germany, May 1993.

13. M. Hueckel. A local visual operator which recognizes edges and lines. *Journal of the ACM*, 20(4):634–647, 1973.

14. D. Huffman. Impossible objects as nonsense sentences. In B. Meltzer and D. Michie, editors, *Machine Intelligence*, volume 6, Edinburgh University Press, 1971.

15. D. Huttenlocher, J. Noh, and W. Rucklidge. Tracking non-rigid objects in complex scenes. In *Proceedings of the 4th Proc. International Conference on Computer Vision*, pages 93–101, Berlin, Germany, May 1993.

16. T. Kanade. Recovery of the three-dimensional shape of an object from a single view. *Artificial Intelligence Journal*, 17:409–460, 1981.

17. D. Lowe. *Perceptual Organization and Visual Recognition*. Kluwer Academic Publishers, 1985.

18. J. Malik. Interpreting Line Drawings of Curved Objects. *The International Journal of Computer Vision*, 1(1):73–103, 1987.

19. L. Morin. *Quelque Contributions des Invariants Projectifs à la Vision par Ordinateur*. PhD thesis, Institute National Polytechnique de Grenoble, LIFIA-IMAG-INRIA Rhone-Alpes, January 1993.

20. J. Mundy and A. Heller. The evolution and testing of a model-based object recognition system. In *Proceedings of the 3rd Proc. International Conference on Computer Vision*, pages 268–282, Osaka, Japan, December 1990.

21. L. Robert and O. Faugeras. Relative 3-D positioning and 3-D convex hull computation from a weakly calibrated stereo pair. *Image and Vision Computing*, 13(3):189–197, 1995.

22. K. Rohr. Recognizing corners by fitting parametric models. *The International Journal of Computer Vision*, 9(3):213–230, 1992.

23. C. Rothwell, J. Mundy, W. Hoffman, and V.-D. Nguyen. Driving vision by topology. In *IEEE International Symposium on Computer Vision*, pages 395–400, November 1995.

24. C. Rothwell. Reasoning about occlusions during hypothesis verification. In Bernard Buxton, editor, *Proceedings of the 4th European Conference on Computer Vision*, Vol. 1, pages 599–609, Cambridge, UK, April 1996.

25. C. Rothwell and J. Stern. Understanding the shape properties of trihedral polyhedra. In Bernard Buxton, editor, *Proceedings of the 4th European Conference on Computer Vision*, Vol. 1, pages 175–185, Cambridge, UK, April 1996.

26. C. Rothwell. *Object recognition through invariant indexing*. Oxford University Science Publications. Oxford University Press, February 1995.

27. S. Sarkar and K. Boyer. Perceptual organization in computer vision: A review and a proposal for a classificatory structure. *IEEE Transactions on Systems, Man, and Cybernetics*, 23:382–399, 1993.

28. R. Shapira and H. Freeman. Computer description of bodies bounded by quadric surfaces from sets of imperfect projections. *IEEE Transactions on Computers*, pages 841–854, 1978.

29. K. Sugihara. *Machine Interpretation of Line Drawings*. MIT Press, 1986.

30. D. Thompson and J. Mundy. Three-dimensional model matching from an unconstrained viewpoint. In *Proceedings of the International Conference on Robotics and Automation*, Raleigh, pages 208–220, 1987.

31. Y. Tsao and K. Fu. Parallel thinning operations for digital binary images. In *Proceedings of the International Conference on Computer Vision and Pattern Recognition*, pages 150–155. 1981.

32. D. Waltz. Understanding line drawings of scenes with shadows. In Patrick H. Winston, editor, *The Psychology of Computer Vision*, pages 19–91. McGraw-Hill, 1975.

33. H. Whitney. Singularities of mappings of Euclidean spaces. I: Mappings of the plane into the plane. *Ann. Math*, 62:374–41, 1955.

34. M. Zerroug and R. Nevatia. Three dimensional part-based descriptions from a real intensity image. In *Proceedings of the ARPA Image Understanding Workshop*, pages 1367–1374. 1994.

35. A. Zisserman, D. Forsyth, J. Mundy, C. Rothwell, J. Liu, and N. Pillow. 3D object recognition using invariance. *Artificial Intelligence Journal*, 78:239–288, 1995.

36. A. Zisserman, J. Mundy, D. Forsyth, J. Liu, N. Pillow, C. Rothwell, and S. Utcke. Class-based grouping in perspective images. In *Proceedings of the 5th Proc. International Conference on Computer Vision*, pages 183–188, Boston, June 1995.

Curvature Based Signatures
for Object Description and Recognition

Elli Angelopoulou, James P. Williams, Lawrence B. Wolff

Computer Vision Laboratory, Department of Computer Science,
The Johns Hopkins University, Baltimore, MD 21218, USA
e-mail: {angelop, jimbo, wolff}@cs.jhu.edu

Abstract. An invariant related to Gaussian curvature at an object point is developed based upon the covariance matrix of photometric values related to surface normals within a local neighborhood about the point. We employ three illumination conditions, two of which are completely unknown. We never need to explicitly know the surface normal at a point. The determinant of the covariance matrix of these three-tuples in the local neighborhood of an object point is shown to be invariant with respect to rotation and translation. A way of combining these determinants to form a signature distribution is formulated that is rotation, translation, and, scale invariant. This signature is shown to be invariant over large ranges of poses of the same objects, while being significantly different between distinctly shaped objects. A new object recognition methodology is proposed by compiling signatures for only a few poses of a given object.

1 Introduction

The recognition of three-dimensional objects using two-dimensional images and the efficient representation of the shape information are of fundamental importance in computer vision and robotics and have a wide variety of applications. A typical approach is to extract a set of high-level features from input images and associate these features with the geometry of the objects in the scene. Such features can vary from simple primitives like 2-D points [15], junctions [32] and 2-D curves [12] to complex structures like deformable models [30, 31], geons [3], generalized cylinders [4, 5, 17] and superquadrics [25, 31].

There is a trade-off between the simplicity of the primitives and the overall complexity of the description/recognition process [6]. A single complex primitive conveys more information about an object than a corresponding simple one. As a result, the simpler the feature, the greater the number of primitives that must be used to

This research was supported in part by ARPA contract F30602-92-C-0191, ARPA grant DAAH04-94-G-0278, AFOSR grant F49620-93-1-0484 and the NSF Young Investigator Award IRI-9357757.

describe the object. This means that more features have to be used in the object descriptors making the object database larger. Also, simple primitives are typically common features that occur in a variety of objects. In order to differentiate among such objects, additional geometric information has to incorporated, adding complexity to the object model. Because simple primitives occur so frequently, multiple matches are not uncommon when attempting to identify an object. The process of eliminating all but one of these matches, typically involves the precise determination of the orientation of the object and its features. This recovery of an object's exact pose is computationaly complex.

On the other hand, complicated primitives place the burden on the process of extracting the features themselves from the images. A small group of such features gives a very precise model of an object. Although fewer primitives need to be stored, comparing such complex features is commensurately difficult. There is also a tendency for these models to be so geometrically exact that they might not allow for minor deformations of the object. Parameterized models [8, 16] provide for this flexibility at the expense of complicating the recognition process by introducing additional complexity in the primitives. Still, some of these models are not powerful enough to encompass fine surface details in their description of the object and its features.

Our method bridges this gap between ease of image-feature extraction and simplicity in the actual object description and identification process, while keeping the object-database size small. Like photometric stereo [33] we use multiple images of the same scene taken under three different illumination conditions. We hold the viewpoint constant under the three illumination conditions. We assume Lambertian reflection. We do not need to know the exact location of the light sources. The only constraint is that one of the light sources be aligned with the optic axis of the camera. This is necessary in order to compensate for the illusion that a uniformly-curved surface bends more when it is foreshortened under orthographic projection. There is no restriction on the placement of the other two light sources other than that they should not be coplanar with the optic axis.

All our computations are performed in photometric space. Our primitive, which is the determinant of the *photometric* covariance matrix, is directly related to the Gaussian curvature of a surface and it is invariant under rigid transformations. We never recover the surface normals themselves, nor do we explicitly compute surface curvature. Our object descriptor condenses the extracted local curvature information in a sequence of at most 20 numbers. This signature is shown to be invariant to pose and scale unless significant new features appear from one pose to the other.

2   A Curvature Based Invariant

The main difference among various smooth objects is their local surface curvature. The light reflectance properties of diffuse objects are directly related to surface orientation [10]. One can extract surface normal orientation and local curvature for such objects through the use of multiple images of the same scene taken from the same viewpoint but under different known, precisely-calibrated illumination conditions (see [11, 13, 14, 33, and 34]). This information can be used for object identification [13]. However, like other recognition systems, this method involves a complex feature extraction process, and relies on the recovery of the normal field of the scene.

We show that multiple images of Lambertian surfaces taken under three unknown but fixed lighting conditions provide enough surface-curvature information to allow for efficient object recognition. Our image acquisition process does not require knowledge of the exact location of the light-source vectors. We compute an invariant based on curvature without ever recovering the surface normals themselves. The computation of this invariant is based on simple image-space (2D) calculations. We never need to solve for any three-dimensional object representation.

2.1 Multiple Illuminations

Lambertian reflection is a good approximation to the way smooth and mildly-rough diffuse surfaces reflect light. According to Lambert's law the observed surface brightness I is proportional to the angle θ between the direction of the incident light-source vector s and the surface normal n. In more detail, $I = L\rho\cos\theta$, where L is the intensity of the point light source and ρ is the albedo, a material dependent coefficient.

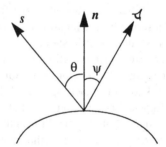

Fig. 1. Diffuse Reflection

We choose to use three non-coplanar light sources. Thus, for each mutually illuminated pixel we have a triplet of intensity values. We treat a 3-tuple of intensity values (I_1, I_2, I_3) as a point in a three-dimensional photometric space. We can assume that both the surface normals and the light-source vectors are of unit length. For each pixel we have three equations of the form:

$$I_i = L_i\rho<s^i, n>$$

where n is the column vector (n_x, n_y, n_z) and s^i is the light-source vector $s^i = (s^i_x, s^i_y, s^i_z)$ for $i = 1, 2, 3$. Thus,

$$I_i = L_i\rho\,(n_x s^i_x + n_y s^i_y + n_z s^i_z) = \rho\,(n_x L_i s^i_x + n_y L_i s^i_y + n_z L_i s^i_z)$$

for $i = 1, 2, 3$. We can write these equations in a matrix form as:

$$I = \rho Sn \tag{1}$$

where I is the column vector (I_1, I_2, I_3), n is the column vector (n_x, n_y, n_z), ρ is the surface albedo, and S is the light source matrix which incorporates the light intensity L_i, for $i = 1, 2, 3$:

$$S = \begin{bmatrix} L_1 s_x^1 & L_1 s_y^1 & L_1 s_z^1 \\ L_2 s_x^2 & L_2 s_y^2 & L_2 s_z^2 \\ L_3 s_x^3 & L_3 s_y^3 & L_3 s_z^3 \end{bmatrix}$$

2.2 DeCov, the Determinant of the Photometric Covariance Matrix

Covariance techniques have been used in object recognition, especially in pose estimation [10, 13]. For rigid objects, the eigenvalues of the covariance matrix of the object's surface normals are invariant under rotation, while its eigenvectors provide the pose of the object. However, as we do not know the incident orientation of the light sources, we do not have enough information to recover the surface normals. We use the covariance matrix of the three surface intensity values (hence *photometric* covariance matrix.) We make our computations in photometric space, where the three axes represent reflectance from the three illumination conditions respectively. Unlike the visible surface normals which are spread in four octants, the intensity values are always positive, i.e. photometric space is one-octant space. This means that four-octant information is compressed in one octant. Rigid transformations on the surface normals result in non-rigid transformations in photometric space. Thus, for the photometric covariance matrix the eigenvalues are not invariant and the eigenvectors do not convey pose information.

DeCov and Curvature. Consider a small surface patch P around a point q and its three sets of intensity values which are created when P is illuminated under three different lighting conditions. The *photometric* covariance matrix C at point q of these intensity values for that patch P provides information about the local surface curvature. When the covariances are taken with respect to the origin, the covariance matrix is:

$$C = \begin{bmatrix} \sum_{p \in P} I_1(p)^2 & \sum_{p \in P} I_1(p) I_2(p) & \sum_{p \in P} I_1(p) I_3(p) \\ \sum_{p \in P} I_1(p) I_2(p) & \sum_{p \in P} I_2(p)^2 & \sum_{p \in P} I_2(p) I_3(p) \\ \sum_{p \in P} I_1(p) I_3(p) & \sum_{p \in P} I_2(p) I_3(p) & \sum_{p \in P} I_3(p)^2 \end{bmatrix}$$

where $I_i(p)$ is the observed intensity value at each point p in the surface patch P, for illumination conditions $i = 1, 2, 3$.

In a more compact way:

$$C = \sum_{p \in P} \left(\begin{bmatrix} I_1(p) \\ I_2(p) \\ I_3(p) \end{bmatrix} \begin{bmatrix} I_1(p) & I_2(p) & I_3(p) \end{bmatrix} \right) = \sum_{p \in P} (I_p (I_p)^T)$$

where I_p is the intensity vector (I_1, I_2, I_3) at point p.
Using equation (1) we can write C as:

$$C = \sum_{p \in P} (\rho_p S n_p (\rho_p S n_p)^T) = S \left(\sum_{p \in P} (\rho_p^2 n_p n_p^T) \right) S^T$$

We define $DeCov$ to be the Determinant of the photometric Covariance matrix C. Matrix C is a positive definite real symmetric matrix, so its determinant $DeCov$ is always positive, $DeCov \geq 0$. For an object of varying albedo, $DeCov$ becomes:

$$DeCov = |C| = \left| S \left(\sum_{p \in P} (\rho_p^2 n_p n_p^T) \right) S^T \right| = |S|^2 \left| \left(\sum_{p \in P} (\rho_p^2 n_p n_p^T) \right) \right| \qquad (2)$$

While, for an object of uniform albedo $DeCov$ is:

$$DeCov = |C| = \left| \rho^2 S \left(\sum_{p \in P} (n_p n_p^T) \right) S^T \right| = \rho^6 |S|^2 \left| \left(\sum_{p \in P} (n_p n_p^T) \right) \right| \qquad (3)$$

Let C' be the covariance matrix of the surface normals of the surface patch P:

$$C' = \sum_{p \in P} (n_p n_p^T)$$

Which means:

$$C' = \begin{bmatrix} \sum_{p \in P} n_x(p)^2 & \sum_{p \in P} n_x(p) n_y(p) & \sum_{p \in P} n_x(p) n_z(p) \\ \sum_{p \in P} n_x(p) n_y(p) & \sum_{p \in P} n_y(p)^2 & \sum_{p \in P} n_y(p) n_z(p) \\ \sum_{p \in P} n_x(p) n_z(p) & \sum_{p \in P} n_y(p) n_z(p) & \sum_{p \in P} n_z(p)^2 \end{bmatrix} \qquad (4)$$

We can then rewrite equation (3) for the $DeCov$ of uniform albedo objects as:

$$DeCov = \rho^6 |S|^2 |C'| \qquad (5)$$

Given a small surface patch $X(x_1, x_2)$ one can use a first order approximation to the Taylor expansion to express the unit surface normal function at some point of that surface as:

$$n(x_1 + u, x_2 + v) \approx n(x_1, x_2) + u\frac{\partial}{\partial x_1}n(x_1, x_2) + v\frac{\partial}{\partial x_2}n(x_1, x_2)$$

Furthermore, the partial derivatives of the surface normals can be expressed in terms of the partial derivatives of the surface itself, through the Weingarten's equations:

$$\frac{\partial n}{\partial x_i} = \sum_j L_j^i \frac{\partial X}{\partial x_i} \tag{6}$$

It is possible to obtain the directions of principal curvature (v_1, v_2) and the principal curvatures (k_1, k_2) without knowing the parameterization of the surface. Once the directions of principal curvature are known, one can parameterize the surface using principal curvature parameterization. In other words, the orthonormal basis $\{e_1, e_2, e_3\} = \{(1, 0, 0), (0, 1, 0), (0, 0, 1)\}$ is aligned with $\{v_1, v_2, n\}$. When the surface is parameterized this way, the Weingarten matrix W is diagonalized and the diagonal elements are the principal curvatures:

$$W = \begin{bmatrix} L_1^1 & L_2^1 \\ L_1^2 & L_2^2 \end{bmatrix} = \begin{bmatrix} k_1 & 0 \\ 0 & k_2 \end{bmatrix}$$

Weingarten's equations (6) are simplified to:

$$\frac{\partial n}{\partial x_i} = k_i \frac{\partial X}{\partial x_i}$$

Thus, one can rewrite the linear approximation of the surface normal function as:

$$n(x_1 + u, x_2 + v) = uk_1\frac{\partial X}{\partial x_1} + vk_2\frac{\partial X}{\partial x_2} + n(x_1, x_2) = uk_1v_1 + vk_2v_2 + n(x_1, x_2)$$

When we are computing the covariance matrix of the surface normals C' in a small neighborhood P we can, without loss of generality, make the following two assumptions: a) the surface patch P is parameterized with respect to the directions of principal curvature; and b) the center of the local coordinate system is at point q, the center of that patch P. Then for each point in that patch, the unit surface normal function is given by:

$$n(u, v) = uk_1v_1 + vk_2v_2 + n(0, 0) = uk_1e_1 + vk_2e_2 + e_3 \tag{7}$$

The unit surface normal function written as a tuple is $n(u, v) = (uk_1, vk_2, 1)$. Thus, the covariance matrix of the surface normals, equation (4), over a small patch $P = \{(u, v) : -i \leq u \leq i, -j \leq v \leq j\}$ about a point q becomes:

$$C' = \begin{bmatrix} \sum\limits_{u,v} u^2 k_1^2 & \sum\limits_{u,v} uv k_1 k_2 & \sum\limits_{u,v} uk_1 \\ \sum\limits_{u,v} uv k_1 k_2 & \sum\limits_{u,v} v^2 k_2^2 & \sum\limits_{u,v} vk_2 \\ \sum\limits_{u,v} uk_1 & \sum\limits_{u,v} vk_2 & 1 \end{bmatrix} = \begin{bmatrix} k_1^2 \sum\limits_{u,v} u^2 & k_1 k_2 \sum\limits_{u,v} uv & k_1 \sum\limits_{u,v} u \\ k_1 k_2 \sum\limits_{u,v} uv & k_2^2 \sum\limits_{u,v} v^2 & k_2 \sum\limits_{u,v} v \\ k_1 \sum\limits_{u,v} u & k_2 \sum\limits_{u,v} v & 1 \end{bmatrix}$$

Since we are summing symmetrically around the origin of the local coordinate system, all the off-diagonal summations equal zero. When we are computing the covariance matrix over neighborhoods of the same size, the remaining summations are constants whose value is determined by the size of the neighborhood we use in the covariance matrix.

Let $a = \sum\limits_{u,v} u^2 = \sum\limits_{u=-i}^{i} u^2$ and $b = \sum\limits_{u,v} v^2 = \sum\limits_{v=-j}^{j} v^2$. Then C' becomes

$$C' = \begin{bmatrix} k_1^2 \sum\limits_{u,v} u^2 & 0 & 0 \\ 0 & k_2^2 \sum\limits_{u,v} v^2 & 0 \\ 0 & 0 & 1 \end{bmatrix} = \begin{bmatrix} k_1^2 a & 0 & 0 \\ 0 & k_2^2 b & 0 \\ 0 & 0 & 1 \end{bmatrix} \tag{8}$$

The determinant of the covariance matrix of the surface normals is:

$$|C'| = k_1^2 k_2^2 ab$$

Gaussian curvature K is defined as the product of the principal curvatures, $K = k_1 k_2$ Thus, the determinant of the covariance matrix of the surface normal is a multiplicative of the square of the Gaussian curvature at that neighborhood.

$$|C'| = cK^2 \tag{9}$$

where $c = ab$ is a constant and it is related to the area of the neighborhood over which the matrix is computed. Respectively, *DeCov*, the determinant of the photometric covariance matrix is a one-to-one function of the square of the Gaussian curvature K^2 at that surface patch:

$$DeCov = \rho^6 |S|^2 |C'| = c\rho^6 |S|^2 K^2 = \beta K^2 \tag{10}$$

where $\beta = c\rho^6 |S|^2$. Even for surfaces of varying albedo, *DeCov* is related to the square of the Gaussian curvature K^2. The difference is that, for such surfaces, *DeCov* incorporates albedo information to the Gaussian curvature measurement. For instance, a Coke-can and a Pepsi-can will give distinct *DeCov* values.

DeCov as an Invariant. *DeCov* is directly related to curvature and surface curvature does not change for rigid objects. As expected *DeCov* is invariant under rigid transformations (rotation, translation) with one noted exception: translation along the optic axis scales the projection of the object on the image plane. Translations parallel to the image plane, which do not involve any scaling of the object's projection on the image plane do not change the intensity images for distant point light sources. This means that the photometric covariance matrix and consequently its determinant does not change.

When an object is rotated, the angles between the surface normals and the light-source direction vectors change, resulting in different brightness values. The photometric covariance matrix for the rotated object becomes:

$$C_R = \sum_{p \in P} (\rho_p SRn_p (\rho_p SRn_p)^T) = SR (\sum_{p \in P} (\rho_p^2 n_p n_p^T)) R^T S^T$$

where R is the applied rotation matrix. However, since $|R| = |R^T| = 1$, *DeCov* remains invariant. For objects of varying albedo we get:

$$DeCov_R = |C_R| = \left| SR (\sum_{p \in P} (\rho_p^2 n_p n_p^T)) R^T S^T \right| = DeCov$$

Similarly, for objects of uniform albedo *DeCov* becomes:

$$DeCov_R = |C_R| = \left| \rho^2 SR (\sum_{p \in P} (n_p n_p^T)) R^T S^T \right| = DeCov$$

The albedo ρ, the light source matrix S, and the Gaussian curvature K are unaffected by scaling. Let the *DeCov* value of the unscaled pose be:

$$DeCov_1 = c_1 \rho^6 |S|^2 K^2$$

where c_1 is related to the area of the unscaled surface patch P. Let the *DeCov* value of the scaled pose be:

$$DeCov_2 = c_2 \rho^6 |S|^2 K^2$$

where c_2 is related to the area of the scaled patch P. All the *DeCov* values are uniformly scaled. The *DeCov* values of the two poses can be placed in the same frame of reference by multiplying each $DeCov_2$ value by the factor (c_1/c_2) :

$$\left(\frac{c_1}{c_2} \right) DeCov_2 = \left(\frac{c_1}{c_2} \right) c_2 \rho^6 |S|^2 K^2 = DeCov_1 \qquad (11)$$

The factor (c_1/c_2) can be computed by dividing the sum of all the $DeCov_1$ values by the sum of all the $DeCov_2$ values:

$$\sum_{\forall q} DeCov_1 = c_1 \rho^6 |S|^2 \sum_{\forall q} K^2 \qquad \sum_{\forall q} DeCov_2 = c_2 \rho^6 |S|^2 \sum_{\forall q} K^2$$

$$\sum_{\forall q} DeCov_1 / \sum_{\forall q} DeCov_2 = c_1/c_2$$

Foreshortening Correction. *DeCov* should be computed over surface patches of uniform area. Unfortunately, our input data is given in terms of the projection of the object on the image plane and not in terms of the actual surface. Although all the pixels in the image have the same area, due to foreshortening, the areas of the corresponding patches on the surface of the object are not the same. The apparent area of a surface patch is less than its actual area, unless the viewing direction is parallel to its surface normal. This phenomenon affects *DeCov* directly. Since *DeCov* is proportional to the square of Gaussian curvature, $DeCov \propto K^2$, the larger the surface area over which we compute *DeCov*, the higher its value. This means that even for objects of uniform curvature, like a sphere, we will be getting higher *DeCov* values at surface patches that are oblique to the camera.

In order to be consistent and compute *DeCov* over surface patches that have the same area, we have to correct for the foreshortening effect. The amount of foreshortening depends on the angle ψ between the viewer and the surface normal (see Fig. 1). If we locate one of the light sources, for example I_1, where the viewer is, then the angle θ_1 between the light source and the surface normal is equal to the angle ψ. According to Lambertian reflection, we have $I_1 = k\cos\theta_1 = k\cos\psi$, where $k = \rho L_1$, and ρ and L_1 are the surface albedo and the intensity of the light source respectively. For objects of uniform albedo, we use a calibration plane that is made of the same material as the object of interest and is oriented so that its surface normal is parallel to the viewing axis ($\psi = 0°$). The intensity value that we get for that plane is $I_1 = k$. We can then recover the amount of foreshortening by dividing the value of I_1 at each pixel by the computed k. For images of varying albedo we can still use the amount $1/I_1$ as a measure of the amount of foreshortening, but in this case our correcting factor is scaled by $1/k$. Once the cosine of the angle ψ (or a multiplicative of it) is recovered, the $m \times m$ neighborhood of pixels around the point q over which *DeCov* is computed is scaled, so that the area of the surface projected into those pixels is constant over the whole image. The following figure shows a pair of *DeCov* error images of a sphere. The brighter the value, the higher the error. The sphere on the left shows the *DeCov* errors caused by foreshortening. On the right is the same sphere after the foreshortening correction is applied.

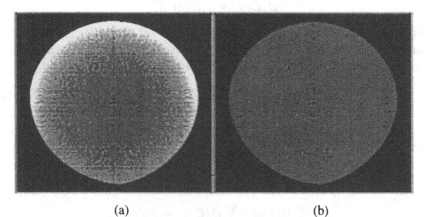

(a)   (b)

Fig. 2. Sphere (a) without and (b) with foreshortening correction

3 Curvature Based Signatures

Our data is a distribution of a large number of *DeCov* values computed over uniform-size surface patches. We group all the *DeCov* values into *G* quantiles. Each quantile contains N/G *DeCov* values, where *N* is the total number of surface points at which *DeCov* is computed. Notice that *N* is *not* the total number of mutually illuminated pixels. *DeCov* is computed over surface patches of uniform area, and it is computed for each minuscule surface patch that corresponds to an unforeshortened pixel. However, due to foreshortening, a pixel does not always correspond to a surface patch of the same area. For example, the *DeCov* value computed at a pixel that corresponds to a surface patch oriented at $\psi = 60°$ to the viewer, maps to $1/\cos\psi = 2$ uniform-sized surface patches, i.e. 2 surface points.

By definition, each quantile *i* holds N/G *DeCov* values which are higher than those stored in quantile *i*-1, and lower than any other *DeCov* value stored in any quantile bigger than *i*. For each quantile *i* we compute the median *DeCov* value m_i and use that as the descriptor for the quantile. We prefer to use median values over the more traditional average values, because medians are less sensitive to outliers. The list $m_1,..,m_G$ of the *G* median values, together with the projected area of the object on the image plane becomes the object's *signature*. The projected area is needed in order to test for possible scaling of the object. As explained in section 2.2, we can use this piece of information to place the two *DeCov* distributions in a common frame of reference. Since each quantile *i* holds a group of *DeCovs* whose values are larger than the values of the *DeCovs* stored in quantile *i*-1, $m_i < m_j$ for $i < j$. Thus, the signature graph, which plots these medians, is always monotonically increasing, but the slope of the graph differs from one object to the other. The measurement used in identifying an object is simply the distance between the medians of corresponding quantiles $|m_{ij} - m_{ik}|$, where *j* and *k* are the two signatures being compared and *i* is the quantile whose medians we are currently comparing.

Grouping the *DeCov* values into quantiles provides for a consistent clustering, which is independent of scaling. By definition, quantiles partition a data set into equally sized groups of data, i.e. each quantile holds the same proportion of data. This is a classification technique which is independent of the size of the data set. In our case, quantiles provide a description that is independent of the number of surface points that constitute the object. Thus, the use of quantiles offers scale independence. Consider an object *O* with total surface area *U* under two different poses. Pose1 is an arbitrary pose and pose2 differs from pose1 by a scaling factor of $1/s$. Assume that each pixel in pose1 maps on an unforeshortened surface patch of area A_1 and that *DeCov* is computed over a surface patch of area A_p. The total number of surface points at which *DeCov* is computed is:

$$N_1 = U/A_1$$

In pose2 each pixel maps on an unforeshortened surface patch of area $A_2 = s^2 A_1$. In this case *DeCov* is computed over a surface patch of area $s^2 A_p$. The total number of surface points for which *DeCov* is computed for pose2 is:

$$N_2 = U/A_2 = U/(s^2 A_1) = N_1/s^2$$

By definition of our signature, each quantile of pose1 holds N_1/G *DeCov* values depicting information for a total surface area of:

$$A_1 N_1 / G$$

Each quantile of pose2 holds N_2/G *DeCov* values for a total surface area of:

$$A_2 N_2 / G = (s^2 A_1)\,(N_1 / s^2)\,/G = A_1 N_1 / G$$

Thus, the total number of *DeCov* values stored in each quantile still represents the same amount of surface area.

4 Experimental Results

4.1 Simulated Data

In the first series of tests we assume an ideal noise-free sensor and Lambertian surfaces. One of the light sources is placed exactly where the camera is located, something that can not be done in a real setup. All point light-sources are of the same intensity and all our objects are made of the same uniform albedo. This setup, which can never be obtained in practice, shows the limit of what can be achieved under the best possible imaging conditions.

Our goal is to test the limitations of *DeCov*. We simulated a variety of objects. Some differ a lot in terms of local curvature, like a sphere, an hourglass shaped vase, and a cylinder. Others have more similar shapes, like a golf-tee (Fig. 3b), a bell and an onion dome. Another object approximates one of the vases (Fig. 3a) which we use in our experiments with real data. Out of rendering convenience, all but one of our synthetic objects are objects of revolution.

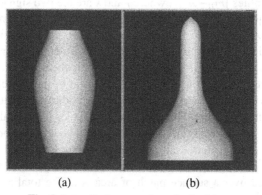

(a)   (b)

Fig. 3. Intensity images of synthetic objects

For consistency, in all our experiments on both synthetic and real data we use a 5×5 template for computing *DeCov*. In our signatures, we use ten quantiles, hence deciles. It is typical for frequency distributions to group the data in a convenient number of classes between five and twenty. Using only five quantiles will cause loss of information. Deciles keep the object descriptor small while providing adequate detail for object identification.

In order to illustrate the relationship between *DeCov* and the square of the Gauss-ian curvature K^2 we create the *DeCov* image of the object. To generate this image we assign a unique color to each adjacent pair of deciles in the signature. We group the quantiles in pairs when color-coding the *DeCov* values, so that we will not overload the images with color variation. The larger the *DeCov*, the darker the pixel. Each grey level value colors 20% of the object's total surface area. Once the coloring scheme is decided, the pixels have to be accordingly colored. Each mutually illumi-nated pixel has its *DeCov* which belongs in one of the G quantiles. We color each pixel according to the color assigned to the quantile in which that pixel's *DeCov* lies.

The dependence of *DeCov* on the square of Gaussian curvature K^2 and its invari-ance is depicted in the *DeCov* images of our synthetic data. The golf tee in Fig. 4 is segmented into "zones" of different curvature. The two big dark regions are areas where the curvature is greatest. A pattern of gradual curvature decrease followed by a symmetric gradual increase can be noticed as one moves from the top of the object to the bottom. This occurs in the flatter part of the golf tee where the object is in tran-sition from the narrow top to the wide bottom.

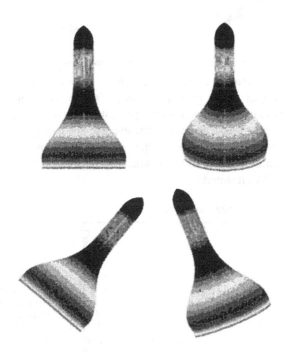

Fig. 4. DeCov images for different poses of a golf tee

Similarly, the vase in Fig. 5 is colored in different zones. We get the highest val-ues around the neck of the vase where we have the highest curvature. Then moving down along the axis of elongation of the vase we get a gradual symmetric increase and then decrease of the *DeCovs* around the area where the vase bulges out. What follows is a lower-curvature region of the vase ending up in a slight inward curvature at the bottom.

Fig. 5. DeCov images for different poses of a vase

The next step is to create the object's signature. This is the actual descriptor used for identifying an object. When matching an object's signature against the database, we look for the minimum average absolute percent difference between corresponding deciles of the signatures. The smaller the average absolute percent difference is between two signatures, the more similar the corresponding poses are. Fig. 6 shows the signatures for different poses of two distinct objects. Notice that signatures of the same object are tightly clustered.

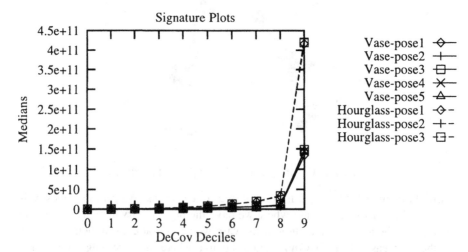

Fig. 6. Clusters of signature plots

Distinct objects create distinct clusters (Fig. 7). Notice the significant variance in the ranges of the median *DeCov* values between the two plots in Fig. 7(a) and Fig. 7(b).

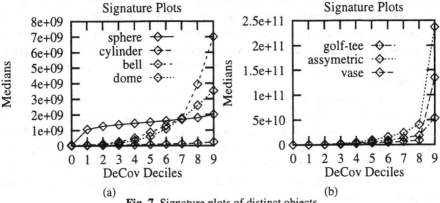

Fig. 7. Signature plots of distinct objects

4.2 Real Data

Our experimental setup is inexpensive and easy to duplicate. We use a Sony XC-77 camera with a 25mm lens. We use three fiber-optic illuminators to create each of our illumination conditions. One of the light sources is placed on top of the camera so that its direction of illumination is roughly parallel to the optic axis. The other two light sources can be placed anywhere as long as the three direction vectors of the light sources are non-coplanar. We have one of the light sources to the right side and one to the left side of the camera. The direction of illumination for each of them forms approximately a 20° angle with the optic axis. None of these light sources needs to be precision mounted. Once the placement of the light-sources is decided, it should remain unchanged with respect to the position of the camera. The whole camera-light-sources apparatus is handled as a single rigid unit. The objects are located at least 30" from the apparatus, which is far relative to the size of the objects. Our object database consists at this point of objects of uniform albedo. They vary from simple objects like a mug, Fig. 8(a), to complex ones like a face mask, Fig. 8(b).

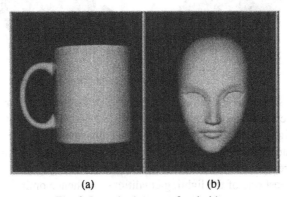

(a)   (b)

Fig. 8. Intensity images of real objects

Since *DeCov* is based on Lambertian reflection it is useful to eliminate specularities. Primary specular reflection produced from a point source can be eliminated by cross-polarization. This involves the placement of one fixed polarizer on each light source and one in front of the camera lens so that the direction of polarization of the polarizer at the camera is orthogonal to the direction of polarization of the polarizers at the light sources. This causes the specular reflection to have a linear polarization that is at 90° to the polarizer on the camera lens. Hence, the specular reflection gets cancelled out.

The relationship between *DeCov* and the square of Gaussian curvature K^2 can be clearly seen in figures 9 and 10. Again, the darker the color, the higher the *DeCov* value. Curvature is evenly distributed over the cylindrical body of the mug. In all the poses the handle clearly stands out as the region of highest curvature. The body contains a small vertical black region due to an unelimated specularity. It is interesting to note that the black line caused by inter-reflection beneath the handle is preserved at the different poses.

Fig. 9. DeCov images for different poses of a mug

The face mask shown in Fig. 10 has a nearly flat forehead and prominent feature ridges. Of particular interest are the sharp eyebrow lines, the dark outlines around the eyes, and the light-colored areas where the eyeballs should be. The eye regions are actually flat and recessed from the rest of the mask. These features are clearly depicted and preserved at the different poses. The white regions near the nose are in shadow for at least one of the lighting conditions and hence omitted from the mutually illuminated region.

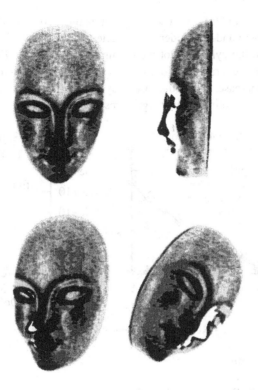

Fig. 10. DeCov images for different poses of a face mask

The signatures of the real data exhibit the same characteristics as the signatures of our simulations. For poses of the same object (Fig. 11) where there is no extreme change in the visible surface normals, the signature plots look very similar, although the clustering is not as tight as it was for the simulated data.

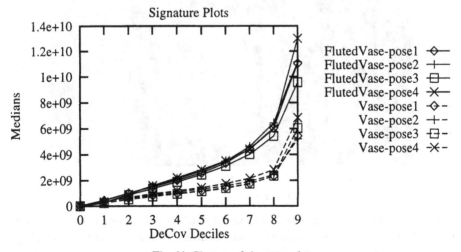

Fig. 11. Clusters of signature plots

As a form of ground truth we use objects of distinct, constant Gaussian curvature, specifically a sphere and a cylinder. As expected, both signature plots are increasing very slowly, with the cylinder plot increasing the slowest (see Fig. 12). As objects become more complex, ranging from various vases to the face mask, the slope of the signature graph increases. As in the case of synthetic data, one can see in Fig. 12 that different objects have quite distinct signatures.

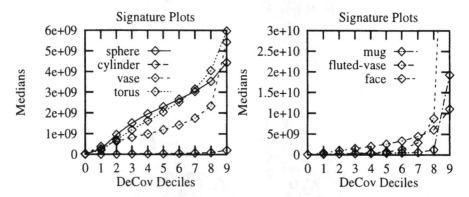

Fig. 12. Signature plots of distinct objects

4.3 Current Limitations

Multiple Poses. *DeCov* is a function of surface features that may not be visible from all viewing angles. For opaque objects there almost always exist degenerate poses in which object features are radically different [2]. Even very symmetric objects, like surfaces of revolution can look quite distinct under different viewing angles.

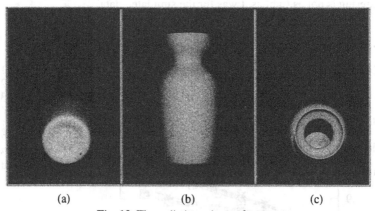

(a)   (b)   (c)

Fig. 13. Three distinct views of a vase

Consider, for example, one of the vases in our database. When one is looking at the bottom of the vase, one sees an almost planar surface with a circular boundary (Fig. 13a). This is the image of an object with an almost zero Gaussian curvature. When the viewing direction is perpendicular to the axis of elongation, one sees a curved surface with a symmetrically curved boundary (Fig. 13b). This is the image of a surface of varying Gaussian curvature. When one is looking down the top of the vase, one sees a concave object with a circular boundary (Fig. 13c). This, too, is an image of a surface of varying Gaussian curvature, but this time, the curvature is significantly different.

Like most of the object recognition systems that do not use complete three dimensional object representations we too have to use multiple poses for each object. In order to recognize an object from any possible viewing angle, we must have a signature for each object-pose that is composed of a completely different set of surface features. The number of poses per object in our database depends on the asymmetry of the object.

The Condition of the Photometric Covariance Matrix. *DeCov* is also limited by the instability inherent to matrices. The covariance matrix for small surface patches that are almost flat (a plane or a small patch of a cylinder with a big diameter) is nearly singular. This makes small uncertainties in the matrix elements, like rounding to the closest integer, or camera noise, to result in large uncertainties in the quantities we compute based on that matrix. Furthermore, the error resulting from these uncertainties increases with scaling. Such matrices are considered unstable, ill-conditioned.

To illustrate that consider the singular matrix V:

$$V = \begin{bmatrix} 1 & 1 & 1 \\ 1 & 1 & 1 \\ 1 & 1 & 1 \end{bmatrix}$$

i.e. its determinant is zero, $|V| = 0$. If some noise is added, which is anticipated when the elements of the matrix are values produced from actual experiments, the matrix becomes non-singular. For example, a nearly singular version of V is:

$$V' = \begin{bmatrix} 1.01 & 1 & 1 \\ 1 & 1 & 1 \\ 1 & 1 & 1.01 \end{bmatrix}$$

Its determinant is now $|V'| = 0.0001$.

Let's multiply this noisy matrix V' by a big scalar like 2,500 giving matrix W. This is actually a small scalar, compared to the values we have in the photometric covariance matrix. (A 5×5 window of intensity values equal to 100 for all three light sources, has covariances equal to 25,000,000.) The new matrix W is:

$$W = \begin{bmatrix} 2525 & 2500 & 2500 \\ 2500 & 2500 & 2500 \\ 2500 & 2500 & 2525 \end{bmatrix}$$

Its determinant is $|W| = 1,562,500$.

A typical measurement for the stability of a square matrix A is the *condition num-*

ber $M(A) = \|A\| \|A^{-1}\|$, where $\|A\|$ is the norm of the matrix. If $M(A) \leq 100$ then A is well-conditioned, if $M(A) \geq 10,000$ then A is ill-conditioned [9, 23]. Depending on the type of norm used, (the infinity norm L_∞, the 1-norm L_1, or the 2-norm L_2), one can compute the M_∞, M_1 or M_2 condition numbers.

However, the condition number is not independent of scaling. Although two matrices A and cA, where c is a scalar, are equally conditioned, $M(A) \neq M(cA)$. Depending on c, one can get a very unrealistic condition number $M(cA)$. A better stability indicator is the *minimum condition number $K(A)$* as defined by Noble [23], which is truly independent of the scaling of A. It allows us to compare different matrices in terms of their conditioning, as it puts them in the same frame of reference. Noble shows that:

$$K_\infty(A) = max(\lambda_i)$$

for $i = 1,..,n$, where n is the size of the matrix A, and λ_i is the eigenvalue of the matrix $A_+ A_+^{-1}$ (or $A_+^{-1} A_+$) if $A_+ A_+^{-1} > 0$ (or if $A_+^{-1} A_+ > 0$). The + subscript means that we replace each element in the original matrix A (or A^{-1}) by its absolute value, $A_+ = [|a_{ij}|]$ and $A_+^{-1} = (A^{-1})_+$ accordingly. Note that $A_+^{-1} \neq (A_+)^{-1}$. If $K_\infty(A) \leq 100$ then A is really well-conditioned, while if $K_\infty(A) \geq 10,000$ then A is truly ill-conditioned [23].

We use the infinity minimum condition number $K_\infty()$ to weight the validity of the results of our object recognizer. In general, a pose is ill-conditioned if the visible surface has Gaussian curvature near zero. This means that the condition number itself can be used in identifying surface patches of near zero Gaussian curvature.

5 Conclusions and Future Research

DeCov is a curvature based invariant which remains constant under rigid transformations. Our object representation technique compiles the dense *DeCov* information down to a signature that is invariant to rotation, translation and scaling. This signature is composed of a short fixed-length sequence of real numbers. The experimental setup is simple. Although different illumination conditions are required, knowledge of the exact location of the light sources is not needed. Surface normals and curvature are never explicitly recovered. *DeCov* is computed directly from the images without the use of any three-dimensional model.

Our object recognition system is flexible and can be easily tuned to individual preferences. Depending on the desired level of surface detail to be used in identifying an object, one can vary the number of quantiles used in the object signature. Additionally, one may choose to use a weighted absolute difference when comparing two signatures. In this way emphasis is placed in specific quantiles. For example, a high weight in the middle quantiles identifies objects mainly by their "average" curvature. Similarly, a small weight in the last quantile discards possible outliers due to specularities and ignores surface details.

We believe that *DeCov* can be further exploited. We are exploring the possibility of making *DeCov* albedo invariant. In that way, one can choose to either recognize an object based on its overall shape and ignore the albedo, or use the varying albedo as additional discriminating information. We are also considering the design of an algorithm which specifies the minimum number of distinct poses that are necessary for reliable object recognition. We are trying to produce a method that doesn't use the intensity values of a light source aligned with the camera. That would result in an unconstrained light-source setup, where the only restriction would be to have non-coplanar direction vectors for the illuminators. Lastly, we are examining the possibility of totally eliminating one of the light sources and computing *DeCov* using only two illumination conditions.

References

1. Beer, F. P. and Johnston, E. R. Jr. *Vector Mechanics for Engineers: Statics and Dynamics.* McGraw-Hill, 1977.
2. Besl, P. J. and Jain, R. C. "Invariant Surface Characteristics for 3D Object Recognition in Range Images." *Computer Vision Graphics and Image Processing*, Vol. 33, 1986. pp. 33-80.
3. Biederman, I. "Human Image Understanding: Recent Research and a Theory." *Computer Vision Graphics and Image Processing: Image Understanding*, Vol. 32, 1985. pp. 27-73.
4. Binford, T. "Visual Perception by Computer." *IEEE Conference on Systems and Control*, December 1971.
5. Brooks, R. "Model-Based 3-D Interpretation of 2-D Images." *IEEE Transactions on Pattern Analysis and Machine Intelligence*, Vol. 5, No. 2, 1983. pp. 140-150.
6. Dickinson, S. J., Pentland, A. P. and Rosenfeld, A. "From Volumes to Views: An Approach to 3-D Object Recognition." *Computer Vision Graphics and Image Processing: Image Understanding*, Vol. 55, No. 2, 1992. pp. 130-154.
7. Faddeev, D. K. and Faddeeva, V. N. *Computational Methods of Linear Algebra.* W.H. Freeman and Co., 1963.
8. Grimson, W. E. L. "Recognition of Object Families Using Parameterized Models." *Proceedings of the First International Conference on Computer Vision*, 1987. pp. 93-100.
9. Hager, W. W. *Applied Numerical Linear Algebra.* Prentice-Hall, 1988.
10. Horn, B. K. P. "Understanding Image Intensities." *Artificial Intelligence*, Vol. 8, No. 2, 1977. pp. 1-31.
11. Horn, B. K. P. and Brooks, M. J. *Shape from Shading.* MIT Press, 1989.
12. Huttenlocher, D. and Ullman S. "Recognizing Solid Objects by Alignment with an Image." *International Journal of Computer Vision*, Vol. 5, 1990. pp. 195-212.
13. Ikeuchi, K., Horn, B. K. P., Nagata, S., Callahan, T. and Feingold O. "Picking up an Object from a Pile of Objects." *Robotics Research: The First International Symposium.* MIT Press, 1984. pp. 139-162.
14. Iwahori, Y., Woodham, R. J. and Bagheri, A. "Principal Components Analysis and Neural Network Implementation of Photometric Stereo." *Proceedings of the Workshop on Physics-Based Modeling in Computer Vision.* IEEE Computer Society Press, 1995. pp. 117-125.
15. Lamdan Y., Shwartz J. and Wolfson H. "On Recognition of 3-D objects from 2-D images." *Proceedings of the IEEE International Conference on Robotics and Automation 1988.* IEEE Computer Society Press, 1988. pp. 1407-1413.
16. Lowe, D. "Fitting Parameterized Three-Dimensional Models to Images." *IEEE Transactions on Pattern Analysis and Machine Intelligence.* Vol. 13, No. 5, 1991. pp. 441-450.

17. Marr, D. and Nishihara, K. "Representation and Recognition of the Spatial Organization of Three-Dimensional Shapes." *Proceedings of the Royal Society of London B*, 1978.

18. Marsden, J. E. and Tromba, J. A. *Vector Calculus*. W.H. Freeman and Co., 1981.

19. Millman, R. S. and Parker, G. D. *Elements of Differential Geometry*. Prentice-Hall, 1977.

20. Mundy, J. L., Zisserman, A. and Forsyth, D. ed. *Applications of Invariance in Computer Vision: Second Joint European-US Workshop Proceedings*. Springer-Verlag, 1994.

21. Myklestad, N. O. Cartesian Tensors: *The Mathematical Language of Engineering*. D. Van Nostrand Company, Inc., 1967.

22. Nalwa, V. S. *A Guided Tour of Computer Vision*. Addison-Wesley Publishing Company, 1993.

23. Noble, B. *Applied Linear Algebra*. Prentice-Hall, 1969.

24. Pentland A. "Perceptual Organization and the Representation of Natural Form." *Artificial Intelligence*, Vol. 28, 1986. pp. 293-331.

25. Pentland A., Sclaroff S., Horowitz B. and Essa, I. "Modal Descriptions for Modeling, Recognition, and Tracking." *Three-Dimensional Object Recognition Systems*. Elsevier Science Publishers B.V., 1993. pp. 423-445.

26. Press, W. H., Flannery, B. P., Teukolsky, S. A. and Vetterling, W. T. *Numerical Recipes in C: The Art of Scientific Computing*. Cambridge University Press, 1988. pp. 353-367.

27. Reiss, T. H. *Recognizing Planar Objects Using Invariant Image Features*. Springer-Verlag, 1993.

28. Shercliff, J. A. *Vector Fields*. Cambridge University Press, 1977.

29. Stein, F. and Medioni, G. "Structural Indexing: Efficient Three Dimensional Object Recognition." *Three-Dimensional Object Recognition Systems*. Elsevier Science Publishers B.V., 1993. pp. 353-373.

30. Terzopoulos, D., Witkin A., and Kaas M. "Symmetry-Seeking Models and 3D Object Recovery." *International Journal of Computer Vision*, Vol. 1, 1987. pp. 211-221.

31. Terzopoulos, D. and Metaxas D. "Dynamic 3D Models with Local and Global Deformations: Deformable Superquadrics." *IEEE Transactions on Pattern Analysis and Machine Intelligence*, Vol. 13, No. 7, 1991. pp. 703-714.

32. Thorpe, C. and Shafer, S. "Correspondence in Line Drawings of Multiple Views of Objects." *Proceedings of the 8th International Joint Conference on Artificial Intelligence*, 1983. pp. 959-965.

33. Woodham, R. J. "Photometric Method for Determining Surface Orientation from Multiple Images." *Optical Engineering*, Vol. 19, No. 1, January/February 1980. pp. 139-144.

34. Woodham, R. J. "Gradient and Curvature from the Photometric-Stereo Method, Including Local Confidence Estimation." *Journal of the Optical Society of America A*, Vol. 11, No. 11, November 1994. pp. 3050-3068.

On 3D Shape Synthesis *

Heung-Yeung Shum, Martial Hebert and Katsushi Ikeuchi

Carnegie Mellon University, Pittsburgh PA 15213, USA

Abstract. We present a novel approach to 3D shape synthesis of closed surfaces. A curved or polyhedral 3D object of genus zero is represented by a curvature distribution on a spherical mesh that has nearly uniform distribution with known connectivity among mesh nodes. This curvature distribution, i.e., the result of forward mapping from shape space to curvature space, is used as the intrinsic shape representation because it is invariant to rigid transformation and scale factor. Furthermore, with regularity constraints on the mesh, the inverse mapping from curvature space to shape space always exists and can be recovered using an iterative method. Therefore, the task of synthesizing a new shape from two known objects becomes one of interpolating the two known curvature distributions, and then mapping the interpolated curvature distribution back to a 3D morph. Using the distance between two curvature distributions, we can quantitatively control the shape synthesis process to yield smooth curvature migration. Experiments show that our method produces smooth and realistic shape morphs.

1 Introduction

A traditional approach to synthesizing 3D shape is to use volume metamorphosis (or volume morphing). Similar to image morphing [1], volume morphing [12] is achieved by first warping two original volumes to new volumes, and then blending both into an intermediate shape, i.e., a morph. A major challenge to a 3D morphing system is automatic feature recognition and matching which are crucial for defining the transformation of the two objects. Unlike image morphing, however, a simple morph map can not be easily established for volume morphing. What is to be morphed between two volumes is not clearly defined. Most 3D morphing systems [4][12] (except [7], which attempts to do automatic feature registration) resort to a good user interface which allows the user to specify feature segments in both objects. However, even when many feature segments are located, smooth transitions between two shapes are not guaranteed

---

* This research is partially sponsored by the Advanced Research Projects Agency under the Department of the Army, Army Research Office under grant number DAAH04-94-G-0006, partially supported by ONR under grant number N00014-95-1-0591, and partially supported by NSF under Contract IRI-9224521. The views and conclusions contained in this document are those of the authors and should not be interpreted as representing the official policies or endorsements, either expressed or implied, of the Department of the Army, or the U.S. government.

because the curvature may not be interpolated properly. Computationally, the volume morphing is very expensive because of the huge amount of data used.

This paper provides a means of synthesizing 3D shapes using surface instead of 3D volume. We restrict our study to shapes of genus zero. We prefer to synthesize shapes of surfaces rather than 3D volumes because surface data is most common in computer vision applications, such as from a light stripe range finder, a laser range finder, or stereo. On the other hand, if volume data is available (e.g., from a CT scan), it can be converted to a surface model using techniques such as "marching cubes [14]". For the purpose of shape synthesis, volume data is perhaps redundant.

We define our 3D shape synthesis problem as follows. Given two 3D shapes A and B, we synthesize a sequence of intermediate shapes (or morphs) which should have the following desirable properties defined in [12]:

1. Realism: the morphs should be realistic objects which have plausible 3D geometry and which retain the essential features of the originals.

2. Smoothness: the change in the sequence of morphs must show a smooth transition between the two originals.

To meet the above requirements, we propose to synthesize a shape in terms of its curvature distribution because curvature is independent of rigid transformations and scalings. However, neither Gaussian curvature nor mean curvature fully captures the essence of 3D shape. People have long searched for proper spherical shape representations using curvatures [6][9][13] for closed shape. For example, from the Extended Gaussian Image, we can reconstruct the original shape if the object is convex. However, to synthesize arbitrary non-convex shape, we have to search for proper representation.

In our work, we use a spherical mesh to represent an object shape, and store local curvature at each node as its intrinsic representation. From this intrinsic representation, we show that, provided that a regularity constraint is introduced, the original shape can be reconstructed up to a scale factor and a rigid transformation. The correspondence between two original shapes can be automatically obtained by minimizing the difference between two curvature distributions as we have shown previously in [5][16]. Then the task of synthesizing a new object from two known objects becomes one of interpolating the two curvature distributions, and then mapping the interpolated curvature distribution back to a 3D shape.

We first illustrate our basic approach on shape synthesis in Section 2. After introducing our spherical shape representation of closed 3D surfaces, Section 3 describes the mapping from shape space to curvature space, while the inverse mapping from curvature space to shape space is discussed in Section 4. Section 5 describes our curvature-based shape synthesis in detail. Correspondence between two shapes is briefly discussed; details of this are given elsewhere [12]. Finally, we show the results of our shape synthesis, and close with a discussion and conclusion.

2 Basic Idea

Our ultimate goal is to generate a sequence of intermediate shapes from two given 3D shapes which can be either closed curved or polyhedral surfaces without holes. We represent our 3D shape using a special spherical coordinate system [5]. A semi-regularly tessellated sphere is deformed so that the meshes sit on the original data points, while the connectivity among the mesh nodes is preserved. After the deformation process, we obtain a spherical representation with local curvature[2] at each mesh node. The local curvature at each node is calculated by its relative position to its neighbors. By enforcing local regularity at each mesh node, we can reconstruct the shape from its curvature distribution up to a scale factor and a rigid transformation.

Our approach to shape synthesis is straightforward: for each original shape, we build its curvature distribution as its intrinsic representation. The correspondence between them is either specified by the user or can be directly computed, e.g., by minimizing the difference between their curvature distributions [16]. The problem of synthesizing a new shape from two known shapes becomes one of interpolating two known curvature distributions. The new shape is reconstructed from the interpolated curvature distribution. This approach is summarized in Figure 1.

An advantage of our shape synthesis using curvatures is that the resulting shapes vary smoothly because the morphs are interpolated according to the original curvature distributions. Volume morphing, on the other hand, can hardly quantitatively specify this kind of shape change. Instead, it has to rely on careful feature selection which provides only qualitative shape change at best. By making explicit use of the curvature information, our method reduces the ambiguity of matching two shapes. This is especially important for smooth curved objects, where features are very difficult to identify manually.

3 From Shape to Curvature Distribution: Forward Mapping

3.1 Spherical Representation of a Closed 3D Surface

To synthesize object shapes, one first has to find appropriate representations of those shapes. A natural discrete representation of a surface is a graph of nodes, or tessellation, such that each node is connected to each of its closest neighbors by an arc of the graph. For example, Szeliski and Tonnesen [17] used oriented particles to model arbitrary surfaces; Witkin and Heckbert [20] also used particles to sample surfaces. We use a special mesh, each node of which has exactly three neighbors. Such a mesh can be constructed as the dual of a triangulation of the surface [5]. Our representation differs from particle systems [17][20] in that

---

[2] We adopt the terms local curvature and curvature distribution here in a different way from the formal definition of curvature such as in [11]. Our local discrete curvature measure was first introduced in [5].

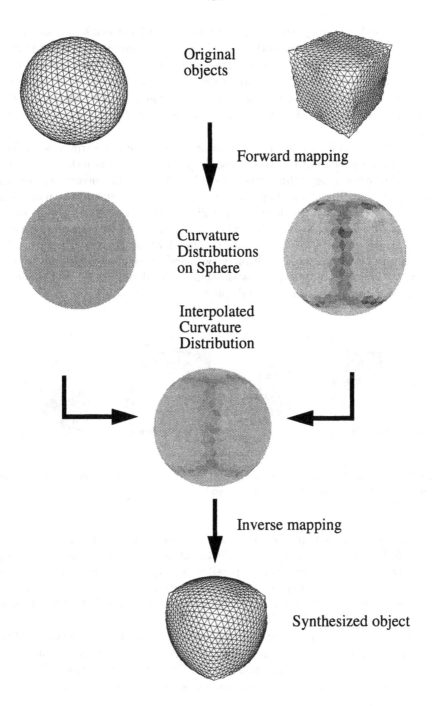

Fig. 1. Shape synthesis in curvature space: an example of a sphere and a hexahedron. The curvature has been color-coded so that darker means bigger positive curvature and lighter means bigger negative curvature.

it is restricted to the spherical topology and has a fixed neighborhood; but it leads to more efficient computation and robust local curvature approximation. To tessellate a unit sphere, we use a standard semi-regular triangulation of the unit sphere constructed by subdividing each triangular face of a 20-face icosahedron into N2 smaller triangles. The final tessellation is built by taking the dual of the 20 N2-face triangulation, yielding a tessellation with the same number of nodes.

In order to obtain a mesh representation for an arbitrary surface, we deform a tessellated surface until it is as close as possible to the object surface. The deformable surface algorithm drives the spherical mesh to converge to the correct object shape by combining forces between the data set and the mesh. Our algorithm originates from the idea of a 2D deformable surface [10][19] and is described in detail in [5]. The deformed surface can accurately represent concave as well as convex surfaces. An example of a free-form object model created using the deformable surface and multiple view merging techniques [15] is shown in Figure 2. The deformation process is robust against data noise and moderate change of parameters such as initial sphere center and radius [15].

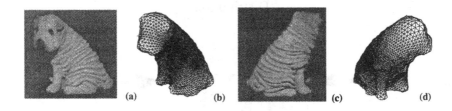

(a)　(b)　(c)　(d)

Fig. 2. An example of a free-form object modeled from a deformable surface: (a) (c) Images of a sharpei; (b) (d) Deformable models of a sharpei.

The key idea of our spherical representation of a surface is to produce meshes in which the density of nodes on the object's surface is nearly uniform[3]. Although perfectly uniform distribution is impossible, a simple local regularity constraint can enforce a very high degree of uniformity across the mesh. We implemented the local regularity constraint in the deformable surface algorithm such that all of the meshes have similar areas [5].

The local regularity constraint is a generalization to three dimensions of the regularity condition on two dimensional discrete curves; this condition simply states that all segments are of equal lengths. The difference between 2D and 3D is that it is always possible to create a uniform discrete curve in 2D, while in 3D

---

[3] Koenderink warned that one has to be very careful of any method that uses the surface area of a polyhedral model (p.597 of [11]). Surface area depends on the way in which triangulations are done. In our previous work, we have shown how areas of different shapes are adjusted before comparison, in particular for partial views [5].

only nearly uniform discrete surfaces can be generated. In practice, the variation of mesh nodes on the surface is on the order of 2% [5].

3.2   3D Local Curvature: An Approximation

After we obtain a nearly uniform surface mesh representation, the next step is to define a measure of curvature that can be computed from the surface representation. Conventional ways of estimating surface curvature, either by locally fitting a surface or by estimating first and second derivatives, are very sensitive to noise. This sensitivity is mainly due to the discrete sampling and, possibly, to noisy data. We introduced in [5] a robust measure of curvature computed at every node from the relative positions of its three neighbors. Our method is robust because all the nodes are at relatively stable positions after the deformation process. The deformable surface process serves as a smoothing operation over the possibly noisy original data. We call this measure of curvature the simplex angle.

The simplex angle α varies between $-\pi$ and π, and is negative if the surface is locally concave, positive if it is convex. Given a configuration of four points, the angle is invariant by rotation, translation, and scaling because it depends only on the relative positions of the points, not on their absolute positions.

The spherical representation can approximate not only free-form objects, but also polyhedral objects. For example, Figure 3 shows an example of a spherical polyhedral approximation of an octahedron with one concave face. Because of the regularity constraint, corners and edges are not represented perfectly. All plane surfaces, however, are well approximated. Different tessellation frequencies result in different approximations.

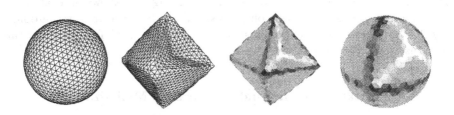

Fig. 3. (a) A spherical tessellation; (b) Deformable surface of an octahedron with a concave dent; (c) Local curvature on each mesh node; (d) Curvature distribution on spherical representation (The curvature on (c) and (d) is negative if it is light, positive if dark, zero if grey).

4 From Curvature Distribution to Shape: Inverse Mapping

4.1 Shape Reconstruction

Now we know how to map object shape to its intrinsic representation defined above, i.e., a curvature distribution on a special spherical coordinate system. But does the inverse mapping exist? In other words, can we reconstruct the shape given its intrinsic representation?

We formulate this reconstruction as an optimization problem which minimizes the curvature difference between the known and reconstructed curvature distributions. The initial shape can be, for example, a sphere where constant curvature is at every mesh node. This minimization problem is, however, complicated because of the nonlinear and coupled nature of the local curvature computation. We devise an iterative method, similar to what has been used in deformable surface extraction, to solve this minimization problem. The motion of each vertex is modeled as a second order dynamics equation,

$$m\frac{d^2 P_i}{dt^2} + k\frac{dP_i}{dt} = F_{int} \tag{1}$$

where m is the mass unit of a node and k is the damping factor. F_{int} is the force which deforms the surface to its assigned curvature distribution. The above equation is integrated over time using finite differences. Starting from an initial shape such as a semi-tessellated sphere, using Euler's method, we can update each mesh node of unit mass by

$$P_i{}^{(t)} = (1 - k)(P_i{}^{(t-1)} - P_i{}^{(t-2)}) + F_{int} \tag{2}$$

Notice that F_{int} has to be updated at each iteration. The reader is referred to [3][5] for the details of how F_{int} is computed. Figure 4 shows an example of the evolution of this reconstruction process. Given its curvature distribution, a toy sharpei is reconstructed from a sphere. We explain in the following how this minimization problem is solved.

4.2 Regularization for the Underconstrained Minimization

The above minimization is unfortunately underconstrained. There are $3n$ unknowns (3D coordinates at each mesh node of the unknown spherical mesh) but only n equations defined by local curvature at each mesh node. The object shape can not be determined without additional constraints. It is well-known that Gaussian curvature and mean curvature can not determine the shape. First fundamental forms are the necessary and sufficient condition for reconstructing shape [11].

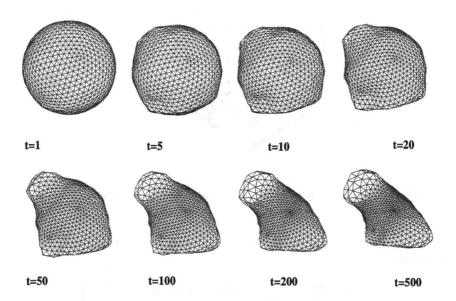

Fig. 4. A sequence of shapes at different steps of deformation from a sphere to a sharpei.

Delingette's Constraint: Metric Parameters One way to add more constraints is to record the relative positioning between each node P and its three neighbors P_1, P_2, and P_3. Delingette [3] defined a so-called metric parameter set $\{\epsilon_1, \epsilon_2, \epsilon_3\}$, such that

$$Q = \epsilon_1 P_1 + \epsilon_2 P_2 + \epsilon_3 P_3 \tag{3}$$

and

$$\epsilon_1 + \epsilon_2 + \epsilon_3 = 1 \tag{4}$$

where Q is the projection of P on the plane $P_1 P_2 P_3$ as shown in Figure 5.

If the shape is known, i.e., if each mesh node P and its three neighbors P_1, P_2, and P_3 are known, the surface normal of the plane $P_1 P_2 P_3$ can be computed as

$$n = \frac{\overline{P_1 P_2} \times \overline{P_1 P_3}}{\|\overline{P_1 P_2} \times \overline{P_1 P_3}\|} \tag{5}$$

From

$$\overline{QP} = \|\overline{QP}\|\overline{n} = (\overline{P_1 P} * \overline{n})\overline{n} \tag{6}$$

and

$$\overline{P_1 Q} = \overline{P_1 P} - \overline{QP} = \epsilon_2 \overline{P_1 P_2} + \epsilon_3 \overline{P_1 P_3}. \tag{7}$$

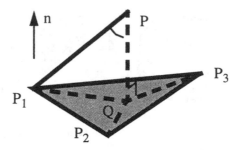

Fig. 5. Metric parameters relating a mesh node with its three neighbors.

We can easily compute the metric parameters from the following 2×2 linear equations

$$\begin{bmatrix} \overline{P_1P_2} \cdot \overline{P_1Q} \\ \overline{P_1P_3} \cdot \overline{P_1Q} \end{bmatrix} = \begin{bmatrix} \|\overline{P_1P_2}\|^2 & \overline{P_1P_2} \cdot \overline{P_1P_3} \\ \overline{P_1P_2} \cdot \overline{P_1P_3} & \|\overline{P_1P_3}\|^2 \end{bmatrix} \times \begin{bmatrix} \epsilon_2 \\ \epsilon_3 \end{bmatrix} \tag{8}$$

The above equation provides us two equations at each mesh node. Thus we have $3n$ sparse nonlinear equations for $3n$ unknowns if we augment our intrinsic representation from $\{\phi_i\}$ to $\{\phi_i, \epsilon_{2i}, \epsilon_{3i}\}$, $i = 0 \ldots n$.

Regularity Constraint: a Means of Regularization The metric parameter augmentation to our intrinsic shape representation may not be necessary. Instead, we introduce a regularity constraint in the shape reconstruction which forces each mesh node to project onto the center of its three neighbors. This regularity constraint implicitly gives a complete intrinsic description $\{\phi_i, 1/3, 1/3\}$, $i = 0 \ldots n$. It shows that our intrinsic representation is sufficient to guarantee that the inverse mapping exists. This regularity constraint is, in spirit, similar to imposing a regularization term for ill-posed problems [18]. Given known local curvature, the expected location of a mesh node can be regularized through its three neighbors. As shown in Figure 6, we can then define F_{int} as the scaled distance between the current mesh node P and its regularized position P^*

$$F_{int} = \alpha \cdot \overline{PP^*}, \alpha = [0, 0.5] \tag{9}$$

Why does this regularity constraint ensure that we reconstruct the correct original shape? The reason is that this same regularity constraint has been enforced in the model extraction process when we construct an object model from range data [5][15]. At the end of the model extraction, the metric parameters converge to the expected value. Figure 7 shows the distribution of metric parameters of the sharpei model and their histograms. It clearly shows that the metric parameters are well regularized around their nominal values of $1/3$. It is also equivalent to that each triangular patch has the same area. Obviously if we

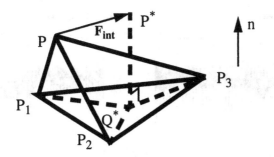

Fig. 6. Regularity and Internal force. $Q^* = (P_1 + P_2 + P_3)/3$.

keep the intrinsic representation but pick an arbitrary set of metric parameters, we will reconstruct another shape which may be close to, but different from, the original shape.

4.3 Examples

The above analysis does not eliminate the local minimum problem associated with our iterative process. Therefore the initial shape plays an important role for convergence to correct shape. To be fair, we start our reconstruction from a semi-tessellated sphere in all experiments. We show two examples of shape reconstruction: a polyhedral object and a free-form object in Figure 8 and Figure 9, respectively.

Both examples show very good reconstruction. The reconstructed shape has a very similar curvature distribution to that of the original one. The relative error in the simplex angle between the original and the reconstructed shapes is less than 1% for each example. However, we do observe an apparent shape discrepancy mainly due to the discretization effect. Because the shape reconstruction is only up to an unknown rigid transformation (rotation and translation) and an unknown scale factor, constant area or constant volume [2] can be enforced during the reconstruction process.

5 Shape Synthesis with Curvature Interpolation

5.1 Shape Correspondence

Given the intrinsic representations of two original shapes, we can compute the correspondence between them based on two important properties of the semi-regularly tessellated sphere: that it has a fixed topology and that each mesh node has exactly three neighbors. We have observed [5] that the correspondence between two shapes is determined once three pairs of nodes are matched. In Figure 10, node P of shape A corresponds to P' of shape B, and the two neighbors of

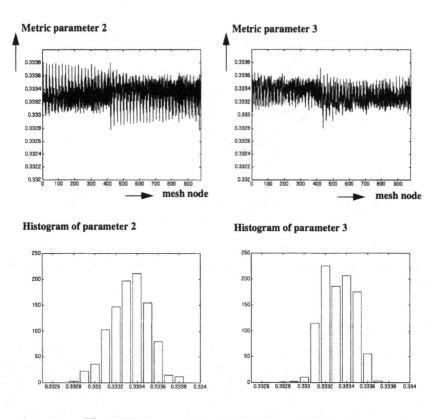

Fig. 7. Metric parameter distributions of a sharpei.

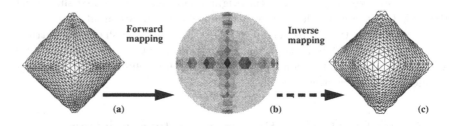

Fig. 8. An example of polyhedral object shape reconstruction (the solid arrow shows the forward mapping from shape to curvature, the dotted arrow shows the inverse mapping from curvature to shape): (a) Deformable surface of an octahedron; (b) Intrinsic representation or its curvature distribution (The curvature is negative if it is light, positive if dark, zero if grey). (c) The reconstructed shape from the curvature distribution (b).

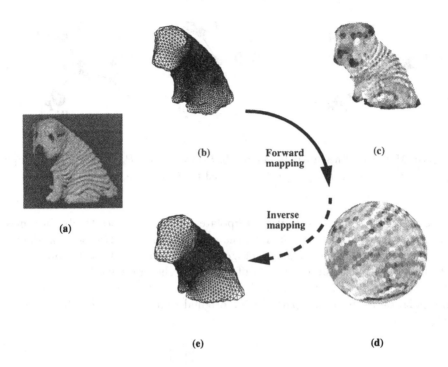

(a)

(b)

Forward mapping

(c)

Inverse mapping

(e)

(d)

Fig. 9. An example of free-form object shape reconstruction (the solid arrow shows the forward mapping from shape to curvature, the dotted arrow shows the inverse mapping from curvature to shape): (a) An image of a sharpei; (b) A sharpei model constructed from range data; (c) Local curvature distribution at each mesh node of a sharpei; (d) Intrinsic representation of a sharpei; (e) Reconstructed sharpei model from its intrinsic representation.

P (P_1, and P_2) are put in correspondence with two of three neighbors (P'_1, P'_2 and P'_3) of P', respectively. Figure 10 shows only 3 valid neighborhood matches, since each node has exactly three neighbors and the connectivity among them is always preserved. Moreover, the number of such correspondences is $3n$ where n is the number of nodes of spherical tessellation [5].

Given the correspondence, the distance between two shapes is defined as the L_2 distance between two curvature distributions [16]. An efficient algorithm has been devised for comparing two shapes using the minimum distance between two shapes, also defined as a shape similarity measure [16].

5.2 Shape Similarity for Interpolation

It is unclear how we can interpolate two shapes unless we know how to compare them. It is very difficult to compare two 3D shapes because of the unknown scale factor and rigid transformation. We have shown [16] that it is possible to quantitatively measure the distance between two shapes using the intrinsic

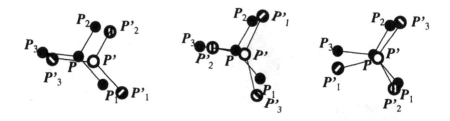

Fig. 10. Matching of neighbors from (P_1, P_2, P_3) to: (a) (P'_1, P'_2, P'_3); (b) (P'_2, P'_3, P'_1); (c) (P'_3, P'_1, P'_2) when P of shape A is matched to P' of shape B.

representation. Thus we can also interpolate these two shapes to obtain a new curvature distribution from the intrinsic representations of two original shapes and their correspondence. An advantage of this approach is that it shows quantitatively how much the morph is different from the originals. For example, at each mesh node of the new mesh C, its curvature can be computed by a linear interpolation of its counterparts in the original shapes A and B. More specifically,

$$\phi_{C_t} = (1 - t)\phi_{A_t} + t\phi_{B_t}, t = [0, 1], i = 1, \ldots, n \qquad (10)$$

where ϕ is the local curvature measure. Alternatively nonlinear cross-dissolve techniques can also be used. For instance, if we use the following interpolation function

$$\phi_{C_t} = (1 - 2t^2)\phi_{A_t} + 2t^2\phi_{B_t}, t = [0, 0.5], i = 1, \ldots, n \qquad (11)$$

$$\phi_{C_t} = 2(1 - t)^2\phi_{A_t} + (1 - 2(1 - t)^2\phi_{B_t}, t = [0.5, 1], i = 1, \ldots, n \qquad (12)$$

to blend curvatures, we can enforce desirable small shape change at initial steps. Figure 11(a) shows linear change of a shape similarity distance between each morph and an original object (a toy sharpei), while (b) shows the distance between each morph and the other original object (a toy pig). Figure 12 shows the nonlinear change of a shape similarity distance. The distance between two shapes is defined as the L_2 norm of the distance between two curvature distributions, which has also been used as a measure of shape similarity in our previous work [16]. It is clear that our synthesis approach is not only consistent with, but also can be controlled by, our metric measure of shape similarity.

5.3 Result and Discussion

Figure 13 shows a sequence of morphs which are synthesized from a toy sharpei and a toy pig, while Figure 14 shows a different view of the same sequence. The models of these free-form objects are constructed from real range images using

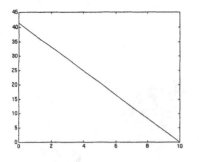 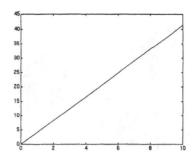

Fig. 11. Linear interpolation of shape distance between the morph sequence and two originals (a) sharpei; (b) pig.

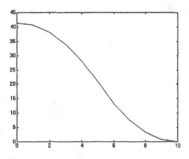 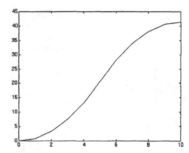

Fig. 12. Nonlinear interpolation of shape distance between the morph sequence and two originals (a) sharpei; (b) pig.

methods described in [15]. The frequency of spherical tessellation is set to *13*, which means that the total number of mesh nodes is *3380*. After the models and their curvature distributions are obtained, we use linear interpolation to generate the intermediate curvature distributions which are then inversely mapped to the morphs. These morphs clearly show a gradual and smooth shape transition between the sharpei and the pig.

Compared with volume morphing, our method is fast and simple to implement. While volume morphing takes more than a day [12] to render a sequence of morphs, our method takes less than an hour to do the shape synthesis. In our experiments, it takes 20 minutes to build deformable models and curvature distribution, and 20 minutes for cross-dissolve curvature interpolations and shape synthesis of a sequence of 10 morphs on a SUN Sparc 20. This does not account for the time taking images and rendering images for the final display. To speed up the iteration process, the morph from the previous step is used to initialize synthesis of the next morphing step.

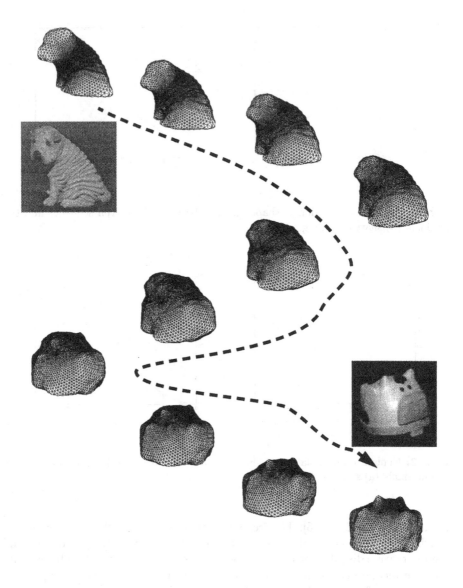

Fig. 13. A morphing sequence between a toy sharpei and a toy pig.

One possible drawback of our approach is that the quality of approximation of a polyhedral or free-form surface depends on the number of patches chosen. Obviously, the larger number of surface patches we use, the better the approximation. Our shape synthesis approach generates good morphing sequence provided that a sufficient number of tessellations is adopted.

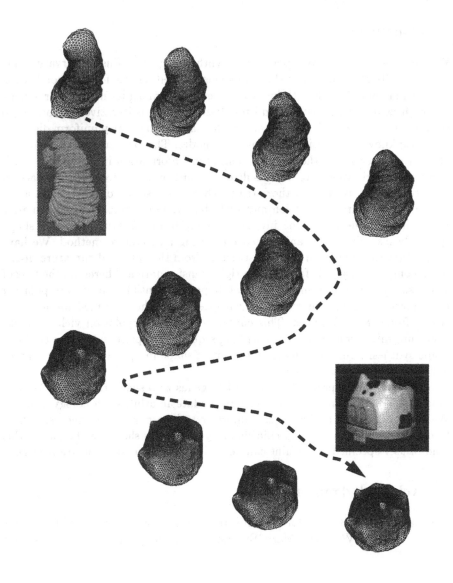

Fig. 14. Another view of the morphing sequence between a toy sharpei and a toy pig.

For the purpose of shape synthesis, we may want to satisfy the user's specification such as manual correspondences, etc. This user-specific requirement can be incorporated into a force F_{ext} such that the (EQ 1) is modified as

$$m\frac{d^2 P_i}{dt^2} + k\frac{dP_i}{dt} = F_{int} + F_{ext} \qquad (13)$$

which is similar to deformable surface model extraction [5] where F_{ext} is dominated by data force.

6 Conclusion

We have described a novel approach to synthesizing shapes using curvature distributions. To store an object shape of genus zero, either free form or polyhedral, we use a spherical mesh with known connectivity among its nodes. Our spherical mesh, which starts with a semi-tessellated sphere, is iteratively deformed to converge to the original object shape while maintaining nearly uniform distribution with known connectivity among mesh nodes. The local curvature computed at each node captures the averaged curvature information in its vicinity. This curvature distribution, i.e., the result of forward mapping from shape space to curvature space, is used as the intrinsic shape representation because it is invariant to rigid transformation and scale factor. Furthermore, with regularity constraints on the mesh, the inverse mapping from curvature space to shape space always exists and can be recovered using an iterative method. We have shown that the shape can be reconstructed, from the spherical curvature distribution only, up to a scale factor and a rigid transformation. Therefore, the task of synthesizing a morph from two original shapes is essentially one of interpolating two known curvature distributions generated from the deformed meshes generated from original shapes. This curvature-based interpolation yields smooth curvature migration along the morphing sequence. Experiments show that our shape synthesis creates realistic and intuitive shape morphs which show a gradual change between two originals.

Currently our approach is restricted to genus zero shape topologies. We can modify the mesh representation to accommodate an arbitrary topology as in [2]. We are also working on better user interface to incorporate the user's specification for more realistic and specific shape synthesis. As shown in the paper, this kind of user-specified constraint can be easily implemented in our framework.

7 Acknowledgement

We thank Henry Rowley for carefully reading this paper and providing many good comments. We thank Marie Elm for proofreading different versions of this paper.

References

1. Beier, T. and Neely, S.: Feature-based Image Metamorphosis. Proc. SIGGRAPH'92, (1992) 35–42
2. Delingette, H., Watanabe, H. and Suenaga, Y.: Simplex Based Animation. Tech report NTT Human Interface Lab. Japan, (1994)
3. Delingette, H.: Simplex Meshes: a General Representations for 3D Shape Reconstruction. INRIA report, **2214** (1994)
4. He, T., Wang, S. and Kaufman, A.: Wavelet-based Volume Morphing. Proc. Visualization'94, (1994) 85–91
5. Hebert, M., Ikeuchi, K. and Delingette, H.: A Spherical Representation for Recognition of Free-form Surfaces. IEEE Trans. PAMI, **17:7** (1995) 681–690

148

6. Horn, B.K.P.: Extended Gaussian Image. Proc. of IEEE, **72:12** (1984) 1671–1686
7. Hughes, J.: Scheduled Fourier Volume Morphing. Proc. SIGGRAPH'92, (1992) 43–46
8. Ikeuchi, K. and Hebert, M.: From EGI to SAI. CMU-CS-95-197, (1995)
9. Kang, S. and Ikeuchi, K.: The Complex EGI: New Representation for 3D Pose Determination. IEEE Trans. PAMI, **15:7** (1993) 707–721
10. Kass, M., Witkin, A. and Terzopoulos, D.: Snakes: Active Contour Models. Int'l. J. Computer Vision, **1:4** (1988) 321–331
11. Koenderink, J.: Solid Shape. The MIT Press, Cambridge, (1990)
12. Lerios, A., Garfinkle, C. and Levoy, M.: Feature-based Volume Metamorphosis. Proc. SIGGRAPH'95, (1995) 449–456
13. Little, J.: Determining Object Attitude for Extended Gaussian Image. Proc. IJCAI, Los Angeles, California, (1985) 960–963
14. Lorensen, W. and Cline, H.: Marching Cubes: A High Resolution 3D Surface Construction Algorithm. Proc. SIGGRAPH'87, (1987) 163–169
15. Shum, H., Hebert, M., Ikeuchi, K. and Reddy, R.: An Integral Approach to Freeform Object Modeling. CMU-CS-95-135, (1995)
16. Shum, H., Hebert, M., Ikeuchi, K.: On 3D Shape Similarity. CMU-CS-95-212, (1995)
17. Szeliski, R. and Tonnesen, D.: Surface Modeling with Oriented Particle Systems. Proc. SIGGRAPH'92, (1992) 185–194
18. Terzopoulos, D.: Regularization of Inverse Visual Problems Involving Discontinuities. IEEE Trans. PAMI, **8:4** (1986) 413–424
19. Terzopoulos, D., Witkin, A. and Kass, M.: Symmetry-Seeking 3D Object Recognition. Int. J. Computer Vision, **1:1** (1987) 211-221
20. Witkin, A. and Heckbert, P.: Using Particles to Sample and Control Implicit Surfaces. Proc. SIGGRAPH'94, (1994) 269–278

Appearance-Based Representations

Shape Constancy in Pictorial Relief [*]

Jan J.Koenderink, Andrea J.van Doorn, Chris Christou
Helmholtz Instituut, Universiteit Utrecht, Buys Ballot Laboratory,
Princetonplein 5, 3584 CC Utrecht, The Netherlands

Joseph S.Lappin
Vanderbilt University, Nashville, TN 37240, USA

Abstract. We measured pictorial relief for a series of pictures of a smooth solid object. The scene was geometrically identical (*i.e.*, the perspective of the same 3D scene) for all pictures, the rendering different. Some of the pictures were monochrome full scale photographs taken under different illuminations of the scene. We also included a silhouette (uniform black on uniform white) and a "cartoon" style rendering (visual contour and key linear features rendered in thin black line on a uniform white ground). Two subjects were naive and started with the silhouette, next did the cartoon, finally the full scale photographs. Another subject had seen the object and did the experiment in the opposite sequence. The silhouette rendering is impoverished, but has considerable relief with much of the basic shape. The cartoon rendering yields well developed pictorial relief, even in the naive subjects. Shading adds only small local details, but different illuminations produce significant alterations of relief. We conclude that *shape constancy* under changes in illumination rules throughout, but that the (small) deviations from true constancy reveal the effect of cues such as shading in a natural setting. Such a "perturbation analysis" appears more promising than either stimulus reduction or cue conflict paradigms.

1 Introduction

Pictures such as photographs lead simultaneously (Pirenne 1970) to the perception of a planar sheet covered with various graytones arranged in a certain order as well as to the perception of a scene in "pictorial space". Pictorial space is quite distinct from the environmental space of the observer though it similarly possesses a 3D character (Gombrich 1982). If pictorial space includes opaque objects the surfaces of these objects appear as 2D barriers to sight into pictorial

[*] An earlier version of this paper was published in *Perception*, volume 5, 1996. This paper is included in this volume by kind permission of Pion Ltd. Jan Koenderink is supported by the Human Frontiers Science Program. Andrea van Doorn is supported by the Netherlands Organization for Scientific Research (NWO). Chris Christou is supported by the ESPRIT Basic Research Action INSIGHT of the European Commission. Joe Lappin visited Utrecht for two months and was supported by the Human Frontiers Science Program.

depth. Such surfaces are "pictorial reliefs". Pictorial reliefs hold some properties in common with surfaces in environmental space, but in many respects the differences outweigh the similarities (Koenderink and van Doorn 1995). Pictorial reliefs are often neither at a well defined distance from the observer, nor do they possess stable Euclidean properties. The *depth dimension* is quite different from the dimensions (Hale 1980) that span frontoparallel planes and indeed incommensurable with these latter dimensions. Where the latter dimensions are "immediately specified" in the visual field of the observer, the depth dimension is an elaborate construction that is highly volatile in the sense that it is variable between observers (Koenderink et al 1992) and depends critically on viewing conditions (Koenderink et al 1994) and the gamut of "depth cues" that the observer can extract from the picture. Changes in these characteristics lead to changes in pictorial relief. "Shape constancy in pictorial relief" is here interpreted in the sense of invariances of pictorial relief over variations of the picture's content. In this paper such changes of content are brought about through manipulation of the *rendering* of a geometrically invariant scene (the same perspective of a single 3D scene, differently rendered).

If one looks monocularly at a monochrome photograph of a familiar scene that is conventionally illuminated, taken from a conventional angle, using a medium focal length, (contact–)printed with a full tonal scale at normal contrast and presented in good light in a frontoparallel position at a distance not too different from the focal length, a satisfactory pictorial relief is almost guaranteed. The boundary conditions are critical (Koenderink and van Doorn 1995; Pirenne 1970). We can change the rendering in many different ways. One is to take photographs after some manipulation of the actual scene. *E.g.*, one displaces the lightsources. Another method is to "retouch" the final print, *e.g.*, one replaces the print with a line tracing of it. In the former case one produces stimuli that contain "ecologically valid" depth cues, in the latter case one produces artificial depth cues, perhaps of the kind used by artists and other draftsmen to evoke pictorial spaces by manual (rather than optical) means (Clifton 1973; Jacobs 1988).

A problem in the study of pictorial relief is that it is extremely vulnerable to *reduction*. Typically, studies that aim to isolate the contribution of a single type of cue to pictorial relief turn out to be effectively sterile (Erens et al 1991, 1993a, 1993b) or result in obvious distortion or inversions (*e.g.*, shape from stereo (Johnston 1991), shape from texture (Todd and Akerstrom 1987), shape from shading (Bülthoff and Mallot 1988)). The knowledge gained from such experiments applies singularly to the set of artificial (because too much reduced) pictures and have no bearing on our understanding of complicated (but perhaps more generic) types of pictures such as photographs. Similar remarks apply to "cue conflict" paradigms.

One may study the influence of a single type of cue, however, *in the context of a nexus of many other cues* in *perturbation studies*. Here one doesn't even try to "isolate" a cue, but varies aspects of the cue parametrically and studies the corresponding changes in the pattern of residuals of the pictorial relief relative to a fiducial baseline. This technique is commonly used in engineering (Truxal 1955) to study the effect of parts on the whole (*e.g.*, a new type of tyre on a car) *in a realistic setting* because it is the only valid way to characterize complex systems. Parametrization is typically easy enough, *e.g.*, one may take photographs of a scene with different positions of the lightsource.

This study is the first phase of a parametric study of the effect of illumination direction on pictorial relief. It turns out that the effects of shading can only be sensibly studied in complicated pictures: If one tries to eliminate cues such as the visual contour, contrast with the background, *etc.* the shading leads to ambiguities that make such studies more or less irrelevant to the understanding of the generic case (Erens et al 1991, 1993a, 1993b). On the other hand the influence of contour on the pictorial relief is so important that it often dominates the shading (Ramachandran 1988). In this initial study we address the issue of the interplay of the silhouette, the contour and various shadings on pictorial relief.

We used the method of reduction for the silhouette and the contour, the method of perturbation for the shading.

2 Methods

2.1 Stimuli

As objects we used mannequins expressly sold for the display of fashion items. These objects can be obtained in a variety of postures of both genders. Many have rather complicated shapes yielding interesting silhouettes from many vantage points and striking shading patterns for typical "Hollywood–type" illuminations (Nurnberg 1948). The mannequin used in this study had a twisted pose (the "figura serpentinata–type" (Lomazzo 1958)) and a semi–gloss white finish. Thus there were no texture cues, but (mainly) strong contour and shading cues.

The mannequin was photographed in a frontal, anterior pose. (See figure 1.) For the shaded pictures we used rather simple illumination schemes (Hunter 1990): A single broad source (halogen bulb in metalized umbrella reflector) provided the main illumination, a large matt white reflection screen filled in the shadows. The background was carefully controlled and promoted a "wrap–around illumination" effect (Hattersley 1979): At the light side of the object it was slightly darker, at the dark side slightly brighter than the object. We used illuminations from the upper left and from the lower right.

In order to photograph a silhouette picture we extinguished the main light source and only illuminated the background. The contrast was raised in Adobe Photoshop© and the picture thresholded to pure black and white.

A "cartoon–type" rendering was produced by tracing variously illuminated shots and picking out the important linear features. The major component is an outline of the silhouette, the next important one is a rendering of the structure of the visual contour (the contour has components in the interior of the silhouette) and a final component involves a few characteristic ridges and ruts that tend to

154

Fig. 1. The stimuli: Silhouette (upper left), cartoon (upper right), illuminated from lower right (lower left), illluminated from upper left (lower right).

show up in the shading from many different lighting setups. The cartoon rendering is equivalent to the type of rendering often used in 19th century linecopy, e.g., Flaxman's popular renderings of classical sculpture (Adams 1974).

All stimuli were converted to grayscale pixmaps that were shown on an 8–bit monitor (although only 1 bit was used for the silhouette and the cartoon renderings). Quality of the full scale pictures was comparable to good postcard size photographs. In the image the original scene radiances are compressed, roughly according to a power function. The compression is very similar to that in generic photographic renderings. Stimuli subtended 18 cm (height) on the screen and were viewed from 57 cm. Stimuli were viewed in a semi dark room, with monocular viewing, using a head and chin rest. We have no indications that monocular versus binocular viewing would make any difference. From the viewing position subjects had no clear sense of light direction in the room, thus possible conflicts with the illumination in picture space were avoided. (This might conceivably be important.)

2.2 Subjects

The experiment was performed by three subjects (each an author). CC is emmetropic, AD is slightly myopic and wears her usual correction, JL is myopic and wears his usual correction. In order to address the possible effect of familiarity subjects CC and JL were carefully kept naive in the sense that they knew neither the object nor the photographs, whereas subject AD knew both the object and all photographs very well. Subjects CC and JL did the experiment in the sequence of increasing richness of optical structure (thus silhouette ⇒ cartoon ⇒ shaded pictures) whereas subject ADfollowed almost the reverse sequence (shaded pictures ⇒ silhouette ⇒ cartoon) and was thus very familiar with the shape. Notice that subjects CC and JL were in no position to guess initially whether the pose was an anterior or posterior one (the silhouette would be the same (Hogarth 1981)), whereas subject AD knew it from the start.

2.3 Paradigm

We use a paradigm that is fast and intuitive for the subjects and that allows us to draw detailed inferences concerning local attitude (slant and tilt) of the pictorial relief. This method has been described in detail elsewhere (Koenderink and van Doorn 1995; Koenderink et al 1992). Briefly, we superimpose a "gauge figure" on the picture and let the subject change the geometry of this probe so as to perceptually "fit" the relief. The gauge figure is an ellipse with a short line segment extended from its center. The subject tries to fit the ellipse (adjusting both its orientation and eccentricity) so as to appear as a circle "painted" on the pictorial surface with the line segment appearing as a surface normal. The ellipse is interpreted as a circle in projection, thus we can convert the parameters of the ellipse to depth gradient or (equivalently) slant and tilt. About 300 points in the picture within the area of the silhouette were probed (one at a time) in random order. This task is very easy. Most subjects take between three quarters of an hour for a session (few seconds per setting), though subject JL took about five times as long. The settings were repeated several times (on different days) to allow us to judge reproduceability.

Various conceptual issues might be raised concerning the paradigm. For instance, since an elliptical figure by itself suggests a slant does this act as a perturbation on pictorial relief? Such issues are inherently difficult (maybe impossible) to answer. However, for the present study we can simply regard the paradigm as an operational method with which to *define* pictorial relief.

2.4 Data evaluation

We apply the usual statistical methods to judge the scatter in the data, then check whether the data conform to any pictorial surface at all. In order to do so the *curl* of the empirically determined "depth gradient" should vanish within the scatter. (Remember that for a smooth function $f(x, y)$ the curl of the gradient is identically zero, this amounts to the equality of mixed partial derivatives $\partial f/\partial x \partial y = \partial f/\partial y \partial x$.) This was the case for all conditions. Then surfaces are fittedto the data. These surfaces are a convenient and intuitive summary of the data. They summarize the deterministic structure of the data completely, *i.e.*, what remains is explained by the session to session variability. We show these results as depth maps with isodepth contours. Each such picture succinctly presents about 600 data items (empirical slant and tilt at 300 points), thus the comparisons are based on 2400 data items (each a mean of up to 9 settings) per subject. Comparisons between conditions are made on the basis of these surfaces (referred to as the "pictorial reliefs").

3 Experiment

3.1 Procedures

Subjects CC and AD performed the task of fitting gauge figures nine times on each of the stimuli, subject JL did the silhouette and cartoon conditions six times, the shaded conditions three times. They finished one stimulus before moving on to the next though in different orders (see methods). In one session each point at which the pictorial surface was probed was visited three times. Different sessions were done on different days. The total experiment lasted several weeks.

3.2 Results

Scatter in the settings For each point at which the pictorial surface attitude was probed we have several independent settings, yielding a point cluster in depth gradient space. Our standard procedure (Koenderink et al 1992) finds a 2D median and discards the two points farthest from this median location as possible outliers. The remaining cluster is used to find a mean location and a covariance ellipse. The covariance ellipse is represented by the principal axes and the orientation. We find that the ellipses are very eccentric, with the major axis aligned with the gradient direction (direction of steepest slope). This means that most of the variance is in the slant whereas the tilt (at least for non–vanishing slants) is rather well defined. The standard deviation in the magnitude of the depth gradient (that is, the tangent of the slant angle) amounts to about 15% of its magnitude. It is perhaps remarkable that the standard deviations are rather similar for the four conditions given the scarcity of cues in especially the silhouette.

Consistency In all cases the data turned out to be almost perfectly consistent. The violations of surface consistency (failure of the curl of the "depth gradient" to vanish identically) are fully explained by the scatter in repeated sessions. This result is perhaps surprising given the fact that the pictorial surface in the case of the silhouette is completely "filled in" (that is: imaginary) whereas in the case of the shaded images there are strong indications of the nature of the physical relief.

Comparison of silhouette and cartoon versus shaded pictures As might have been expected from such a reduced picture the results are not very articulate for the case of the *silhouette*. (See figures 2, 3 and 4.)There are interesting and revealing differences between the three subjects though. Subject AD was familiar with the picture, she knew that the view was an anterior one. That she used this foreknowledge is evident from the fact that her pictorial relief clearly shows the left breast of the mannequin though the silhouette shows no indication of it and indeed the view might as well have been a posterior one with no breast showing at all. This detail is "imaginary relief", rather than "pictorial relief". In subject CC's (who was fully naive) relief this left breast is missing, whereas in subject JL's (also naive) it is present, though much less prominent than in the fuller renderings. In AD's result we see the twist of the posture, which is completely missing in CC and JL's results. Thus it cannot be doubted that the subjects were able to use familiarity cues and in the case of subject AD with astonishing effect. In her case the high resemblance to the actual shape is obvious and the availability of additional cues would only raise any correlation marginally. In the case of the naive observers CC and JL, the relief is not very articulate (ventral region, posture of legs) and apparently interpolated from the outline. In the case of JL the left breast seems to be "invented" as a mirror image of the right one. Interestingly, subjects CC and JL both perceived the silhouette only as an anterior view.

For all subjects the relief for the *cartoon* picture is highly articulate and indeed hardly less so than that for the fully shaded gray tone pictures. This shows that even for the naive observers the visual contour with a few linear details suffices to capture most of the structure of the shape. The differences between the silhouette case and the cartoon rendering are noticeable for subjects CC and JL, less so for subject AD who already did quite well on the silhouette given

Subject CC

Silhouette
Cartoon

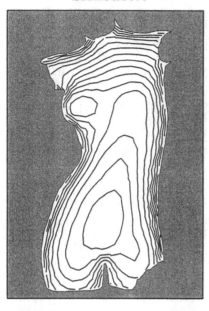
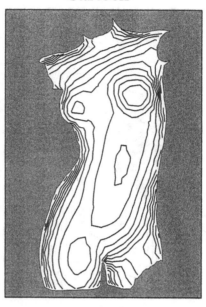

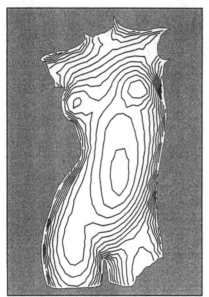
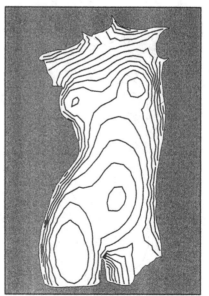

**Illumination
from upper left**
**Illumination
from lower right**

Fig. 2. The pictorial relief for subject CC for the various renderings.

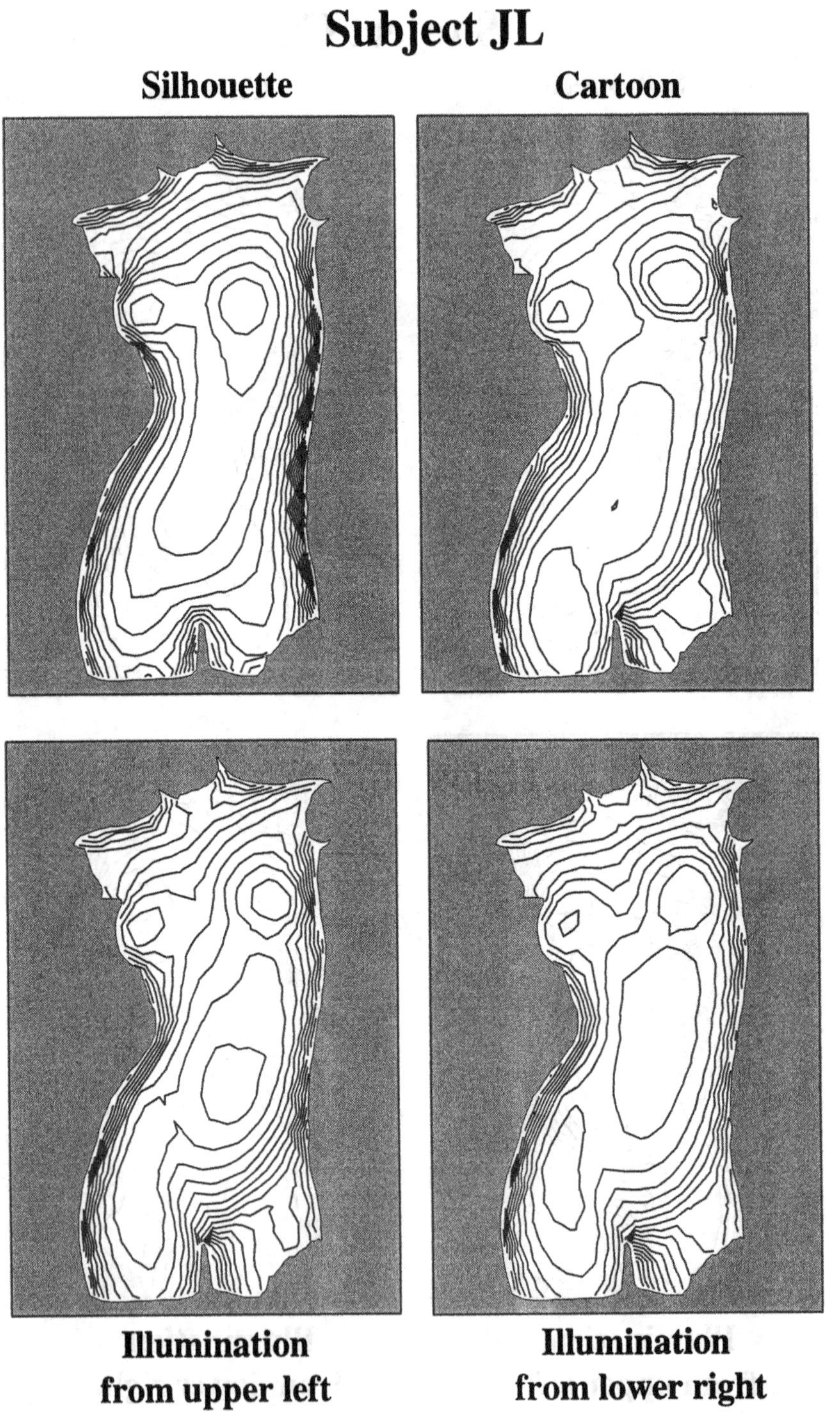

Subject JL

Silhouette   **Cartoon**

**Illumination
from upper left**   **Illumination
from lower right**

Fig. 3. The pictorial relief for subject JL for the various renderings.

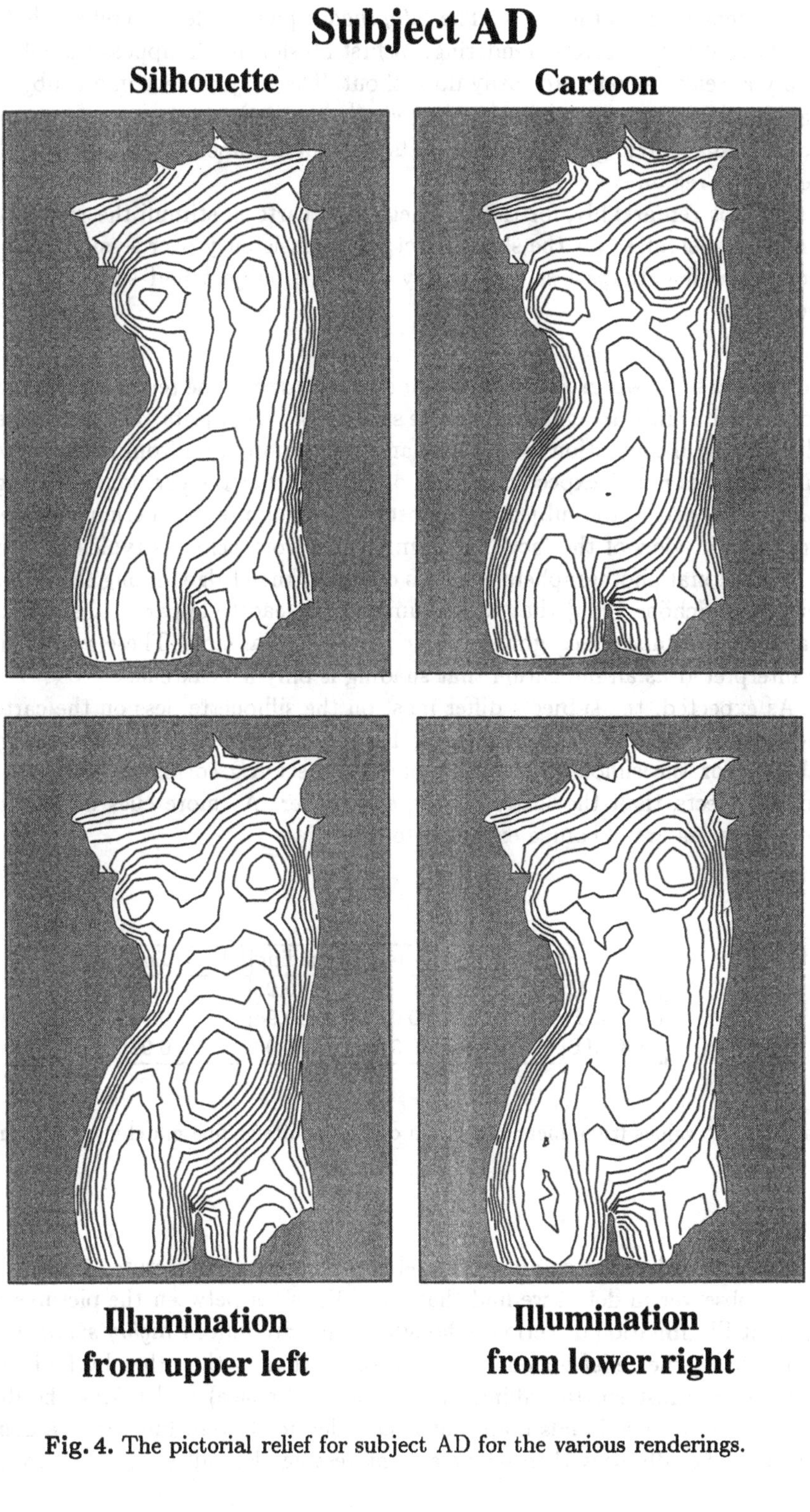

Subject AD

Silhouette

Cartoon

Illumination
from upper left

Illumination
from lower right

Fig. 4. The pictorial relief for subject AD for the various renderings.

her foreknowledge of the shape. The *fully shaded* pictures lead to reliefs that add little to that of the cartoon rendering. At first blush it might appear that shading is only a weak cue one can easily do without. The major effect in all subjects is that the shaded pictures capture the twist in the body, the counter–rotation of shoulder and pelvic girdle. Other effects are small shape corrections, mainly in the ventral region.

Perhaps remarkably, for the silhouette and cartoon stimuli the results were just as consistent as for the shaded pictures. Consistency over sessions was also good, though perhaps less remarkably after the subject had seen the picture once.

Differences between differently shaded pictures The reliefs for the shaded pictures are highly articulate and quite similar for all subjects. The differences in relief for the illumination from below and the illumination from above are only minor. This is not unexpected in view of the fact that the cartoon rendering already suffices to let the subjects get most of the structure, but perhaps somewhat surprising in view of the fact that illumination from below is typically avoided by professional photographers *etc.* It is common knowledge among them (Nurnberg 1948; Schöne 1954) that such illuminations lead to apparent distortions of shape and diminished recognition of *e.g.*, even familiar faces. These results might be interpreted as an indication that shading is only a weak cue.

As expected, the subjects differ most on the silhouette, less on the cartoon, and least on the shaded picture (table 1). (Correlations were done on the mean values.) For the silhouette the non–naive subject AD correlates less with the naive subjects than the latter among each other. As more cues are available ideosyncracies are of course expected to count less.

subjects	silhouette	cartoon	LR shading	UL shading
AD-CC	0.85	0.95	0.98	0.87
AD-JL	0.83	0.91	0.94	0.94
CC-JL	0.94	0.91	0.96	0.92

Table 1. R^2 values from linear regression of depths after affine matching. Comparison of subjects.

However, when we compare the relief for the shaded pictures for any particular observer in detail we find that the differences between the pictures with different illumination direction—though small—are indeed *highly significant*. If we try to fit the relief for one illumination to that of the other by finding the best linear transform (an affinity in 3D pictorial space) and taking the difference we obtain a residuals map that is not identically zero but shows a definite pattern. The affinities themselves are interesting: For all subjects we find an overall slant of about 8° about a similar axis (table 2). Thus this slant is a major effect, apparently depending on the difference of illumination direction. An analysis of residuals reveals that (again for all subjects) all slopes facing the source come forward, those facing away from the source recede. This yields an example of measurable perturbations due to a parametric change in the scene.

subject	slant	tilt	original R^2	R^2 after correction
AD	8.5°	−65°	0.71	0.93
CC	7.1°	−93°	0.39	0.91
JL	9.4°	−30°	0.57	0.94

Table 2. R^2 values from linear regression and slant, tilt and R^2 values from affine regression of depths from shaded images (illumination from upper left and from lower right).

4 Conclusions

We have studied the pictorial reliefs due to pictures of a single object in a single view (constant scene geometry for all stimuli) as a function of the rendering. Two of the pictures were full (gray-)scale pictures of photographic quality and differed only in the way the scene was illuminated: One from below and from the right, the other from above and from the left. The other pictures were prepared by a process of reduction. In a cartoon rendering only linear features (*e.g.*, the visual contour) were indicated as black lines on a uniform white ground. In a silhouette rendering the object was only visible as a uniform black blob on a uniform white ground.

We find that the cartoon rendering leads to a fully developed pictorial relief and that full shading adds relatively little in a quantitative sense, whereas the silhouette rendering leads to impoverished relief for naive observers and to a reasonably articulate but inaccurate relief for an observer who was familiar with both the object and the gray scale pictures. We also find that the difference between the shaded pictures is small but significant. In the first approximation "shape constancy" rules. But the significant differences mean that the influence of shading can be studied as a change of the relief as the rendering is parametrically perturbed.

To us this seems an important methodological observation: If one tries to *isolate* the shading as a depth cue nothing is left because the reduction has gone too far and one obtains ambiguous results (Erens et al 1991). Subjects feel uneasy in judging relief of shaded images without contours; it certainly doesn't feel "natural". If one tries to study shading in a more natural context one finds that shading seems to add very little because most of the relief is already determined

by other cues such as the visual contour. This does not mean that the shading
cue is irrelevant though, because a change in the shading pattern leads to small
but very significant changes of the pictorial relief. Such changes will be missed
when only a single instance of shading is studied. The way to study cues such
as shading appears to be to study pictorial relief in families of realistic pictures
in which certain parameters that pertain to the shading (*e.g.*, the direction of
illumination) are systematically perturbed.

More generally, it appears that many of the optical structures that lead to
pictorial (or visual, because it is unlikely that these observations apply merely
to pictures) relief are most effective in a context where a rich nexus of cues
is available. Reduction of the stimulus (or, for that matter, "conflicting cues")
typically leads to merely artificial results. The more promising way to study the
efficacy of optical structures (not so much in the sense of leading to veridical
results as in the sense of leading to confidendence in the judgement of a single
observer and concordance between judgements of several observers) seems to be
a perturbation analysis. Here we have shown that such an approach is indeed
viable.

References

1. Bülthoff H B, Mallot H O, 1988 "Integration of depth modules: stereo and
   shading" *J.Opt.Soc.Am.* **A5** 1749–1757
2. Clifton J, 1973 *The eye of the artist* (Westport, Conn.: North Light Publishers)
3. Erens R G F, Kappers A M L, Koenderink J J, 1991 "Limits on the perception
   of local shape from shading", in *Studies in Perception and Action* Eds P J Beek,
   R J Bootsma, P C W van Wieringen (Amsterdam) pp. 65-71
4. Erens R G F, Kappers A M L, Koenderink J J, 1993a "Perception of local shape
   from shading" *Perception & Psychophysics* **54** (2) 145–156
5. Erens R G F, Kappers A M L, Koenderink J J, 1993b "Estimating local shape
   from shading in the presence of global shading" *Perception &
   Psychophysics* **54** (3) 334–342
6. Adams R M, 1974 *The roman stamp. Frame and facade in some forms of
   neo-classicism* (Berkeley: University of California Press)
7. Gombrich E H, 1982 *The image and the eye, further studies in the psychology of
   pictorial representation* (Oxford: Phaidon)
8. Hale N C, 1980 *Abstraction in Art and Nature* (New York: Watson–Guptill)
9. Hattersley R, 1979 *Photographic lighting, learning to see* (Englewood Cliffs,
   New Jersey: Prentice–Hall, Inc.)
10. Hogarth B, 1981 *Dynamic light and shade* (New York: Watson–Guptill)
11. Hunter F, Fuqua P *Light, Science and Magic, an introduction to photographic
    lighting* (Boston: Focal Press)
12. Jacobs T S, 1988 *Light for the artist* (New York: Watson–Guptill)
13. Johnston E E, 1991 "Systematic distortions of shape from stereopsis"
    Vis.Res. **31** 1351–1360
14. Koenderink J J, Doorn A J van, 1995 "Relief: Pictorial and otherwise" *Image
    and Vision Computing* **13** (5) 321–334
15. Koenderink J J, Doorn A J van, Kappers A M L, 1992 "Surface perception in
    pictures" *Perception and Psychophysics* **32** 487–496

16. Koenderink J J, Doorn A J van, Kappers A M L, 1994 "On so–called paradoxical monocular stereoscopy" *Perception* **23** 583–594
17. Lomazzo P, 1958, in: *A documentary history of art* Vol.II, Ed. E G Holt (Garden City, New York: Doubleday)
18. Nurnberg W, 1948 *Lighting for portraiture* (London: The Focal Press)
19. Pirenne M H, 1970 *Optics, Painting and photography* (Cambridge: Cambridge University Press)
20. Ramachandran V, 1988 "Shape from shading" *Nature* **331** Januari 14th
21. Schöne W, 1954 *Über das Licht in der Malerei* (Berlin: Gebr.Mann Verlag)
22. Todd J J, Akerstrom R A, 1987 *JEP: Human Perc. and Performance* **13(2)** 242–255
23. Truxal J G, 1955 *Automatic feedback control system synthesis* (New York: McGraw-Hill)

Dimensionality of Illumination Manifolds in Appearance Matching

Shree K. Nayar[1]   and Hiroshi Murase[2]

[1] Department of Computer Science,
Columbia University,
New York, N.Y. 10027, USA

[2] NTT Basic Research Laboratory,
Morinosato Wakamiya, Atsugi-shi,
Kanagawa 243-01, Japan

Abstract. Appearance matching was recently demonstrated as a robust and efficient approach to 3D object recognition and pose estimation. Each object is represented as a continuous appearance manifold in a low-dimensional subspace parametrized by object pose and illumination direction. Here, the structural properties of appearance manifolds are analyzed with the aim of making appearance representation efficient in off-line computation, storage requirements, and on-line recognition time. In particular, the effect of illumination on the structure of the appearance manifold is studied. For an ideal diffuse surface of arbitrary texture, the appearance manifold is linear and of dimensionality 3. This enables the construction of the entire illumination manifold from just three images of the object taken using linearly independent light sources. This result is shown to hold even for illumination by multiple light sources and for concave surfaces that exhibit interreflections. Finally, a simple but efficient algorithm is presented that uses just three manifold points for recognizing images taken under novel illuminations.

1   Introduction

Appearance matching techniques are fast becoming popular in machine vision. Recent applications include face recognition [20] and the real-time recognition of complex 3D objects [7]. A representation of object appearance called the *parametric eigenspace* has resulted from this work [6]. For a given vision application, a visual workspace is first defined as the range of visual appearances that result from varying the parameters of the task. This workspace is sampled to obtain an image set that is used to compute a low-dimensional linear subspace [15], called the eigenspace, in which the visual workspace is represented by one or more parametrized manifolds. During recognition, novel images are projected to the eigenspace. The closest manifold and the exact location of the closest point on the manifold reveal the task parameters.

The parametric eigenspace representation has found several applications. These include learning [7] and real-time recognition of 3D objects [6], positioning and tracking of 3D objects by a robot manipulator [11], and illumination planning for robust object recognition [8]. Recently, a recognition system with 100 complex objects in its database was developed that is solely based on appearance matching [9]. The sheer efficiency of appearance matching enables the system to accomplish both recognition and pose estimation in real time using no more than a standard workstation equipped with an image sensor (see Figure 1).

Fig. 1. An automated recognition system with 100 objects in its database. A complete recognition and pose estimation cycle takes less than 1 second on a Sun SPARC workstation [9].

In the context of large systems, the primary bottleneck in appearance matching has turned out to be the learning stage which includes the acquisition of large image sets, the computation of eigenspaces from large covariance matrices, and the construction of parametric appearance manifolds. In object recognition, each object is represented as a separate manifold in eigenspace that is parametrized by pose and illumination. The efficiency of the learning stage is determined by the number of sample images needed to compute an accurate appearance manifold. This brings us to the following question: What is the smallest number of images needed for constructing the appearance manifold for any given object?

The answer to the above question lies in the structural properties of appearance manifolds. The structure of an object's manifold is closely related to its geometric and reflectance properties. In special cases, such as solids of high symmetry and solids of revolution, one can make concrete statements regarding the dimensionality of the manifold. For instance, given a fixed illumination direction and viewpoint, the manifold for a sphere of uniform reflectance is simply a point since the sphere appears the same in all its poses. This unfortunately is an extreme instance of little practical value. Under perspective projection, the relation between object shape and manifold structure is complex to say the least. A general expression that relates object pose to manifold structure would be much to hope for.

In contrast, the function space associated with object reflectance is more concise and hence conducive to analysis. Our work draws on the previous work of Petrov [17], Shashua [18][19], Hallinan [5], Nayar and Murase [10] and Epstien et al. [3][3]. It is possible to establish, under certain reflectance assumptions, a closed-form relationship between illumination parameters and manifold structure. Given that the eigenspaces we use are linear subspaces, the class of linear reflectance functions [17][18] is of particular interest to us. For this reflectance class, the structure of the illumination manifold is completely determined from a small number of samples of the manifold. In particular, for Lambertian surfaces of arbitrary texture, the entire illumination manifold can be constructed from just three images taken using known illuminants. Alternatively, the dimensionality of the illumination manifold is exactly 3. This result becomes intuitive when one considers photometric stereo [21], where three light sources provide all the information required to estimate the albedo and the unit normal at each surface point. We use the above bound on the manifold dimensionality to show that novel images of the object can be recognized from just three projections on the illumination manifold in eigenspace without the explicit construction of the manifold.

In addition, the validity of the above results for illumination by multiple sources and in the presence of interreflections caused by concave surfaces is demonstrated. With respect to interreflections, it was shown in [14] that a concave Lambertian surface of arbitrary texture behaves exactly like a Lambertian one without interreflections, but with a different set of surface normals and albedo values. The underlying assumption is that all points on the surface are visible and illuminated, i.e. no self-occlusions and self-shadows. We have used this notion of a pseudo surface to show that the dimensionality of the illumination manifold in eigenspace is in fact preserved in the case of interreflections.

For ideal diffuse objects, these results have direct implications on the efficiency of both learning and recognition, as they dramatically reduce the number of images needed for appearance representation. These results for diffuse surfaces stem from the observation [18] that the image of a diffuse object under any illumination can be expressed as a linear combination of images taken using three independent basis illuminants. Such a linear combination does not generally exist for objects with nonlinear reflectance functions. For instance, a pure specular object would produce only strong highlights for each of the basis illuminants. The highlights produced by a novel source cannot in general be expressed as a linear combination of basis images. In fact, it is hard to envision non-trivial upper bounds on the dimensionality of a vector space containing illumination manifolds for the class of nonlinear reflectance functions.

---

[3] Some of the results presented in this paper overlap with those of Belhumeur et al. [1] and Belhumeur and Kriegman [2]. These works and the one reported in this paper were conducted more or less concurrently.

2 Linear Reflectance Models

We will assume throughout our presentation that all surface properties and image brightness values correspond to a single wavelength, λ. To accomodate the general case of colored surfaces and colored illuminants, we assume that all brightness values are measured using narrow-band filters, say, narrow-band red, green, and blue filters. This ensures that the wavelength of light is, in effect, fixed for any given color band.

A linear reflectance function may be written as:

$$e(\mathbf{x}) \; = \; \mathbf{p}(\mathbf{x}) \cdot \mathbf{q} \tag{1}$$

where, $e(\mathbf{x})$ is the image irradiance or intensity at point $\mathbf{x}$, $\mathbf{p}(\mathbf{x})$ represents local surface properties, and $\mathbf{q}$ is an arbitrary vector that could depend on the illumination and the viewpoint of the observer. If $Dim(\mathbf{p}) = Dim(\mathbf{q}) = k$, we have a k-order linear reflectance model (see [17][18] for details).

When is the linear reflectance model valid in practice? In general, reflectance functions can be viewed as the combination of surface (specular) and body (diffuse) components [13]. Surface reflection is a nonlinear function of viewpoint. In contrast, the body component is relatively less viewpoint dependent. Though the dependence can be significant in the case of surfaces with high macroscopic roughness (see [16]), many man-made objects (for instance, those with matte paints) as well as some natural surfaces can be approximated by a linear model. The most popular and widely used approximation is the Lambertian model where surface radiance, and hence also image brightness, depend only on the irradiance of the surface and not the observer's viewpoint. The image brightness of a Lambertian surface element illuminated by a point light source is:

$$e(\mathbf{x}) \; = \; \mathbf{n}(\mathbf{x}) \cdot \mathbf{s} \tag{2}$$

Here, $\mathbf{n}(\mathbf{x}) = \rho(\mathbf{x})\,\hat{\mathbf{n}}(\mathbf{x})$ is the normal vector, where ρ is the local surface albedo and $\hat{\mathbf{n}}$ is the local unit surface normal. Similarly, $\mathbf{s} = b\,\hat{\mathbf{s}}$ is the source vector, where b represents the intensity of the source and $\hat{\mathbf{s}}$ is a unit vector in the direction of the source. We are assuming here that the source is distant and hence its direction is independent of the location of the surface point in the scene. From the above expression, we see that the Lambertian model is a linear one of order 3.

3 Images as Linear Combinations

Given that $\mathbf{s}$ is a constant three-dimensional vector, it is clear that three non-coplanar source vectors, say $\{\mathbf{a}_1, \mathbf{a}_2, \mathbf{a}_3\}$, can be used as a basis to represent any source vector. Any arbitrary source vector is simply a linear combination of the three basis vectors:

$$\mathbf{s} \; = \; \alpha_1\,\mathbf{a}_1 + \alpha_2\,\mathbf{a}_2 + \alpha_3\,\mathbf{a}_3 \tag{3}$$

Let us define $\Lambda = [\mathbf{a}_1\ \mathbf{a}_2\ \mathbf{a}_3]$ as the basis source matrix. Then, the coefficient vector $\Phi = [\alpha_1\ \alpha_2\ \alpha_3]^{\mathrm{T}}$ for any given source can be determined as:

$$\Phi = \Lambda^{-1}\mathbf{s} \tag{4}$$

The image brightness values of a Lambertian surface point due to the three basis illuminants are:

$$
\begin{aligned}
e_1(\mathbf{x}) &= \mathbf{n}(\mathbf{x}) \cdot \mathbf{a}_1 \\
e_2(\mathbf{x}) &= \mathbf{n}(\mathbf{x}) \cdot \mathbf{a}_2 \\
e_3(\mathbf{x}) &= \mathbf{n}(\mathbf{x}) \cdot \mathbf{a}_3
\end{aligned}
\tag{5}
$$

From (3) and (5), the brightness of the surface point due to a novel source is:

$$
\begin{aligned}
e(\mathbf{x}) &= \mathbf{n}(\mathbf{x}) \cdot \mathbf{s} \\
&= \alpha_1 \mathbf{n}(\mathbf{x}) \cdot \mathbf{a}_1 + \alpha_2 \mathbf{n}(\mathbf{x}) \cdot \mathbf{a}_2 + \alpha_3 \mathbf{n}(\mathbf{x}) \cdot \mathbf{a}_3
\end{aligned}
\tag{6}
$$

If we define $\Gamma = [\,e_1,\ e_2,\ e_3\,]^{\mathrm{T}}$, we have:

$$e(\mathbf{x}) = \Gamma^{\mathrm{T}}(\mathbf{x})\Phi \tag{7}$$

Note the similarity between the linear combination in the above expression and the one in (3). The brightness due to any novel source is the same linear combination of the basis brightness values as the novel source is of the basis illuminants. This result may not seem obvious at first glance. However, it turns intuitive when one notes that the three brightness values in $\Gamma(\mathbf{x})$ and the three corresponding sources Λ contain all the information required to estimate the albedo $\rho(\mathbf{x})$ and the unit normal vector $\hat{\mathbf{n}}(\mathbf{x})$ as done in the case of photometric stereo [21]. It is therefore not surprising that the three brightness values corresponding to the basis illuminants can be used to predict brightness for any desired source vector.

Note that (7) holds for all points on the imaged object. Hence, if the basis illuminants and the novel source are distant and are visible to all observed points on the object, the image $\mathbf{I}$ of the object under a novel illumination can be expressed as a linear combination of its three images $\Pi = [\,\mathbf{I}_1, \mathbf{I}_2, \mathbf{I}_3\,]$ due to the basis illuminants.

$$\mathbf{I} = \Pi\,\Phi \tag{8}$$

The above linear combination holds true irrespective of the texture (albedo variation) of the surface. It appears that the above results have been arrived at independently by several investigators (see Petrov [17], Shashua [18][19], Hallinan [5], Nayar and Murase [10] and Epstien et al. [3]). The linear combination of (8) was initially used in its explicit form for the analysis of surface color in [17] as well as specularity detection and photometric recognition in [18].

The main assumptions used thus far are; (a) the surface is illuminated by a single distant point source that is visible to all imaged points, and (b) the surface is convex and consequently image brightness is due to only direct source illumination and not interreflections. In subsequent sections, we will see that these two assumptions can be relaxed quite a bit, and expression (8) does in fact hold true for multiple sources and concave surfaces with interreflections.

4 Multiple Novel Sources

Consider a Lambertian surface simultaneously illuminated by R point sources. In this case, the image brightness of a surface point will be the sum of the contributions of individual sources:

$$e(\mathbf{x}) = \sum_{r=1}^{R} \mathbf{n}(\mathbf{x}) \cdot \mathbf{s}_r$$

$$= \mathbf{n}(\mathbf{x}) \cdot \sum_{r=1}^{R} \mathbf{s}_r \ = \ \mathbf{n}(\mathbf{x}) \cdot \tilde{\mathbf{s}} \tag{9}$$

Here, $\tilde{\mathbf{s}}$ serves are a single *effective source* and is simply the average, or center of mass, of the set of individual source vectors. In the above derivation, the individual sources need not be point sources. They could be extended sources with arbitrary radiance functions and still each be replaced by an effective point source. The above result is well-known and has surfaced in various guises in previous work. Its implication is that, for Lambertian reflectance, any number of novel sources that are all visible to the entire imaged surface can be viewed as a single effective point source $\tilde{\mathbf{s}}$. Hence, the linear combination of (8) holds true for multiple novel sources; the coefficient vector Φ is simply that of $\tilde{\mathbf{s}}$.

5 Interreflections

Next, let us consider the case of a concave Lambertian surface with arbitrary texture. We assume that each of the basis illuminants, and subsequent novel sources, are visible to all points on the observed surface. The image brightness of each surface point in this case is due to not only the source but also the contributions of other points on the surface that are visible to it. Though, in general, an infinite number of interreflections occur between any two mutually visible surface points, the brightness of a surface point can be expressed as the sum of the brightness due to direct source illumination and contributions due to the *final* radiance values of all surface points visible to it. Thus, for any given wavelength λ of incident light, the brightness image of the surface is:

$$\mathbf{I} \ = \ \mathbf{I}_s + \mathbf{P\,K\,I} \tag{10}$$

where, $\mathbf{I}_s$ is the image due to direct source illumination, $\mathbf{P}$ is the albedo matrix whose diagonal elements are the albedo values of individual infinitesimal elements on the surface, and $\mathbf{K}$ is the interreflection kernel that captures the relative geometric configurations of pairs of surface elements.

In [14], the above brightness equation was analyzed to show that the concave surface behaves exactly like a Lambertian one without interreflections, but with a different set of surface normals and albedo values. The underlying assumption is that all points on the surface are visible and illuminated, i.e. no self-occlusions and self-shadows. This apparent surface is called the *pseudo surface*. The relation between the pseudo surface and the actual surface was found to be:

$$\mathbf{F}_p \ = \ [\mathcal{I} - \mathbf{P\,K}\,]^{-1}\,\mathbf{F} \tag{11}$$

Here, the matrix $\mathbf{F}_p = [\mathbf{n}_p{}^1, \mathbf{n}_p{}^2, ..., \mathbf{n}_p{}^n]$ represents the pseudo surface and is composed of the pseudo normal vectors of all n visible surface elements. The true surface is given by the matrix $\mathbf{F} = [\mathbf{n}^1, \mathbf{n}^2, ..., \mathbf{n}^n]$.

This result implies that, under any given illumination that satisfies the assumptions stated above, the image of a concave surface $\mathbf{F}$ with all its interreflections exactly equals the image of its corresponding pseudo surface $\mathbf{F}_p$ without interreflections. As a result, all the linear combinations derived in the previous sections hold true for concave Lambertian surfaces of arbitrary texture. In general, the pseudo surface varies with the wavelength of incident light. This dependence on wavelength vanishes in the cases on gray textured surfaces illuminated by white-light sources. In the general case of colored surfaces and colored illuminants, the linear combination of (8) remains valid when the images are taken using narrow-band filters, say, narrow-band red, green, and blue filters [12][4].

6   Multispectral Images

If the application involves the use of a color image sensor, as stated earlier, it is assumed that each color band is obtained using a narrow-band spectral filter. Then, the intensities of the three basis illuminants, irrespective of their spectral distributions, are fixed for each of the narrow bands. In other words, all the results derived thus far remain valid in each band. Appearance manifolds for an object can then be constructed independently for each band (as in [9]) and recognition is deemed successful if all bands of a novel image are found to match the same object in the appearance database.

Alternatively, an image vector can be constructed by concatenating the multiple bands of the color image. Appearance representation in subspaces is invariant to the order of concatenation since this order only alters the order of values in the principle vectors (dimensions) of the subspace [15]. However, since the linear combinations of the previous sections can be expected to differ between bands, a concatenated vector cannot be assumed to represent any single linear combination. Here, some of the results related to color-rank in [17] could lead to interesting results.

7   Illumination Manifold in Eigenspace

We are now equipped to analyze the dimensionality of illumination manifolds in eigenspaces. An eigenspace $\mathcal{E}$ is an image subspace that is typically computed using the Karhunen-Loéve transform [15]. The bases of $\mathcal{E}$ are then the normalized eigenvectors of the covariance matrix computed from an image set, one that typically includes images of a large number of objects taken at different poses and illumination conditions during a learning (or training) stage [7]. Suppose the dimensionality of the eigenspace is d. The eigenvectors $e_s, s = 1, 2, ...d$ are those with the largest eigenvalues, λ_s, of the covariance matrix, such that, $\lambda_1 \geq \lambda_2 \geq .... \geq \lambda_d$. Our main concern here is the analysis of the illumination manifold of an object, i.e. the projections in eigenspace $\mathcal{E}$ of images of an object taken under different source directions, for a fixed pose.

Scale and brightness normalizations are applied to all object images before they are either used to construct eigenspace representations or to recognize novel object images [7]. The scale normalization ensures that both appearance representation and recognition are invariant to the magnification of the imaging system under weak-perspective projection. The brightness normalization is used to achieve invariance to the intensity of illumination. As a result of brightness normalization, all images lie on a unit ball in a high but finite dimensional Hilbert space.

First, let us define the normalized basis images as $i_1 = I_1/m_1$, $i_2 = I_2/m_2$, and $i_3 = I_3/m_3$, where, $m_i = \| I_i \|$. Further, we define the magnitude matrix M to be a 3×3 diagonal matrix with the m_i as its three diagonal elements, and the normalized image matrix as $N = [i_1 \, i_2 \, i_3]$. Using (8), a novel image can be written as:

$$I = \alpha_1 m_1 i_1 + \alpha_2 m_2 i_2 + \alpha_3 m_3 i_3$$
$$= N M \Phi \tag{12}$$

The magnitude of the novel image is related to its source coefficient vector as:

$$m = \sqrt{\Phi^T N^T M^T M N \Phi} \tag{13}$$

The normalized novel image can now be expressed as a combination of the three normalized basis images:

$$i = \frac{1}{m} N M \Phi \tag{14}$$

Now, the projections of the three normalized basis images in a d-dimensional eigenspace are:

$$g_1 = [e_1 \, e_2 \, ... \, e_d]^T i_1$$
$$g_2 = [e_1 \, e_2 \, ... \, e_d]^T i_2$$
$$g_3 = [e_1 \, e_2 \, ... \, e_d]^T i_3 \tag{15}$$

Likewise, the projection of a novel image is:

$$g = [e_1 \, e_2 \, ... \, e_d]^T i \tag{16}$$

If we define the basis projection matrix as $G = [g_1 \, g_2 \, g_3]$, the above expression and (14) yield:

$$g = \frac{1}{m} G M \Phi \tag{17}$$

Therefore, given the three eigenspace projections corresponding to the basis illuminants, we can determine the projection for any novel source s from its coefficient vector Φ. If the basis and novel images are not normalized, the illumination manifold spans a linear subspace $\mathcal{G} = g(s)$ whose dimensionality is 3 for a Lambertian surface of arbitrary texture, irrespective of the dimensionality d of the eigenspace used. When the basis and novel images are brightness normalized, the illumination manifold spans a nonlinear subspace of dimensionality

2. In this case, the nonlinearity arises from the normalization procedure, and the reduction in dimensionality by one results from all normalized images being constrained to lie on a unit ball in the Hilbert space.

The above results imply that we do not need to take a large number of images by sampling the entire illumination space for each pose of each object as done in [7]. For any given object pose, the entire illumination manifold can be constructed from just the three basis projections.

It is worth reiterating that the above results are valid only for Lambertian surfaces and not much can be stated regarding the dimensionality of the illumination manifold for surfaces with nonlinear reflectance functions. This becomes intuitive when one considers the extreme case of a pure specular object that only produces highlights for each of the basis sources. In this case, the image produced by any novel source cannot in general be expressed as a linear combination of any number of basis images. Fortunately, there does exist a class of real-world objects that closely approximate Lambertian reflectance, and for such objects the above results prove useful as shown in the following sections.

8 Recognition of Novel Object Images

We have assumed thus far that the eigenspace $\mathcal{E}$ is known a-priori. In practice, such an eigenspace is computed from a large image set obtained by varying pose and illumination in small increments [7]. Given the above results, it is possible to dramatically reduce the number of images that need to be taken during learning. We now need to vary object pose in small increments and take only three images for each pose corresponding to the independent basis illuminants[4]. All of the acquired images are used to compute $\mathcal{E}$. It is assumed here that $\mathcal{E}$ is less sensitive to illumination variations than pose variations. This is typically the case when dealing with objects of complex shape and textural properties (see [6]).

Next, all images of an object are projected to eigenspace. The expression in (17) can be used to compute eigenspace projections corresponding to any desired number of source directions. In [7], the projections due to both pose and illumination variations are interpolated and the resulting manifold is densely resampled. The resulting points are stored in a database and serve as a discrete appearance representation of the object. Given a novel image, a segmentation algorithm is used to extract object regions. Each object region is normalized in scale and brightness and projected to eigenspace. A nearest neighbor algorithm is then used to identify the object, its pose and the illumination.

The idea of storing each object as a large number of densely resampled manifold points is practical only when the number of objects is small. Below, we present an algorithm that can be used to find the closest point on an illumination manifold directly from the three basis projections without ever constructing

---

[4] The basis illuminants must be chosen with some care to ensure that they are representative of appearance variations due to illumination. For instance, three sources that form a compact cluster in the physical world, even if independent, may not in practice be able to accurately predict appearances due to sources that are distant from the cluster.

the illumination manifold. The algorithm is based on the observation that since the illumination manifold can be expressed in terms of three basis projections, this expression can in turn be used to determine if a novel eigenspace projection lies on the manifold.

Given a scale normalized novel object image $\mathbf{I}'$, it is first normalized in brightness to get $\mathbf{i}' = \mathbf{I}'/m'$, where $m' = \| \mathbf{I}' \|$. If the novel image does lie on a particular illumination manifold, its projection $\mathbf{g}'$ in eigenspace must satisfy:

$$\mathbf{g}' = \frac{1}{m'} \mathbf{G M} \Phi \tag{18}$$

This expression gives us d equation with just 3 unknowns, namely, the source coefficients in Φ. Given that $10 \leq d \leq 30$ in most previous applications of parametric eigenspaces [6] [9], what we have above is an overdetermined linear system that is easily solved to obtain an estimate of the source coefficients $\tilde{\Phi}$.

Since the object in the novel image is unknown, the estimate $\tilde{\Phi}$ may or may not correspond to a point on the illumination manifold. A simple test can be employed to verify the validity of $\tilde{\Phi}$ by checking if the eigenspace projection of the novel image matches the projection corresponding to $\tilde{\Phi}$. To this end, we define the error measure:

$$\varepsilon = \| \mathbf{g}' - \frac{1}{\tilde{m}} \mathbf{G M} \tilde{\Phi} \| \tag{19}$$

where, the magnitude $\tilde{m}$ is computed by using $\tilde{\Phi}$ in equation (13). In theory, if the novel image does belong to the illumination manifold in question, we have $\varepsilon = 0$. In practice, a threshold is applied to ε to determine if the novel projection is close enough to the illumination manifold to be assumed to belong to it. We therefore have a simple and efficient algorithm that uses only three basis projections on an illumination manifold to check if a novel projection belongs to the manifold.

9 Experiments

Figure 2 shows 12 images of a complex object taken under known illumination directions. The object is more or less diffuse in reflectance, has surface patches with different albedo values, and includes concavities that cause interreflections. All the experiments were conducted in an eigenspace that was precomputed for a large set of objects. The source directions are expressed as (θ, ϕ), the azimuth and polar angles subtended by the source in an object centered coordinate frame with its z-axis pointing towards the sensor. Three of the 12 images, namely images 2, 9, and 12, were used to determine the structure of the illumination manifold. The directions of these basis illuminants are $(-9.1°, 81.0°)$, $(17.7°, 81.3°)$, and $(0.0°, 64.4°)$, respectively. The remaining 9 images in Figure 2 were used to test the accuracy of the illumination manifold. Figure 3 shows projections of the 9 images in eigenspace. For display, the projections are shown in a 2D subspace of a 10D eigenspace. The known source directions for the 9 test images were

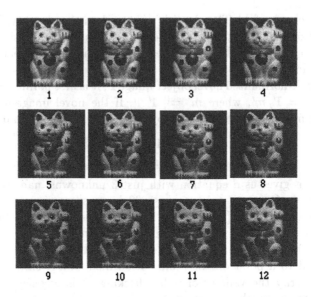

Fig. 2. The image set used in the experiments. The object is primarily diffuse in reflectance, has patches with unknown albedo values, and includes concavities that produce interreflections. Image 2, 9, and 12 correspond to the basis illuminants and were used to determine the structure of the illumination manifold. The remaining 9 images were used to test the accuracies of the theoretical manifold and recognition.

used in expression (17) to determine theoretical predictions of their locations on the illumination manifold. These projections are shown in Figure 4. We see that the actual projections of the 9 images and their theoretical predictions are is strong agreement, demonstrating the accuracy of the illumination manifold derived from just three basis projections. The slight discrepancies in the plots could have resulted from several factors; the object is not purely Lambertian, includes a few specular patches, and produces self-shadows that are not accounted for in the theory.

The second experiment involves the recognition of novel images and the estimation of illumination direction using just three basis projections. Table 1 compares the actual illumination directions used to take the test images with the directions estimated using the algorithm described in section 8. We see that the estimates are very accurate. Also shown in the table are the distances in eigenspace between predicted and actual projections.

Acknowledgements

The authors would like to thank Brian Funt of Simon Fraser University for references and discussions on the use of illumination spaces in color analysis, and Simon Baker of Columbia University for his detailed comments on an early draft that have helped improve the paper. This research was conducted at the

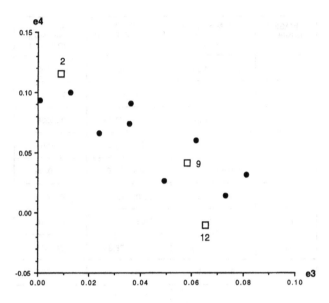

Fig. 3. Projections of the basis images 2, 9, and 12 (shown as boxes) and the remaining 9 test images (shown as dots) in a 2D subspace of a 10D eigenspace.

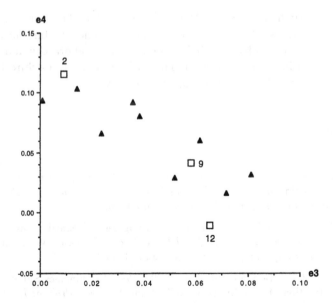

Fig. 4. Projections of the three basis images 2, 9, and 12 (shown as boxes) and theoretical projections of the 9 test images (shown as traingles) determined directly from their known source directions (using expression 17). These projections can be compared with the actual image projections in Figure 3 to evaluate the accuracy of the theoretical illumination manifold. Corresponding projections in the two plots appear in similar positions with respect to the basis projections.

IMAGE NUMBER	ACTUAL SOURCE DIRECTION (θ, ϕ)	ESTIMATED SOURCE DIRECTION (θ, ϕ)	ERROR IN EIGENSPACE PROJECTION
1	(-9.1 , 64.6)	(-8.2 , 63.3)	0.00003
3	(9.1 , 64.6)	(9.6 , 66.3)	0.00033
4	(17.7 , 65.4)	(19.5 , 69.3)	0.00236
5	(-9.1 , 72.5)	(-7.5 , 72.2)	0.00011
6	(0.0 , 72.3)	(-0.8 , 72.6)	0.00029
7	(9.1 , 72.5)	(7.8 , 73.6)	0.00022
8	(17.7 , 73.1)	(15.5 , 75.8)	0.00010
10	(0.0 , 80.9)	(-0.8 , 78.3)	0.00011
11	(9.1 , 81.0)	(9.5 , 78.9)	0.00005

Table 1. Results of illumination direction estimation using just three basis projections in eigenspace.

NTT Basic Research Laboratory, Atsugi, Japan, and at the Center for Research in Intelligent Systems, Department of Computer Science, Columbia University. It was supported in part by the NTT Basic Research Laboratory and in part by an NSF National Investigator Award. The authors thank Ken-ichiro Ishii of the NTT Basic Research Lab. for his support of this work.

References

1. P. J. Belhumeur, A. L. Yuille, and R. Epstein, "Learning and recognizing objects using illumination subspaces," *Proc. of ECCV Workshop on Object Representation in Computer Vision*, Cambridge, U.K., April 1996.

2. P. N. Belhumeur and D. Kriegman, "Learning and recognizing objects using illumination subspaces," *Proc. of IEEE Conf. on Computer Vision and Pattern Recognition*, (to appear), San Fransisco, June 1996.

3. R. Epstein, P. W. Hallinan, and A. L. Yuille, "5±2 Eigenimages Suffice: An Empirical Investigation of Low-Dimensional Lighting Models," *Proc. of IEEE Workshop on Physics Based Modeling in Computer Vision*, pp. 108-116, Boston, June 1995.

4. B. V. Funt and M. S. Drew, "Color Space Analsyis of Mutual Illumination," *IEEE Transactions on Pattern Analysis and Machine Intelligence*, Vol. 15, No. 12, pp. 1319-1325, December 1993.

5. P. W. Hallinan, "A Low-dimensional Lighting Representation of Human Faces for Arbitrary Lighting Conditions," *Proc. of IEEE Conf. on Computer Vision and Pattern Recognition*, pp. 995-999, June 1994.

6. H. Murase and S. K. Nayar, "Visual Learning and Recognition of 3D Objects from Appearance," *International Journal of Computer Vision,* Vol. 14, No. 1, pp. 5-24, January, 1995.

7. H. Murase and S. K. Nayar, "Learning and Recognition of 3D Objects from Appearance," *Proc. of IEEE Workshop on Qualitative Vision,* pp. 39-50, June 1993.

8. H. Murase and S. K. Nayar, "Illumination Planning for Object Recognition in Structured Environments," *Proc. of IEEE Conf. on Computer Vision and Pattern Recognition,* Seattle, pp. 31-38, June 1994.

9. S. K. Nayar, S. A. Nene, and H. Murase, "Real-Time 100 Object Recognition System," *Proc. of ARPA Image Understanding Workshop,* Palm Springs, February 1996. Also, to appear in *Proc. of IEEE Intl. Conf. on Robotics and Automation,* Minnesota, April 1996.

10. S. K. Nayar and H. Murase, "On the Dimensionality of Illumination Manifolds in Eigenspace," CUCS-021-94, Technical Report, Department of Computer Science, Columbia University, New York, August 1994.

11. S. K. Nayar, H. Murase, and S. A. Nene, "Learning, Positioning, and Tracking Visual Appearance," *Proc. of IEEE Intl. Conf. on Robotics and Automation,* San Diego, May 1994.

12. S. K. Nayar and Y. Gong, "Colored Interreflections and Shape Recovery," *Proc. of DARPA Image Understanding Workshop,* San Diego, CA, 1992.

13. S. K. Nayar, K. Ikeuchi, and T. Kanade, "Surface Reflection: Geometrical and Physical Perspectives," *IEEE Transactions on Pattern Analysis and Machine Intelligence,* Vol. 13, No. 7, pp. 611-634, July 1991.

14. S. K. Nayar, K. Ikeuchi, and T. Kanade, "Shape from Interreflections," *International Journal of Computer Vision,* Vol. 2, No. 3, pp. 173-195, 1991.

15. E. Oja, *Subspace methods of Pattern Recognition,* Research Studies Press, Hertfordshire, 1983.

16. M. Oren and S. K. Nayar, "Generalization of the Lambertian Model and Implications for Machine Vision," *International Journal of Computer Vision,* Vol. 14, No. 2-3, pp. 227-251, April, 1995.

17. A. P. Petrov, "Color and Grassman-Cayley coordinates of shape," in *Human Vision, Visual Processing and Digital Display II,* SPIE Proc., Vol. 1453, pp. 342-352, 1991.

18. A. Shashua, "On Photometric Issues in 3D Visual Recognition from a Single 2D Image," Technical Report, Artificial Intelligence Lab., MIT, 1993. Also, to appear in the *International Journal of Computer Vision.*

19. A. Shashua, "Illumination and View Position in 3D Visual Recognition," *Proc. of Neural Information Processing,* pp. 404-411, December 1991.

20. M. A. Turk and A. P. Pentland, "Face Recognition Using Eigenfaces," *Proc. of IEEE Conference on Computer Vision and Pattern Recognition,* pp. 586-591, June 1991.

21. R. J. Woodham, "Photometric method for determining surface orientation from multiple images," *Optical Engineering,* Vol. 19, pp. 139-144, 1980.

Learning Object Representations from Lighting Variations

R. Epstein[1], A. L. Yuille[2], and P. N. Belhumeur[3]

[1] Division of Applied Sciences, Harvard University, Cambridge MA, 02138
[2] Smith-Kettlewell Eye Research Institute, San Francisco, CA 94115.
[3] Department of Electrical Engineering, Yale University, New Haven, CT 06520-8267

Abstract. Realistic representation of objects requires models which can synthesize the image of an object under all possible viewing conditions. We propose to learn these models from examples. Methods for learning surface geometry and albedo from one or more images under fixed posed and varying lighting conditions are described. Singular value decomposition (SVD) is used to determine shape, albedo, and lighting conditions up to an unknown 3×3 matrix, which is sufficient for recognition. The use of class-specific knowledge and the integrability constraint to determine this matrix is explored. We show that when the integrability constraint is applied to objects with varying albedo it leads to an ambiguity in depth estimation similar to the bas relief ambiguity. The integrability constraint, however, is useful for resolving ambiguities which arise in current photometric theories.

1 Introduction

The image of an object depends on many imaging factors such as lighting conditions, viewpoint, articulation and geometric deformations of the object, albedo of the object, and whether it is partially occluded by other objects. It is therefore necessary to design object representations which capture all the image variations caused by these factors. Such representations can then be used for object detection and recognition.

We believe that realistic representations of objects will require models which can *synthesize* the image of an object for all possible values of the imaging factors. Such an approach has long been advocated by people influenced by Bayesian probability theory [8]. This approach has some similarities to "appearance based models" [23] but, as we will argue in the next section, there are some important differences.[4]

We propose to learn these representations from examples. Learning from examples allows us the possibility of representing objects which are too complicated for current modelling systems. If statistical techniques are used, this allows us to concentrate on the important characteristics of the data and ignore unimportant details. For example, aspect graphs [18] give an elegant way of characterizing the

---

[4] Alternative approaches, such as extracting invariant features, [22] may only be applicable to limited classes of objects such as industrial parts.

different views of objects. But for many objects, they are difficult to calculate and hard to use. By contrast, the statistical methods used in [23] are able to recognize certain objects from different viewpoints using simpler techniques. It seems therefore, that some of the complexities of the aspect graph representation are unnecessary, at least for some classes of objects.

We argue that it is important to model the variations of all the factors affecting the image *independently* and *explicitly*. This will allow the object representations to be more general, suitable for more complicated objects, and more easy to generalize to new instances. For example, the appearance based matching algorithm of Murase and Nayar [23] is highly successful within its choosen domain of simple rigid objects but its avoidance of geometric and reflectance models means that it could be fooled by simply repainting one of the learned objects. The new (repainted) object would then have to be learnt again, requiring a costly training procedure. Similar problems would arise if the object is allowed to deform geometrically.

Modelling variations explicitly also makes it easy to incorporate prior knowledge about the object class into the learning procedure. If the object class is known, and explicit models are used, then far less training data will be needed. It appears that humans can make use of this type of class specific knowledge in order to generalize rapidly from one instance of an object [21]. In related work, we are exploring whether our models can account for these and other psychophysical experiments.

In this paper, therefore, we will describe methods for learning the geometry and reflectance functions of objects from one of more images of the object. We assume fixed pose but vary the lighting conditions[5]. For this paper we assume Lambertian reflectance functions with non-constant albedo, but we are currently generalizing our work to other types of relectance models.

We describe mechanisms for learning the shape, reflectance, and albedo of an object with or without the use of class specific knowledge. In particular, we make use of the surface integrability constraint and discover a close relation between the bas relief ambiguity and integrability. We illustrate the usefulness of our representations by synthesizing images. In related work, Belhumeur and Kriegman [2] characterize the set of images that can be generated by using Lambertian models, of the type we learn here, and give further examples of image synthesis.

Our approach makes use of singular value decomposition (SVD) which has previous been applied to the related problem of photometric stereo by Hayakawa [13]. For Lambertian sources with a single illuminant, SVD allows one to estimate shape, albedo, and lighting conditions up to an unknown 3×3 constant matrix, which we call the $\mathbf{A}$ matrix. We observe that the stated assumptions in [13] only determines $\mathbf{A}$ up to an unknown rotation matrix. It can be shown [30] this assumption is valid for certain types of surfaces but will be incorrect for others. However, we demonstrate that a variety of general purpose and/or class specific assumptions, including surface integrability, can be used to determine the $\mathbf{A}$

---

[5] An extension to variable pose is described in Epstein and Yuille (in preparation).

matrix uniquely. Moreover, it can be shown [2] that the set of allowable images of the object (from fixed viewpoint) can be determined *without* knowing **A**.

2   Appearance Based Models and Image Synthesis

To set our work in context, it is important to describe how it relates to other work on image synthesis and the influential work on appearance based models [23].

Appearance based models (ABM's) of objects are learned by applying principal component analysis (PCA) to a representative dataset of images of an object. For certain classes of objects, this produces a low-dimensional subspace which captures most of the variance of the dataset. The object can then be represented by a manifold defined in this low-dimensional space. The position of the image on this manifold will depend on the lighting and viewpoint conditions. An input image, or subpart of an image, can be matched to the appearance manifold and hence recognized. This approach is extremely successful within specific domains.

It is interesting to contrast ABM's with image synthesis models of the type that we use in this paper. Our approach requires specifying a representation for the object and an imaging model. The representation model should be flexible enough to deal with all the variations described previously – due to lighting, articulation, geometric deformations, etc. The imaging model enables us to synthesize an image of the object. The representation and imaging models are learnt by statistical techniques from samples of the data.

Synthesis models and ABM's are similar in two important respects. Firstly, unlike many (most) current object recognition systems, they do not first extract sparse features, such as edges, from the image (see [9]). However, the word "appearance" in ABM's is slightly misleading because the ABM's only model the appearance of the object within the low dimensional subspace. They ignore all image variations that project outside this subspace. The synthesis models, by contrast, generate all possible image variations. Secondly, both synthesis models and ABM's are statistical with their models being generated by the data. This makes them more robust with respect to noise which can destroy more deterministic modelling approaches such as geometric invariants [22].

From our viewpoint, however, the ABM's are limited because they do not represent variables like shape and lighting explicitly. It is straightforward to adapt synthesis models to take into account geometrical deformations or to add paint onto the surface of an object. But an ABM would have to learn all such changes from scratch. Similar problems would also apply in the related eigenface approach [27] where the eigenfaces combine albedo, lighting, and geometrical changes, but represent none of them explicity. Like ABM's this approach involves projecting the image onto a low-dimensional space and ignoring anything that lies outside this space.

The ABM's have gone a long way in demonstrating the advantages of using much richer descriptions than simply sparse features like edges and corners for recognition. Still, a drawback of these approaches is that in order to recognize an object seen from a particular pose and under a particular illumination, they must have previously seen the object under the same conditions. Yet, if one tries to enumerate all possible poses and permutes these with all possible illumination conditions, things get out of hand quite quickly. Fortunately, this brute-force approach to modeling, which requires observing objects under the full range of parametric variation, is unnecessary since appearance can usually be predicted from a modest number of images.

Indeed, both eigenfaces and ABM's can be considered to be feature based methods where the features are extracted by applying linear filters determined by PCA. It can be argued [3] that if the goal is discrimination between objects, rather than representation, then better linear filters can be used based on Fisher's linear discriminant. PCA projects into the subspace which captures most of the variance between objects. By contrast, Fisher's linear discriminant [7] projects into the subspace which maximizes the variation between different objects. This can be illustrated by considering applying both techniques to a set of faces in which a small subclass of people have glasses. The PCA approach would tend to project onto a subspace which ignores the glasses (because they appear in two few samples to significantly affect the variance). By contrast, Fisher's linear discriminant would project into a subspace which included the glasses because they would be powerful cues for distinguishing between people.

A more explicit way of modelling faces occurs in [4] where the eigenfaces are considered to be principal components of the albedoes of faces. Two-dimensional geometrical distortions are applied to allow for changes in viewpoint and expression. These deformations occur by warping a set of feature points, corresponding to the facial features, and interpolating the warp over the rest of the face.

Lighting variations are also handled explicity by a related model by Hallinan [12] which is able to recognize faces under highly variable lighting conditions and to distinguish reliably between faces and non-faces. Lighting variations are represented by a linear combination of lighting basis images obtained from PCA. To model geometric changes, Hallinan [12] uses two-dimensional image warps. Though this not an explicit model of surface geometry, it can be shown that the spatial warps correspond to warps of the surface normal vectors of the underlying three dimensional shape [29]. It is therefore straightforward to recompute the surfaces from the warps. Hallinan's lighting models were the starting point for this current work and we will return to them later in the paper.

Another model, that uses image synthesis and explicit representations is the face recognition system reported in [1]. This face model uses three dimensional geometry and a Lambertian imaging model. By using a dataface of face geometry, obtained by laser scanning, a strong prior distribution for the shape of faces is obtained. Using this prior the three dimensional geometry of the face can be estimated from a single image. However, the types of geometric models used in this system are somewhat limited and only apply to objects made of single parts, such as faces. For objects with several articulating parts more sophisticated geometrical models should be used, perhaps of the type described in [31].

3   The Lambertian Model and Lighting Basis Functions

Suppose we pick an object and fix its pose and articulation. Then the principle of superposition ensures that the set of images of the object, as the lighting varies, lies within a linear space[6]. How does this obervation relate to reflectance function models of image formation?

The most used reflectance model is the Lambertian model [14] which is often written as:

$$I(x, y) = a(x, y)\mathbf{n}(x, y) \cdot \mathbf{s} \equiv \mathbf{b}(x, y) \cdot \mathbf{s}, \tag{1}$$

where $a(x, y)$ is the albedo of the object, $\mathbf{n}(x, y)$ is its surface normal, $\mathbf{b}(x, y) \equiv a(x, y)\mathbf{n}(x, y)$ and $\mathbf{s}$ is the light source direction (the light is assumed to be at infinity). If this equation applies then it is clear [26],[28],[25] [20], that the space of images of the object, as the light source direction changes, spans a three dimensional subspace. In other words, any image of the object can be expressed as:

$$I(x, y) = \sum_{i=1}^{3} \alpha_i b_i(x, y), \tag{2}$$

for some coefficients $\{\alpha_i\}$, where i labels the vector components. This is a linear subspace model of image formation.

Equation (1), however, has several limitations. It ignores attached shadows (where $\mathbf{b}(x, y) \cdot \mathbf{s} \leq 0$), cast shadows, and partial or hidden shadows (where there are several light sources and the light from some of them are shadowed). It also ignores interreflections. When these effects are taken into account, the dimensionality of the image space rises enormously [2]. Moreover, the model ignores specularities and will break down if the light source is close to the object. These limitations mean that caution is necessary when using this model.

Alternatively, motivated by the principle of superposition, one can try to analyze the empirical structure of the set of possible images . In a series of empirical studies [11], [5] principal component analysis (PCA) was used to analyze the space of images generated by one object at fixed pose with varying lighting conditions. The lighting conditions were sampled evenly on the view hemisphere, so the dataset included extreme lighting configurations. The experimental results showed that 5 ± 2 eigenvalues were typically enough to account for most of the variance. For faces, the percentage of variance covered by the first five eignevalues was approximately 90 %. For objects which were highly specular (such as a helmet) or with many shadows (such as an artificial parrot) the percentage decreased. Nevertheless, the specularities and shadows, though perceptually very saliant, contributed little to the variance. In addition, Hallinan [11] showed that if *different* faces were aligned geometrically, using affine transformations, then the first five eigenvalues still captured approximately 90 % of the variance.

These results meant that for each object we could approximate the image space by a linear combination of the first five eignevectors or *lighting basis functions*. In other words an image of the object, under fixed viewpoint, could be expressed as:

---

[6] In fact it can shown to lie within a convex cone inside this linear space [2]

$$I_M(\mathbf{x}; \{\alpha_i\}) = \sum_{i=1}^{5} \alpha_i B_i(\mathbf{x}), \tag{3}$$

where the $\{B_i(.)\}$ are the lighting basis functions (i.e. the first five principal components), and the $\{\alpha_i\}$ are the coefficients (which depend on the specific lighting conditions).

If this number of coefficients is set equal to three then this would be similar to the Lambertian linear model, see equation (2). Indeed it was observed that the first three lighting basis functions usually corresponded to the image lit from in front, from the side, and from above. This is explained in [30].

The empirical linear subspace model, see equation (3), was used by Hallinan [12] to successfully model lighting variation. Such models are attractive but they do have several limitations: albedo and shading information is combined indiscriminantly and there is no explicit 3-D model. (Although, under certain circumstances [29] it does allow recovery of the three-dimensional shape.)

For reasons described above, we would prefer a more explicit representation based on three-dimensional shape and albedo. We argue, therefore, that the success of the linear subspace results suggest that Lambertian models are a good approximation to a number of real objects. Indeed, it was *conjectured* [5] that the first three principal components of this space correspond to Lambertian illumination of the object and higher order principal components dealt with specularities and sharp shadows.

4 Learning the Models

Our approach consists of learning models of the objects – their surface geometry and albedo – using variants of the Lambertian model which make it robust to shadows and specularities. This is done with four different schemes.

Suppose we have a set of images of an object illuminated by M different point light sources. We denote these light sources by $\{s(\mu) : \mu = 1, ..., M\}$. The resulting images are represented by $\{I(p, \mu) : \mu = 1, ..., M \;\; p = 1, ..., P\}$ where the index p labels the pixels of the image (these pixels lie on a two dimensional grid but it is convenient to represent them as a vector).

Our first scheme assumes that we have multiple images of the object[7] and the light sources are known. This is of least interest since it is a strong assumption and corresponds to standard photometric stereo [26, 28, 14, 17], though with nonconstant albedo. We investigated this scheme mainly to test the Lambertian asumptions about our data. We concluded that the model is a good approximation though robust techniques are needed to reduce the influence of shadows and specularities.

If, however, there are multiple unknown light sources then we show that SVD can be applied (see also [13]) to simultaneously estimate the surface geometry and albedo up to a 3×3 linear transformation, the $\mathbf{A}$ matrix. This transformation

---

[7] Fixed pose and varying illumination.

arises due to an ambiguity in the Lambertian equation (1). This is because for any arbitrary invertible linear transformation $\mathbf{A}$:

$$\mathbf{b} \cdot \mathbf{s} = \mathbf{b}^T \mathbf{s} = \mathbf{b}^T \mathbf{A}\mathbf{A}^{-1}\mathbf{s}. \tag{4}$$

Our second learning scheme, follows from this result and the proof in [2] that the set of images of the object are *independent* of the precise value of $\mathbf{A}$ provided the objects are viewed from front on. This means that it unecessary to estimate $\mathbf{A}$. Our second scheme, therefore consists merely of applying SVD to the input data and thereby generating the light cone representation described in [2].

For our third learning scheme, we demonstrate that the $\mathbf{A}$ matrix can be recovered by using the surface integrability constraint and the assumption that we either have an image of the object under ambient lighting, or that the sampling set of lighting conditions allows us to generate one. We compare our assumptions to those of [13] and prove that his method relies on an, unstated, assumption about the dataset which will often not be valid. In addition, we describe a new perceptual ambiguity related to the integrability constraint. This scheme results in the full albedo and three-dimensional shape of the object.

In our fourth learning scheme, we consider the use of prior knowledge about the class of the viewed object. We demonstrate that $\mathbf{A}$ can be learnt by merely assuming that we know a prototype object of that class. Not suprising, if the object class is known then fewer images are needed to learn the object model. This seems to agree with current psychophysical results [21].

4.1 Learning the Models with known light source direction

Suppose we assume that the light source vectors $\{\mathbf{s}(\mu) : \mu = 1, ..., M\}$ are known. This is true for our dataset because the images have been gathered under controlled conditions.

We can formulate estimating shape and albedo as a least squares optimization problem:

$$E[b; V] = \sum_{\mu,p} V(p, \mu)\{I(p, \mu) - \sum_i b_i(p)s_i(\mu)\}^2 \tag{5}$$

where $V(p, \mu)$ is a binary indicator function whose value is 1 if point p is not in shadow, or have a specularity, under lighting condition μ, and is zero otherwise.

The arguments of the energy function – b, s, V – represent the sets $\{\mathbf{b}(p) : p = 1, ..., P\}$; $\{\mathbf{s}(\mu) : \mu = 1, ..., M\}$, and $\{V(p, \mu) : p = 1, ..., P \quad \mu = 1, ..., M\}$ respectively.

We observe that the energy can be written as the sum of P independent energies $E_p[\mathbf{b}(p), \{V(p, \mu) : \mu = 1, ..., M\}] = \sum_\mu V(p, \mu)\{I(p, \mu) - \sum_i b_i(p)s_i(\mu)\}^2$. These energy functions E_p $(p = 1, ..., P)$ are all quadratic in b and so they can be minimized by linear algebra provided the V are specified. This allows us to estimate the surface normal and albedo at all points p independently.

We first assume that there are no specularities or shadows, in other words we set $V(p, \mu) = 1$, $\forall p, \mu$. This gives the results shown in figures (1, 2). This is equivalent to the photometric stereo techniques described in [26], [28], [14].

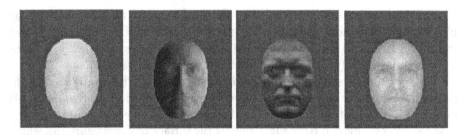

Fig. 1. The albedo and normals estimated directly assuming known light source directions and *without* using robust techniques to remove specularities and shadows. The first three images are the z, x, and y components of the surface normal respectively. The rightmost image is the albedo. Observe that the estimated albedo appears to get darker near the boundaries of the face causing the albedo image to appear to be non-flat. This is due to failure to treat the shadows correctly.

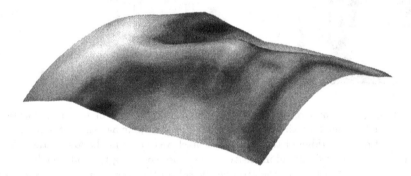

Fig. 2. The surface computed from the normals in the previous figure. The face appears flattened. This is because the algorithm's failure to remove shadows means that it underestimates the albedo in shaded regions and correspondingly makes the surface flatter.

These results are reasonable but close inspection shows that the estimated albedo becomes darker towards the boundaries of the face, see figure (1), and the shape of the face is flattened, see figure (2). This is because the algorithm knows nothing about shadows and tries to model them as regions of dark albedo. This in turn causes the shape to appear too frontoparallel. We conclude that the object is approximately Lambertian but that it also has shadows and specularities.

We observe, however, that specularities are bright, shadows are dark, and a point will tend to be in shadow or specular only for a limited set of lighting directions. Thus if we histogram the intensity values at a single image point, as it is illuminated from many directions, the brightest and darkest points will tend to be specularities and shadows[8].

---

[8] Ideally perfect shadows would have zero intensity, but our light sources are not true point sources and there was some ambient light present when our database was collected.

Thus we can remove most of the effects of shadows and specularities by plotting the histogram, see figure (3), and set $V = 0$ for the bottom $\alpha_1\%$ and top $\alpha_2\%$. If α_1 and α_2 are sufficiently large (say 30%) then we set $V = 0$ for the remaining data (which we now assume is purely Lambertian).

We now minimize the E_p again using linear algebra. The results are significantly improved, see figures (4, 5). Observe that the albedo image in figure (4) appears to be much flatter, suggesting that we have removed much of the effects of the shadows. This is further supported by the surface plot, see figure (5), which is no longer foreshortened – compare with figure (2). Thus eliminating the shadows by pruning the histogram gives us significantly more uniform albedoes on the skin and a more accurately estimated shape.

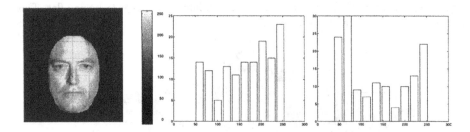

Fig. 3. Histograms for two pixels on the bridge and the side of the nose, locations shown in the left image by the two black dots. The middle image shows the histogram for the pixel on the bridge of the nose. This pixel was never in shadow so there is no peak in the histogram for small intensity values (corresponding to shadows). The right image shows the histogram for the pixel on the side of the nose. This pixel was often in shadow and so its histogram has a peak at low intensity values. Note that background ambient illumination prevents the shadows from being perfectly dark.

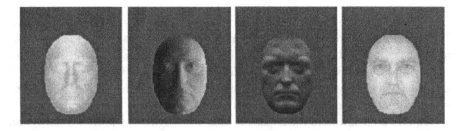

Fig. 4. The normals and albedo calculated directly. Residuals at low intensity values < 70 have been removed.

Alternatively, instead of eliminating the top $\alpha_2\%$ and the bottom $\alpha_1\%$ of $\{I(p, \mu) : \mu = 1, ..., M\}$ we could instead eliminate all intensities below a *shadow threshold* and above a *specularity threshold*. Or, we could do residual analysis

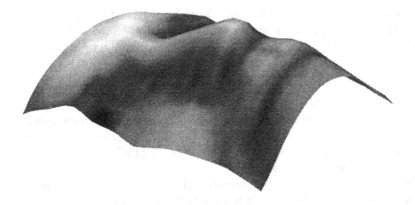

Fig. 5. Surface computed from normals above.

Fig. 6. This figure shows the extent to which each pixel was thresholded. Pixel brightness corresponds to the number of images in which the pixel was over threshold. Hence, dark pixels were thresholded out (considered in shadow) more. Observe that points will low albedos, such as the irises of the eyes, are overrepresented.

to check whether the intensities thrown away correspond to true shadows or specularities. We can use our estimate $b^*(p)$ to *predict* what the intensities would be for those cases. Those light source configurations for which the predictions agree with the observed intensities are no longer assumed to be due to shadows, or specularities, and so are used to make a second estimate of $b(p)$. This process can be repeated.

4.2 Light Source Direction Unknown: Using SVD to estimate surface properties and light source directions up to a linear transformation

It is unrealistic to assume that the light source directions will be given. Thus we need a method which can estimate them and the surface properties simultaneously. In other words, we need to minimize the energy function $E[b, s] = \sum_{\mu,p} \{I(p,\mu) - \sum_{i=1}^{3} b_i(p)s_i(\mu)\}^2$ as a function of b *and* s. Fortunately minimization of this function, up a linear transform, can be done using singular value decomposition (SVD). This has been first applied to photometric stereo in [13].

Observe that the intensities $\{I(\mu, p)\}$ can be expressed as a $M \times P$ matrix $\mathbf{J}$. Similarly we can express the surface properties $\{b_i(p)\}$ as a $P \times 3$ matrix $\mathbf{B}$ and the light sources $\{s_i(\mu)\}$ as a $3 \times M$ matrix $\mathbf{S}$. SVD implies that we can write $\mathbf{J}$ as:

$$\mathbf{J} = \mathbf{U}\Sigma\mathbf{V}^T, \qquad (6)$$

where Σ is a diagonal matrix whose elements are the square roots of the eigenvalues of $\mathbf{J}\mathbf{J}^T$ (or equivalently of $\mathbf{J}^T\mathbf{J}$). The columns of $\mathbf{U}$ correspond to the normalized eigenvectors of the matrix $\mathbf{J}^T\mathbf{J}$. The ordering of these columns corresponds to the ordering of the eigenvalues in Σ. Similarly, the columns of $\mathbf{V}$ correspond to the eigenvectors of $\mathbf{J}\mathbf{J}^T$.

If our image formation model is correct then there will only be three nonzero eigenvalues of $\mathbf{J}\mathbf{J}^T$ and so Σ will have only three nonzero elements. We do not expect this to be true for our dataset because of shadows, specularities, and noise. But SVD is guaranteed to gives us the best least squares solution in any case. Thus the biggest three eigenvalues of Σ, and the corresponding columns of $\mathbf{U}$ and $\mathbf{V}$ represent the Lambertian part of the reflectance function of these objects. We define the vectors $\{\mathbf{f}(\mu) : \mu = 1, ..., M\}$ to be the first three columns of $\mathbf{U}$ and the $\{\mathbf{e}(p) : p = 1, ..., P\}$ to be the first three columns of $\mathbf{V}$.

This assumption enables us to use SVD to solve for $\mathbf{B}$ and $\mathbf{S}$ up to a linear transformation. The solution is:

$$\mathbf{s}(\mu) = \mathbf{P}\mathbf{f}(\mu), \ \forall \ \mu,$$
$$\mathbf{b}(p) = \mathbf{Q}\mathbf{e}(p), \ \forall \ p, \qquad (7)$$

where $\mathbf{P}$ and $\mathbf{Q}$ are 3×3 matrices which are constrained to satisfy $\mathbf{P}^T\mathbf{Q} = \Sigma_3$, where Σ_3 is the 3×3 diagonal matrix containing the square roots of the biggest three eigenvalues of $\mathbf{J}\mathbf{J}^T$. There is an ambiguity $\mathbf{P} \mapsto \mathbf{A}\mathbf{P}$, $\mathbf{Q} \mapsto \mathbf{A}^{-1}{}^T\mathbf{Q}$ where $\mathbf{A}$ is an arbitrary invertible matrix.

This means we can determine $\{\mathbf{s}\}$ and $\{\mathbf{b}\}$ up a linear transform. It can be shown [2] that this is sufficient to recognize objects from front-on under arbitrary illumination. To verify that these linear subspaces are correct we use our knowledge of the light source directions to determine the $\mathbf{P}$ and $\mathbf{Q}$ matrices (i.e. we use least squares to solve $\mathbf{s}(\mu) = \mathbf{P}\mathbf{f}(\mu)$, $\forall \ \mu$ for $\mathbf{P}$.) The resulting albedos and surface normals are shown in figure (7). The results are similar to those obtained by using knowledge of the light source directions directly. They appear slightly better than the results without residuals, figure (1), and slightly worse than the results with residuals, figure (4).

4.3 Estimating the linear transformations

We would like, however, to estimate the true geometry and albedo because this would enable us to predict how the object changes as the viewpoint varies (and to deal with cast shadows). The next subsection discusses ways to use additional information can be used to determine the linear transformation and hence to determine the surface albedo and shape.

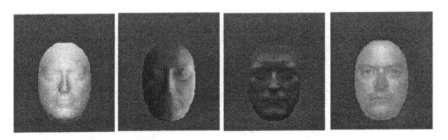

Fig. 7. Normals and albedo calculated directly from SVD using known light source directions to estimate the linear transformations.

Objects of Unknown Class Suppose we have an object of unknown class and we wish to determine the **A** matrix.

One plausible assumption is that we have an estimate of the object's albedo. This might consist of an additional image of the object taken under ambient lighting conditions[9]. Alternatively we can assume that the light source directions sample the view hemisphere and so, by taking the mean of our dataset we get an approximation to an ambient image of the object. It should be emphasized that this estimated albedo need only be very approximate.

We use the mean of our dataset to estimate the albedo. This means that, using Equation (7), for each point p in the image we have a constraint on the linear transformations:

$$a(p)^2 = \mathbf{e}^T(p)\mathbf{P}^T\mathbf{P}\mathbf{e}(p), \ \forall \ p = 1, ..., P. \tag{8}$$

We impose these constraints using a least squares goodness of fit criterion. This can be solved using SVD to estimate $\mathbf{P}^T\mathbf{P}$. This yields $\mathbf{P}^T\mathbf{P} = \mathbf{W}\mathbf{M}\mathbf{W}^T$, where $\mathbf{M}$ is diagonal. We then estimate $\mathbf{P}^* = \mathbf{M}^{1/2}\mathbf{W}$ which is correct up to rotation.

We note that Hayakawa assumes that this rotation matrix is the identity [13]. It can be shown, however, that this is not always the case. Indeed, see [30], it can be shown to hold if the matrices $\sum_p b_i(p)b_j(p)$ and $\sum_\mu s_i(\mu)s_j(\mu)$ are both diagonal. But, for example, it does not hold if $\sum_\mu s_i(\mu)s_j(\mu)$ is diagonal but $\sum_p b_i(p)b_j(p)$ is not. The condition that these matrices are both diagonal can be traced to symmetry assumptions about the dataset. It is straightforward to generate situations for which they fail.

Fortunately, however, this rotation ambiguity can be cured by using the surface integrability constraint, see section 5. The results shown in figures (8,9) are consistent with integrability.

Objects of Known Class We can use knowledge about the class of the object to determine the linear transformations $\mathbf{P}$ and $\mathbf{Q}$, and hence determine the surface properties and the light sources uniquely.

---

[9] Recall that the image of an object under ambient lighting conditions is given by the albedo [14]

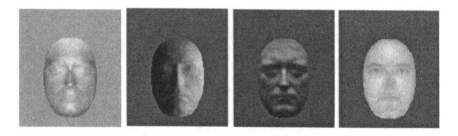

Fig. 8. Normals and albedo calculated from eigenvectors. We used the mean of the dataset as an initial estimate of albedo. The matrix $P^T P$ is then calculated from $a^2(x) = e^T(x)P^T Pe(x)\forall x$. SVD on $P^T P$ gives $P^T P = W * M * W^T$, M diagonal. We then take, as an estimate of P, $P^* = sqrt(M) * U$. This is correct up to rotation. In the above results, we take the rotation matrix to be the identity and check consistency with integrability.

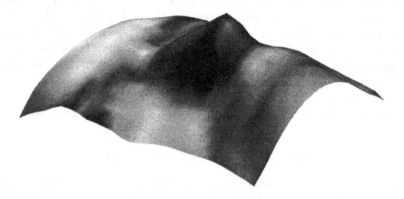

Fig. 9. Surface computed from normals above.

To do this all we need is a $\mathbf{b}(p)$ vector from a prototype member of the class. For example, we assume that we know $\mathbf{b}_{Pr}(p)$ for a prototype face Pr. Then when we get the data for a new face image we will estimate its $\mathbf{P}$ and $\mathbf{Q}$ matrices by assuming that it has the same surface properties as the prototype. Thus we estimate $\mathbf{P}$ by minimizing:

$$\sum_p |\mathbf{b}_{Pr}(p) - \mathbf{P}e(p)|^2, \tag{9}$$

where the $e(p)$ are computed from the new dataset. We are minimizing a quadratic function of $\mathbf{P}$ so the result, $\mathbf{P}^*$, can be obtained by linear algebra.

We now solve for the surface properties using:

$$\mathbf{b}(p) = \mathbf{P}^*e(p), \quad \forall\, p. \tag{10}$$

Observe that the prototype is used merely in conjunction with the dataset to solve for the 3×3 matrix $\mathbf{P}$. Our results demonstrate that the surface properties computed using this assumption are good.

Fig. 10. Normals and albedos calculated for a new subject using the results shown in figure 7 as a prototype.

This result has used prior knowledge about object class in the simplest possible form – a prototype model. More sophisticated class knowledge, such as a prior probability distribution for shapes and albedoes, would lead to improved results.

5 Surface Integrability

The surface integrability constraint requires that the normal vectors are consistent with a surface (for a discussion, see [15].) It puts restrictions on the set of normals vectors but it is not sufficient to determine the surface uniquely. We will show that for Lambertian objects with unknown albedo this leads to an ambiguity including scaling in depth.

The unit normals $\mathbf{n}(\mathbf{x}) = (n_1(\mathbf{x}), n_2(\mathbf{x}), n_3(\mathbf{x}))$ of a surface must obey the following surface integrability constraint to ensure that they form a consistent surface:

$$\frac{\partial}{\partial y}\left(\frac{n_1(\mathbf{x})}{n_3(\mathbf{x})}\right) = \frac{\partial}{\partial x}\left(\frac{n_2(\mathbf{x})}{n_3(\mathbf{x})}\right). \tag{11}$$

This constraint is a necessary and sufficient condition and can be derived from the fact that any surface can be locally parameterized as $z = f(x, y)$ with normals of form:

$$\mathbf{n}(\mathbf{x}) = \frac{1}{\{\nabla f \cdot \nabla f + 1\}^{(1/2)}}(f_x, f_y, -1). \tag{12}$$

It is straightforward to see that the vector $\mathbf{b}(\mathbf{x}) = a(\mathbf{x})\mathbf{n}(\mathbf{x})$ also satisfies the same constraint – i.e. we can replace (n_1/n_3) and (n_2/n_3) by (b_1/b_3) and (b_2/b_3) in the constraint equations.

Now recall that the linear algebra in the previous section determined the $\mathbf{b}(\mathbf{x})$ up to an unknown linear transformation determined by the $\mathbf{P}$ matrix.

The surface integrability constraint will partially determine the $\mathbf{P}$ matrix. It is straightforward to show, and to verify, that the only linear transformations which preserve the integrability constraint are:

$$b_1(\mathbf{x}) \mapsto \lambda b_1(\mathbf{x}) + \mu b_3(\mathbf{x}),$$

$$b_2(\mathbf{x}) \mapsto \lambda b_2(\mathbf{x}) + \nu b_3(\mathbf{x}),$$
$$b_3(\mathbf{x}) \mapsto \rho b_3(\mathbf{x}). \tag{13}$$

Observe that there is a constant scaling factor in this transformation which can never be determined (a dark surface lit with a bright light is indistinguishable from light surface lit by a dark light) so we could set $\rho = 1$ without loss of generality.

If the $\mathbf{A}$ matrix is known up to a rotation ambiguity, as in section 4.3, then integrability determines the remaining part of the transformation.

Moreover, if the albedo is known to be constant, then the class of transformations are reduced to the well known convex/concave (or light up/light down) ambiguity well known in the psychophysics literature. This is because the requirement that $\mathbf{b}(\mathbf{x})$ has constant magnitude (independent of $\mathbf{x}$) puts further restrictions on the transformation.

Thus for objects with unknown albedo, we get a class of perceptual ambiguities corresponding to the transformations given in equation (13). To understand these ambiguities we let the transformed surface be represented by $z = \bar{f}(x, y)$. It is straightforward calculus to see that:

$$\bar{f}(x, y) = \lambda f(x, y) + \mu x + \nu y. \tag{14}$$

In other words, the ambiguity consistent with the integrability constraint consists of scaling the depth by a factor λ and adding a planar surface $z = \mu x + \nu y$. Interestingly, it has been reported [19] that humans appear to differ in their judgement of shape from shading by a scaling in the z direction. This connection is being explored in our current work.

Thus we see that the integrability constraints reduces the ambiguity in reconstructing the surface but it does not eliminate it altogether. To solve the problem uniquely we must impose additional constraints.

6 Learning an Object from a Single View

In previous sections we developed methods for learning object models assuming that we have multiple images of the object. In practice, however, we may only have one image of each object. Moreover, it is important to know how much we can learn about an object from a single image.

A single image, however, gives us little information about the object. Recall that, assuming Lambertian models, we can express the image as $I(\mathbf{x}) = \mathbf{b}(\mathbf{x}) \cdot \mathbf{s}$ where $\mathbf{b}(\mathbf{x})$ and $\mathbf{s}$ are unknown. This equation, without additional assumptions, is not sufficient to determine $\mathbf{b}(\mathbf{x})$ and $\mathbf{s}$[10]. To make progress we must use knowledge about the class of the object. One way to do this would be to do statistics on the class of objects to develop a prior distribution for them[1]. Instead we will determine techniques for learning object models making as few assumptions

---

[10] Current shape from shading algorithms usually assume known light source and constant albedo.

as possible about the object class. Our assumptions are: (i) a prototype model, $\mathbf{b}^p(\mathbf{x})$, for the class, and (ii) symmetry assumptions about the object.

For faces the symmetry assumption is valid and we can select a prototype head from our database. It is convenient to use as a prototype one of our previously learnt models shown in figures (8,9). The algorithm proceeds in several stages.

Stage I. We use the prototype model to estimate the light source direction. More precisely, we solve for:

$$s^* = \arg\min_s \int d\mathbf{x}\,|I(\mathbf{x}) - \mathbf{b}^p(\mathbf{x}) \cdot \mathbf{s}|^2. \tag{15}$$

Stage II. The symmetry assumption. We assume that the object is symmetric across the y-axis at $x = 0$. This means that we can express the model as:

$$(b_1(x,y), b_2(x,y), b_3(x,y)) = (h_1(x,y), h_2(x,y), h_3(x,y)), \quad x \geq 0,$$
$$(b_1(x,y), b_2(x,y), b_3(x,y)) = (-h_1(-x,y), h_2(-x,y), h_3(-x,y)), \quad x \leq 0, \tag{16}$$

where $(h_1(x,y), h_2(x,y), h_3(x,y))$ represents the right half of the face.

By using the image of the left and the right part of the face we can observe $s_1 h_1(x,y) + s_2 h_2(x,y) + s_3 h_3(x,y)$ and $-s_1 h_1(x,y) + s_2 h_2(x,y) + s_3 h_3(x,y)$. Therefore, using the fact that we know s from Stage I, we know $s_1 h_1(x,y)$ and $s_2 h_2(x,y) + s_3 h_3(x,y)$. Thus we know two components of $\mathbf{h}(x,y)$. It remains to determine the third component $-s_3 h_2(x,y) + s_2 h_3(x,y)$. Of course, this requires that neither $s_1 = 0$ nor $s_2 = s_3 = 0$. So the lighting cannot be purely front-on or purely from the x-direction.

Stage III. To determine the third component $- -s_3 h_2(x,y) + s_2 h_3(x,y)$ – we make use of the integrability constraint and, if necessary, the prior model. The integrability constraint is:

$$\frac{\partial}{\partial x}\frac{h_2(x,y)}{h_3(x,y)} = \frac{\partial}{\partial y}\frac{h_1(x,y)}{h_3(x,y)}, \quad \forall x,y. \tag{17}$$

Multiplying this equation by $h_3^2(x,y)$ and expanding it gives:

$$h_3(x,y)\frac{\partial}{\partial x}h_2(x,y) - h_2(x,y)\frac{\partial}{\partial x}h_3(x,y) = h_3(x,y)\frac{\partial}{\partial y}h_1(x,y) - h_1(x,y)\frac{\partial}{\partial y}h_3(x,y). \tag{18}$$

We define two new vectors $p_2(x,y)$ (known) and $p_3(x,y)$ (unknown) by:

$$p_2(x,y) = \frac{s_2 h_2(x,y) + s_3 h_3(x,y)}{(s_2^2 + s_3^2)}, \quad p_3(x,y) = \frac{-s_3 h_2(x,y) + s_2 h_3(x,y)}{(s_2^2 + s_3^2)},$$
$$h_2(x,y) = s_2 p_2(x,y) - s_3 p_3(x,y), \quad h_3(x,y) = s_3 p_2(x,y) + s_2 p_3(x,y). \tag{19}$$

Then we express integrability by defining a function $K(x,y)$:

$$K(x,y) = (s_2^2 + s_3^2)p_3(x,y)\frac{\partial p_2(x,y)}{\partial x} - (s_2^2 + s_3^2)p_2(x,y)\frac{\partial p_3(x,y)}{\partial x}$$
$$-(s_3 p_2(x,y) + s_2 p_3(x,y))\frac{\partial h_1(x,y)}{\partial y} + h_1(x,y)\frac{\partial(s_3 p_2(x,y) + s_2 p_3(x,y))}{\partial y}, \tag{20}$$

and requiring that $K(x,y) = 0 \ \forall \ (x,y)$.

Observe that this constraint is linear in the unknown variable $p_3(x,y)$ and we have one constraint for each position (x,y). Thus there may be sufficient information in these constraints to determine $p_3(x,y)$ uniquely, although possibly there are some linear dependencies between the constraints which would prevent uniqueness. It therefore seems wise to impose these constraints by least squares – i.e. write a quadratic cost function for $p_3(x,y)$ by summing the squares of $K(x,y)$ over (x,y) – *and* add an additional prior term. This gives an energy function:

$$E[P_3] = \int d\mathbf{x} K^2(\mathbf{x}) + \lambda \int d\mathbf{x} \{p_3(x,y) - \frac{1}{(s_2^2 + s_3^2)}(-s_3 h_2^p(x,y) + s_2 h_3^p(x,y))\}^2,$$

(21)

where λ is a constant and $h_2^p(x,y), h_3^p(x,y)$ are the y and z components of the prototype model for the right half of the face.

This completes the three stages. Results are shown in figures (11, 12,13).

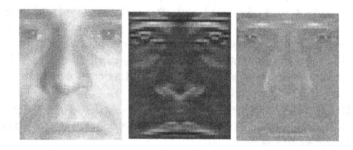

Fig. 11. Left – the original input image. Center – the estimate of p3. Right – the estimate of the albedo.

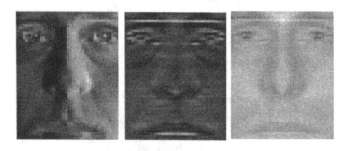

Fig. 12. Estimated b vectors of the face.

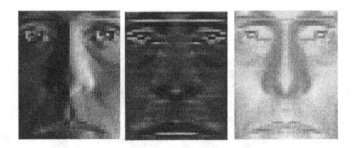

Fig. 13. Estimated normals of the face.

7 Object Synthesis

This section briefly shows how to peform recognition by using our learned object models to synthesize images. The methods used are described in [2] which includes further examples.

We first learned the illumination subspace for each face in the database, by determining b^* up to the $\mathbf{A}$ matrix. We then presented the algorithm with input images of the faces in the database seen under different lighting conditions. The algorithm estimates the best lighting conditions for generating the input assuming a Lambertian model – this is done by finding s^* to minimize $\sum_{x,y}\{I(x,y) - \mathbf{b}^*(x,y) \cdot \mathbf{s}\}^2$ – and then synthesizes the image using s^*. The algorithm appeared to have no problem in estimating the correct lighting and in synthesizing an image similar to the input, even if the input image was taken under novel lighting conditions and included shadows and specularities, see figures (14,15,16).

Figure (14) shows some of the images used to construct the model. Figure (15) shows four of the input images to the system and figure (16) shows the result of using the algorithm to obtain synthesized images closest to the corresponding inputs. Observe that the synthesized images are similar except for certain shadows and specularities which cannot be synthesized using a purely Lambertian model. Although these shadows and specularities are perceptually salient, they are small in the least squares sense and do not prevent the light sources from being estimated accurately.

8 Conclusion

This paper developed a variety of techniques for learning models of the 3D shape and albedoes of objects. We demonstrated, using the dataset of faces constructed in [12], that the resulting models were fairly accurate and that they could be used to synthesize images of objects under arbitrary lighting conditions.

Our four learning schemes used different amounts of knowledge about the light source distribution and the object class. The first learning scheme assumed knowledge of light source directions and was equivalent to standard photometric

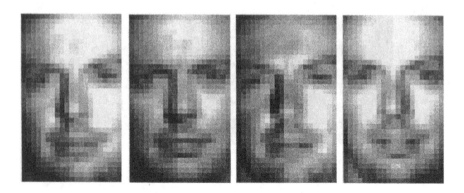

Fig. 14. Five of the original images used to construct the basis.

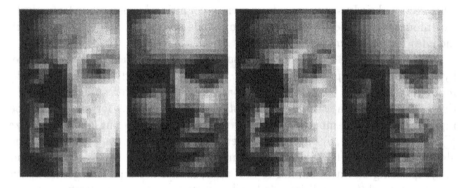

Fig. 15. Four of the input images used to test the the fitting algorithm.

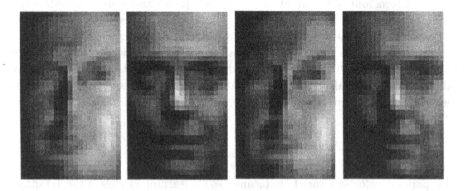

Fig. 16. The synthesized images corresponding to the input images in the previous figure. They are found by first estimating the best principal light source direction and then reconstructing the best fit. Note that estimate of the best light source direction is found only up to an arbitrary invertible linear transformation.

stereo. The remaining three schemes used SVD to estimate light source directions, albedo, and shape up a linear transformation $\mathbf{A}$. We discussed why it was uneccessary to know $\mathbf{A}$ in order to construct the light cone representation [2]. We also described ways to estimate $\mathbf{A}$ using surface integrability and/or prior knowledge about the object class.

While exploring surface integrability, we found an additional ambiguity in depth estimation which might be related to experimental findings by Koenderink [19]. This is being explored in current work.

We observed that surface integrability could be used to resolve an ambiguity in the SVD approach to photometric stereo [13] and described cases in which Hayawara's model would fail. Our work therefore has relevance to photometric stereo.

In addition, it has been applied to allowing for lighting variations of a moving object and hence improving tracking devices [10]. We are currently working on other applications and attempting to generalize to other reflectance functions.

9 Acknowledgments*

Support was provided by NSF Grant IRI 92-23676 and ARPA/ONR Contract N00014-95-1-1022. We thank David Mumford, Dan Kersten, David Kriegman and Mike Tarr for helpful discussions. Comments from three anonymous referees were also appreciated.

References

1. J.J. Atick, P.A. Griffin and A. N. Redlich. Statistical Approach to Shape from Shading. Preprint. Computational Neuroscience Laboratory. The Rockefeller University. New York. NY. 1995.
2. P. Belhumeur and D. Kriegman. "What is the set of images of an object under all lighting conditions?". In *Proceedings of Conference on Computer Vision and Pattern Recognition*. San Francisco. CA. 1996.
3. P.N. Belhumeur, J.P. Hespanha, and D.J. Kriegman. "Eignefaces vs. Fisherfaces: Recognition using Class Specific Linear Projection". In *Proceedings of ECCV*. Cambridge, England. 1996.
4. A. Lanitis, C.J. Taylor and T.F. Cootes. A Unified approach to Coding and Interpreting Face Images. In *Proceedings of ICCV*. Boston. MA. 1995.
5. R. Epstein, P.W. Hallinan and A.L. Yuille. "5 ± Eigenimages Suffice: An Empirical Investigation of Low-Dimensional Lighting Models". In *Proceedings of IEEE WORKSHOP ON PHYSICS-BASED MODELING IN COMPUTER VISION*. 1995.
6. R. Epstein and A.L. Yuille. In preparation. 1996.
7. R.A. Fisher. The use of multiple measures in taxonomic problems. *Ann. Eugenics*, 7: pp 179-188. 1936.
8. U. Grenander, Y. Chow, and D.M. Keenan. **Hands: A Pattern Theroetic Study of Biological Shapes**. New York: Springer-Verlag. 1991.
9. W.E.L. Grimson. **Object Recognition by Computer.** MIT Press. Cambridge, MA. 1990.

10. G. Hager and P.N. Belhumeur. In *Proc ECCV*. Cambridge, England. 1996.

11. P.W. Hallinan. "A low-dimensional lighting representation of human faces for arbitrary lighting conditions". In. *Proc. IEEE Conf. on Comp. Vision and Patt. Recog.*, pp 995-999. 1994.

12. P.W. Hallinan. **A Deformable Model for Face Recognition under Arbitrary Lighting Conditions.** PhD Thesis. Division of Applied Sciences. Harvard University. 1995.

13. K. Hayakawa. "Photometric Stereo under a light source with arbitrary motion". *Journal of the Optical Society of America* A, 11(11). 1994.

14. B.K.P. Horn. **Computer Vision.** MIT Press, Cambridge, Mass. 1986.

15. B.K.P. Horn and M. J. Brooks, Eds. **Shape from Shading.** Cambridge MA, MIT Press, 1989.

16. P.J. Huber. **Robust Statistics.** John Wiley and Sons. New York. 1980.

17. Y. Iwahori, R.J. Woodham and A Bagheri "Principal Components Analysis and Neural Network Implementation of Photometric Stereo". In *Proceedings of the IEEE Workshop on Physics-Based Modeling in Computer Vision.* pp117-125 (1995).

18. J.J. Koenderink and A.J. van Doorn. "The internal representation of solid shape with respect to vision". *Biological Cybernetics*, Vol. 32. pp 211-216. 1979.

19. J.J. Koenderink, A.J. van Doorn, A.M.L. Kappers "Surface Perception in Pictures".*Perception and Psychophysics*, vol 52, pp487-496. 1992.

20. Y. Moses, Y. Adini, and S. Ullman. "Face Recognition: The problem of compensating for changes in the illumination direction". In *European Conf. on Comp. Vision.*, pp 286-296. 1994.

21. Y. Moses, S. Ullman, and S. Edelman. "Generalization to Novel Images in Upright and Inverted Faces". Preprint. Dept. of Applied Mathematics and Computer Science. The Weizmann Institute of Science. Israel. 1995.

22. J.L. Mundy and A. Zisserman (eds.) **Geometric Invariance in Computer Vision.** MIT Press. Cambridge, MA. 1992.

23. H. Murase and S. Nayar. "Visual learning and recognition of 3-D objects from appearance". *Int. Journal of Computer Vision.* 14. pp 5-24. 1995.

24. S.K. Nayar, K. Ikeuchi and T. Kanade "Surface reflections: physical and geometric perspectives" *IEEE trans. on Pattern Analysis and Machine Intelligence*, vol 13 p611-634. 1991.

25. A. Shashua. **Geometry and Photometry in 3D Visual Recognition.** PhD Thesis. MIT. 1992.

26. W. Silver. *Determining Shape and Reflectance Using Multiple Images.* PhD Thesis. MIT, Cambridge, MA. 1980.

27. M. Turk and A. Pentland. "Faces recognition using eigenfaces". In *Proceedings of IEEE Conf. on Comp. Vision and Pattern Recognition.* pp 586-591. 1991.

28. R. Woodham. "Analyszing Images of Curved Surfaces". *Artificial Intelligence*, 17, pp 117-140. 1981.

29. A.L. Yuille, M. Ferraro, and T. Zhang. "Shape from Warping". Submitted to CVGIP. 1996.

30. A.L. Yuille. "A Mathematical Analysis of SVD applied to lighting estimation and image warping". Preprint. Smith-Kettlewell Eye Research Institute. San Francisco. 1996.

31. Song Chun Zhu and A.L. Yuille. "A Flexible Object Recognition and Modelling System". In *Proceedings of the International Conference on Computer Vision.* Boston 1995.

Learning Appearance Models for Object Recognition

Arthur R. Pope[1] and David G. Lowe[2]

[1] David Sarnoff Research Center, CN 5300, Princeton, NJ 08543–5300
[2] Dept. of Computer Science, University of British Columbia, Vancouver, B.C., Canada V6T 1Z4

Abstract. We describe how to model the appearance of an object using multiple views, learn such a model from training images, and recognize objects with it. The model uses probability distributions to characterize the significance, position, and intrinsic measurements of various discrete features of appearance; it also describes topological relations among features. The features and their distributions are learned from training images depicting the modeled object. A matching procedure, combining qualities of both alignment and graph subisomorphism methods, uses feature uncertainty information recorded by the model to guide the search for a match between model and image. Experiments show the method capable of learning to recognize complex objects in cluttered images, acquiring models that represent those objects using relatively few views.

1 Introduction

The multiple-view object recognition approach models an object with a series of views, each describing the object's appearance over a small range of viewing conditions. Systems adopting this approach generally assume that all features of a model view have the same likelihood of being detected and the same positional uncertainty, perhaps because better models are difficult to obtain [8, p. 182]. Clearly, though, features differ in incidence, localization accuracy, and stability. A system that learns models from example images can directly measure these differences. This paper describes how to represent feature uncertainty in a multiple-view model, learn such models from training images, and recognize objects with them.

Information about feature uncertainty can help guide the matching process that underlies recognition. Features whose presence is most strongly correlated with that of the object can be given priority during matching; features best localized can contribute most to an estimate of the object's position; and features whose positions vary most can be sought over the largest image neighborhoods. Our matching method, based on both iterative alignment and graph matching, achieves these goals. We hypothesize initial pairings between model and image features, use them to estimate an aligning transformation, use the transformation to evaluate and choose additional pairings, and so on, pairing as many features as possible. The transformation estimate includes an estimate of its uncertainty derived from the uncertainties of the paired model and image features. Potential feature pairings are evaluated using the transformation, its uncertainty, and topological relations among features so that the least ambiguous pairings are adopted earliest, constraining later pairings. The method is called *probabilistic alignment* to emphasize its use of uncertainty information.

Two processes are involved in learning a multiple-view model from training images (Fig. 1). First, the training images must be clustered into groups that correspond to distinct views of the object, with the goal that there be as many groups as necessary, but no more. Second, each group's members must be generalized to form a model view characterizing the most representative features of that group's images. Our method couples these two processes in such a way that clustering decisions consider how well the resulting groups can be generalized, and how well those generalizations describe the training images. The multiple-view model produced thus achieves a balance between the number of views it contains, and the descriptive accuracy of those views.

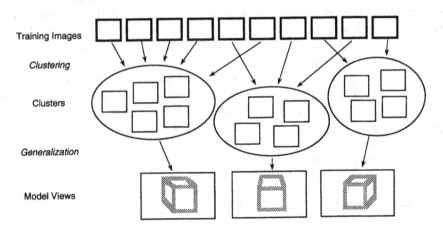

Fig. 1. Learning a multiple-view model from training images requires a clustering of the training images and a generalization of each cluster's contents.

2 Related Research

2.1 Use of Uncertainty Information in Matching

Iterative alignment has been used with a Kalman filter to estimate transformations from feature pairings in both 2D–2D matching [1] and 2D–3D matching [12]. Besides being efficient, this allows feature position uncertainty to determine transformation uncertainty, which in turn is useful in predicting feature positions in order to rate additional feature pairings [12]. However, this (partial) least-squares approach can only represent uncertainty in either image or model features, not both; total least squares can represent both, but may not be accurate in predicting feature positions from the estimated transformation [22, p. 5]. Most have chosen to represent image feature uncertainty; we have chosen to emphasize model feature uncertainty, which in our case carries the most useful information.

Some recognition methods are based on matching attributed graphs in which nodes and arcs represent features and their relations, and attributes record measurements. PREMIO [6] uses Gaussian distributions characterizing the expected number of feature and relation matches, and the expected deviation of attributes from their norms, to define a graph similarity measure that guides a fast, heuristic search for matches; all features of one model view, however, share common distributions. In view description networks [5], attribute distributions are determined by regularly sampling idealized features.

Whereas graph matching enforces topological and geometric relations among groups of features—e.g., ensuring that model line segments sharing a common junction are paired with image line segments sharing a similar junction—alignment enforces the viewpoint consistency constraint [13]. By combining these two approaches, we gain advantages from employing all constraints.

Recognition methods that search transformation space by accumulating votes may use feature uncertainty to weight votes (e.g., [17], although they assume the same uncertainty for all features). Methods that avoid tesselating the space [4, 7] have required the use of bounded error models of feature uncertainty to achieve their high efficiency. However, in our situation, where models are learned from positive training examples only, there is no way to determine error bounds; we use Gaussian error models instead. Empirical evidence [23, ch. 3] supports this choice, at least for some features.

2.2 Learning Appearance Models

Some approaches model an object as a subspace within a large space of possible appearances, and use principal components analysis to obtain a concise description of the particular subspace occupied by a given set of training examples (e.g., [21, 14]). However, applications of this approach have used global appearance representations, such as entire images, and thus they have not supported recognition of occluded objects.

Connell and Brady [9] have described a system that learns an appearance model of a 2-D object (or one view of a 3-D object), using structures of localized features. The system incorporates many interesting ideas. They use graphs to represent the part/whole and adjacency relations among object regions described by smoothed local symmetries (ribbon shapes). An attribute of a region, such as its elongation or curvature, is encoded symbolically by the presence or absence of additional graph nodes according to a Gray code. A structural learning procedure forms a model graph from multiple example graphs, most commonly by deleting any nodes not shared by all graphs (the well-known dropping rule for generalization). Similarity between two graphs is measured by a purely syntactic measure: simply by counting the nodes they share. Consequently, this system accords equal importance to all features, and it uses a somewhat arbitrary metric for comparing attribute values.

Learning a multiple-view model from real images requires some means of comparing and clustering appearances. Although several researchers have clustered images rendered from CAD models and thus avoided the feature correspondence problem, only a few have clustered real images. Among them, Gros [11] measures the similarity of an image pair as the proportion of matching shape features, whereas Seibert and Waxman [20] use a vector clustering algorithm with fixed-length vectors encoding global appearance. Our method, in comparison, uses a clustering measure based on objective performance criteria (accuracy and efficiency), and an appearance representation less affected by occlusion.

3 Method

Representations used for images, models, and transformations are described in Sects. 3.1 and 3.2. A match, comprising a set of feature pairings and an aligning transformation, is rated by the measure described in 3.3. One component of this measure estimates the probability that two features match given their respective position distributions and an aligning transformation; it is described in 3.4; other components have been described previously [15]. The method of estimating a transformation from feature pairings is described in 3.6. A matching procedure, described in 3.5, uses the match quality measure and transformation estimator to match model features with image features.

The matching procedure is used both to learn a model from training images and to recognize a modeled object in a scene. The learning procedure is described in 3.7. Recognition combines the matching procedure with an indexing procedure for selecting likely model views from a model database, and a verification procedure for deciding whether a match presents sufficient evidence that an object is present. Suitable indexing and verification methods have been described elsewhere (e.g., [2, 19]), and will not be discussed here.

A more complete description of the entire approach may be found in [16].

3.1 Image and Model Representations

An image is represented by a graph with nodes denoting features and arcs denoting abstraction and composition relations among them. A feature may, for example, be a segment of intensity edge, a particular arrangement of such segments, the response of a corner detector, or a region of uniform color. A typical image is described by many features of various types, scales, and degrees of abstraction, some found by low-level detectors, others by grouping.

Formally, an image graph G is a tuple $\langle F, R \rangle$ where F is a set of image features and R is a relation over elements of F. A feature $f_k \in F$ is a tuple $\langle t_k, \mathbf{a}_k, \mathbf{b}_k, \mathbf{C}_k \rangle$; t_k is the feature's type, $\mathbf{b}_k$ and $\mathbf{C}_k$ are the mean and covariance of its image position, and $\mathbf{a}_k$ is a vector of descriptive attributes (e.g., the curvature of a circular arc, the interior angle of a junction). An element of R, $\langle k, l_1, \ldots, l_n \rangle$, indicates that feature k groups or abstracts features l_1 through l_n.

An object is modeled by a series of model views. A model view is represented by a graph similar to an image graph, but one that includes information for estimating the probability that a feature will be found in various positions and with various attributes. It describes, for each model feature, a distribution of where that feature may be expected to be found once the model and image have been satisfactorily aligned by a transformation.

Formally, a model graph $\bar{G}$ is a tuple $\langle \bar{F}, \bar{R}, \bar{m} \rangle$, where $\bar{F}$ is a set of model features, $\bar{R}$ is a relation over elements of $\bar{F}$, and $\bar{m}$ is the number of training images used to produce $\bar{G}$. A feature $\bar{f}_j \in \bar{F}$ is a tuple $\langle \bar{t}_j, \bar{m}_j, \bar{A}_j, \bar{B}_j \rangle$; $\bar{t}_j$ is the feature's type, $\bar{m}_j$ is the number of training images in which the feature was observed, and $\bar{A}_j$ and $\bar{B}_j$ are the sequences of attribute vectors and positions drawn from those training images. The mean and covariance matrix of $\bar{B}_j$ are denoted $\mathbf{b}_j$ and $\mathbf{C}_j$. $\bar{R}$ is defined similarly to R.

3.2 Coordinate Systems

Feature positions are specified by 2D location, orientation, and scale. Image features are located in an *image coordinate system* of pixel rows and columns. Model features are located in a *model coordinate system* shared by all features within a model graph. Two schemes are used:

$xy\theta s$ The feature's location is represented by $[x\ y]$, its orientation by θ, and its scale by s.

$xyuv$ The feature's location is represented by $[x\ y]$. Its orientation and scale are represented by the orientation and length of the 2D vector $[u\ v]$.

We will prefer the $xy\theta s$ scheme for measuring feature positions and the $xyuv$ scheme for aligning features in the course of matching a model with an image. They are related by $\theta = tan^{-1}(v/u)$ and $s = \sqrt{u^2 + v^2}$. Where necessary, superscripts $^{xy\theta s}$ and ^{xyuv} indicate which scheme is in use.

A 2D similarity transformation T is used to align features.[3] Fortunately, the $xyuv$ scheme allows T to be estimated from feature pairings by solving a system of linear equations.[4] The transformation of image position $\mathbf{b}_k = [x_k\ y_k\ u_k\ v_k]$ involving a rotation by θ_t, a scaling by s_t, and a translation by $[x_t\ y_t]$ (in that order), has two linear formulations, both used here:

$$T(\mathbf{b}_k) = \begin{bmatrix} 1 & 0 & x_k & -y_k \\ 0 & 1 & y_k & x_k \\ 0 & 0 & u_k & -v_k \\ 0 & 0 & v_k & u_k \end{bmatrix} \begin{bmatrix} x_t \\ y_t \\ u_t \\ v_t \end{bmatrix} = \mathbf{A}_k \mathbf{b}_t \text{ and}$$

---

[3] There is an analogous formulation using affine transformations with advantages only in modeling 3D planar objects.

[4] Ayache and Faugeras [1], among others, have also used this formulation to express the transformation as a linear operation.

$$T(\mathbf{b}_k) = \begin{bmatrix} u_t & -v_t & 0 & 0 \\ v_t & u_t & 0 & 0 \\ 0 & 0 & u_t & -v_t \\ 0 & 0 & v_t & u_t \end{bmatrix} \begin{bmatrix} x_k \\ y_k \\ u_k \\ v_k \end{bmatrix} + \begin{bmatrix} x_t \\ y_t \\ 0 \\ 0 \end{bmatrix} = \mathbf{A}_t \mathbf{b}_k + \mathbf{x}_t \ .$$

3.3   Match Quality Measure

A match is a consistent set of pairings between some model and image features, plus a transformation closely aligning paired features. We seek a match that maximizes both the number of features paired and the similarity of paired features. Our match quality measure quantifying these goals extends that reported in [15] to include an evaluation of how well the transformation aligns features.

Pairings are represented by $E = \langle e_1, e_2, \ldots \rangle$, where $e_j = k$ if model feature j matches image feature k, and $e_j = \perp$ if it matches nothing. H denotes the hypothesis that the modeled view of the object is present in the image. Match quality is associated with the probability of H given E and T, which Bayes' theorem lets us write as

$$P(H \mid E, T) = \frac{P(E \mid T, H) \, P(T \mid H)}{P(E \wedge T)} P(H) \ . \tag{1}$$

There is no practical way to represent the high-dimensional, joint probability functions $P(E \mid T, H)$ and $P(E \wedge T)$ so we approximate them by adopting simplifying assumptions of feature independence. The joint probabilities are decomposed into products of low-dimensional, marginal probability functions, one per feature:

$$P(H \mid E, T) \approx \prod_j \frac{P(e_j \mid T, H)}{P(e_j)} \frac{P(T \mid H)}{P(T)} P(H) \ . \tag{2}$$

The measure is defined using log-probabilities to simplify calculations. Moreover, all positions of a modeled view within an image are assumed equally likely, so $P(T \mid H) = P(T)$. With these simplifications the measure becomes

$$g(E, T) = \log P(H) + \sum_j \log P(e_j \mid T, H) - \sum_j \log P(e_j) \ .$$

$P(H)$, the prior probability that the object as modeled is present in the image, can be estimated from the proportion of training images used to construct the model. The remaining terms are described using the following notation for random events: $\tilde{e}_j = k$, the event that model feature j matches image feature k; $\tilde{e}_j = \perp$, the event that it matches nothing; $\tilde{\mathbf{a}}_j = \mathbf{a}$, the event that it matches a feature whose attributes are $\mathbf{a}$; and $\tilde{\mathbf{b}}_j = \mathbf{b}$, the event that it matches a feature whose position, in model coordinates, is $\mathbf{b}$.

There are two cases to consider in estimating the conditional probability, $P(e_j \mid T, H)$, for a model feature j.

1. When j is unmatched, this probability is estimated by considering how often j was found during training. We use a Bayesian estimator, a uniform prior, and the $\bar{m}$ and $\bar{m}_j$ statistics recorded by the model:

$$P(\tilde{e}_j = \perp | T, H) = 1 - P(\tilde{e}_j \neq \perp | T, H) \approx 1 - \frac{\bar{m}_j + 1}{\bar{m} + 2} . \qquad (3)$$

2. When j is matched to image feature k, this probability is estimated by considering how often j matched an image feature during training, and how the attributes and position of k compare with those of previously matching features:

$$P(\tilde{e}_j = k | T, H) \approx P(\tilde{e}_j \neq \perp | T, H) \, P(\tilde{a}_j = a_k | \tilde{e}_j \neq \perp, H)$$
$$P(\tilde{b}_j = T(b_k) | \tilde{e}_j \neq \perp, T, H) . \qquad (4)$$

$P(\tilde{e}_j \neq \perp)$ is estimated as in (3). $P(\tilde{a}_j = a_k)$ is estimated using the series of attribute vectors $\bar{A}_j$ recorded with model feature j, and a non-parametric density estimator described in [15]. Estimation of $P(\tilde{b}_j = T(b_k))$, the probability that model feature j will match an image feature at position b_k with transformation T, is described in Sect. 3.4.[5]

Estimates of the prior probabilities are based, in part, on measurements from a collection of images typical of those in which the object will be sought. From this collection we obtain prior probabilities of encountering various types of features with various attribute values. Prior distributions for feature positions assume a uniform distribution throughout a bounded region of model coordinate space.

3.4 Estimating Feature Match Probability

The probability that a model and image feature match depends, in part, on their positions and on the aligning transformation. This dependency is represented by the $P(\tilde{b}_j = T(b_k) | ...)$ term in (4). To estimate it, we transform the image feature's position into model coordinates, and then compare it with the model feature's position (Fig. 2). This comparison considers the uncertainties of the positions and transformation, which are characterized by Gaussian pdfs.

Image feature k's position is reported by its feature detector as a Gaussian pdf in $xy\theta s$ image coordinates with mean $b_k^{xy\theta s}$ and covariance matrix $C_k^{xy\theta s}$. To allow its transformation into model coordinates, this pdf is re-expressed in $xyuv$ image coordinates using an approximation adequate for small θ and s variances. The approximating pdf has a mean, b_k^{xyuv}, at the same position as $b_k^{xy\theta s}$, and a

---

[5] For simplicity, our notation does not distinguish probability mass and probability density. $P(\tilde{e}_j)$ is a mass because $\tilde{e}_j$ assumes discrete values, whereas $P(\tilde{a}_j)$ and $P(\tilde{b}_j)$ are densities because $\tilde{a}_j$ and $\tilde{b}_j$ are continuous. But since (2) divides each conditional probability mass by a prior probability mass, and each conditional probability density by a prior probability density, here we can safely neglect the distinction.

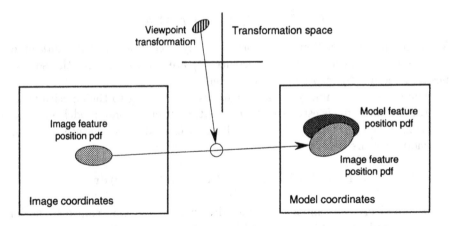

Fig. 2. An aligning transformation maps an image feature's position to model coordinates, where it is compared with a model feature's position to yield one component of the probability that the image and model features match.

covariance matrix $\mathbf{C}_k^{xyuv}$ that aligns the Gaussian envelope radially, away from the $[u\,v]$ origin:

$$\mathbf{b}_k^{xyuv} = [x_k \;\; y_k \;\; s_k \cos\theta_k \;\; s_k \sin\theta_k] \text{ and}$$

$$\mathbf{C}_k^{xyuv} = \mathbf{R} \begin{bmatrix} \sigma_l^2 & 0 & 0 & 0 \\ 0 & \sigma_l^2 & 0 & 0 \\ 0 & 0 & \sigma_s^2 & 0 \\ 0 & 0 & 0 & \sigma_\theta^2 \end{bmatrix} \mathbf{R}^{\mathrm{T}} \;,$$

$$\text{where } \mathbf{R} = \begin{bmatrix} 1 & 0 & 0 & 0 \\ 0 & 1 & 0 & 0 \\ 0 & 0 & \cos\theta_k & -\sin\theta_k \\ 0 & 0 & \sin\theta_k & \cos\theta_k \end{bmatrix}$$

and σ_l^2, σ_s^2 and σ_θ^2 are the variances in image feature position, scale and orientation estimates.

T is characterized by a Gaussian pdf over $[x_t\,y_t\,u_t\,v_t]$ vectors, with mean $\mathbf{t}$ and covariance $\mathbf{C}_t$ estimated from feature pairings as described in Sect. 3.6. Using it to transform the image feature position from $xyuv$ image to model coordinates again requires an approximation. If we would disregard the uncertainty in T, we would obtain a Gaussian pdf in model coordinates with mean $\mathbf{A}_k\mathbf{t}$ and covariance $\mathbf{A}_t\mathbf{C}_k\mathbf{A}_t^{\mathrm{T}}$. Alternatively, disregarding the uncertainty in k's position gives a Gaussian pdf in model coordinates with mean $\mathbf{A}_k\mathbf{t}$ and covariance $\mathbf{A}_k\mathbf{C}_t\mathbf{A}_k^{\mathrm{T}}$. With Gaussian pdfs for both feature position and transformation, however, the transformed position's pdf is not of Gaussian form. At best we can approximate it as such, which we do with a mean and covariance given in $xyuv$ coordinates by

$$\mathbf{b}_{kt}^{xyuv} = \mathbf{A}_k\mathbf{t} \text{ and}$$

$$\mathbf{C}_{kt}^{xyuv} = \mathbf{A}_t \mathbf{C}_k^{xyuv} \mathbf{A}_t^{\mathrm{T}} + \mathbf{A}_k \mathbf{C}_t \mathbf{A}_k^{\mathrm{T}} .$$

Model feature j's position is also described by a Gaussian pdf in $xyuv$ model coordinates. Its mean $\mathbf{b}_j$ and covariance $\mathbf{C}_j$ are estimated from the series of position vectors $\bar{B}_j$ recorded by the model.[6]

The desired probability—that j matches k according to their positions and the transformation—is estimated by integrating, over all $xyuv$ model coordinate positions $\mathbf{r}$, the probability that both the transformed image feature is at $\mathbf{r}$ and the model feature matches something at $\mathbf{r}$:

$$P(\tilde{\mathbf{b}}_j = T(\mathbf{b}_k) \mid \ldots) = \int_{\mathbf{r}} P(\tilde{\mathbf{r}}_j = \mathbf{r}) \, P(\tilde{\mathbf{r}}_{kt} = \mathbf{r}) \, d\mathbf{r} .$$

Here $\tilde{\mathbf{r}}_j$ and $\tilde{\mathbf{r}}_{kt}$ are random variables drawn from the Gaussian distributions $N(\mathbf{b}_j, \mathbf{C}_j)$ and $N(\mathbf{b}_{kt}, \mathbf{C}_{kt})$. It would be costly to evaluate this integral by sampling it at various $\mathbf{r}$, but fortunately the integral can be rewritten as a Gaussian since it is essentially one component in a convolution of two Gaussians:

$$P(\tilde{\mathbf{b}}_j = T(\mathbf{b}_k) \mid \ldots) = G(\mathbf{b}_j - \mathbf{b}_{kt}, \mathbf{C}_j + \mathbf{C}_{kt}) ,$$

where $G(\mathbf{x}, \mathbf{C})$ is a Gaussian with zero mean and covariance $\mathbf{C}$. In this form, the desired probability is easily computed.

3.5 Matching Procedure

Matches between a model graph and an image graph are identified by a process that combines iterative alignment and graph matching. First, possible pairings of higher-level features are ranked according to the contribution each would make to the match quality measure. The pairing $\langle j, k \rangle$ receives the rating

$$g_j(k) = \max_T \log P(\tilde{e}_j = k \mid T, H) - \log P(\tilde{e}_j = k) , \qquad (5)$$

favoring pairings where j has a high likelihood of matching, j and k have similar attributes, and the transformation estimate obtained by aligning j and k has low variance. The maximum over T is easily computed because $P(\tilde{e}_j = k \mid T, H)$ is a Gaussian in T.

Alignments are attempted from the highest-ranked pairings. Each estimates a transformation from the initial pairing, and then proceeds by repeatedly identifying additional consistent pairings, adopting the best, and updating the transformation estimate with them until the match quality measure cannot be improved further. Consistency is judged with respect to previously adopted pairings and the relations recorded by graph arcs. Again, a pairing is rated according to its match quality measure contribution:

$$g_j(k; E, T) = \log P(\tilde{e}_j = k \mid T, H) - \log P(\tilde{e}_j = k) .$$

---

[6] When $\bar{B}_j$ contains too few samples for a reliable estimate of $\mathbf{C}_j$, the estimate that $\bar{B}_j$ yields is blended with another determined by system parameters. Also, minimum variances are imposed on $\mathbf{C}_j$ to overcome situations where $\bar{B}_j$ has zero variance in some dimension.

This favors the same qualities as (5), while also favoring pairings aligned closely by the estimated transformation. To postpone ambiguous choices, highly-ranked but conflicting pairings of a common feature are downgraded.

As alignments yield matches, the best is retained and its match quality measure provides a threshold for cutting off subsequent alignments. Alignments are attempted until a match is found meeting acceptance criteria (e.g., a minimum fraction of edges matched) or resource limits are reached.

3.6 Estimating Aligning Transformation

From a series of feature pairings, an aligning transformation is estimated by finding the least-squares solution to a system of linear equations. Each pairing $\langle j, k \rangle$ contributes to the system the equations

$$\mathbf{U}_j^{-1} \mathbf{A}_k \mathbf{t} = \mathbf{U}_j^{-1} \mathbf{b}_j + \tilde{\mathbf{e}} \ .$$

$\mathbf{A}_k$ is the matrix representation of image feature k's mean position, $\mathbf{t} = [x_t y_t u_t v_t]$ is the transformation estimate, and $\mathbf{b}_j$ is model feature j's mean position. $\mathbf{U}_j$ is the upper triangular square root of j's position covariance (i.e., $\mathbf{C}_j = \mathbf{U}_j \mathbf{U}_j^T$); it weights both sides of the equation so that the residual error $\tilde{\mathbf{e}}$ has unit variance.

A recursive estimator solves the system, efficiently updating the transformation estimate as pairings are adopted. We use the square root information filter (SRIF) [3] form of the Kalman filter for its numerical stability, and its efficiency with batched measurements. The SRIF works by updating the square root of the information matrix, which is the inverse of the estimate's covariance matrix. The initial square root, $\mathbf{R}_1$, and state vector, $\mathbf{z}_1$, are obtained from the first pairing $\langle j, k \rangle$ by

$$\mathbf{R}_1 = \mathbf{U}_j^{-1} \mathbf{A}_k \quad \text{and} \quad \mathbf{z}_1 = \mathbf{U}_j^{-1} \mathbf{b}_j \ .$$

With each subsequent pairing $\langle j, k \rangle$, the estimate is updated by triangularizing a matrix composed of the previous estimate and data from the new pairing:

$$\begin{bmatrix} \mathbf{R}_{i-1} & \mathbf{z}_{i-1} \\ \mathbf{U}_j^{-1}\mathbf{A}_k & \mathbf{U}_j^{-1}\mathbf{b}_j \end{bmatrix} \overset{\triangle}{\rightarrow} \begin{bmatrix} \mathbf{R}_i & \mathbf{z}_i \\ 0 & \mathbf{e}_i \end{bmatrix} \ .$$

When needed, the transformation and its covariance are obtained from the triangular $\mathbf{R}_i$ by back substitution:

$$\mathbf{t}_i = \mathbf{R}_i^{-1} \mathbf{z}_i \quad \text{and} \quad \mathbf{C}_{t_i} = \mathbf{R}_i^{-1} \mathbf{R}_i^{-T} \ .$$

3.7 Model Learning Procedure

The learning procedure assembles one or more model graphs from a series of training images showing various views of an object. To do this, it clusters the training images into groups and constructs model graphs generalizing the contents of each group (Fig. 1). We shall describe first the clustering procedure, and then the generalization procedure, which the clustering procedure invokes repeatedly.

We use $\mathcal{X}$ to denote the series of training images for one object. During learning, the object's model $\mathcal{M}$ consists of a series of clusters $\mathcal{X}_i \subseteq \mathcal{X}$, each with an associated model graph $\bar{G}_i$. Once learning is complete, only the model graphs must be retained to support recognition.

Clustering Training Images. An incremental conceptual clustering algorithm is used to create clusters among the training images. Clustering is incremental in that, as each training image is acquired, it is assigned to an existing cluster or used to form a new one. Like other conceptual clustering algorithms (e.g., COBWEB [10]), the algorithm uses a global measure of overall clustering quality to guide clustering decisions. This measure is chosen to promote and balance two somewhat-conflicting qualities. On one hand, it favors clusterings that result in simple, concise, and efficient models, while on the other hand, it favors clusterings whose resulting model graphs accurately characterize (or match) the training images.

The minimum description length principle [18] is used to quantify and balance these two qualities. The principle suggests that the learning procedure choose a model that minimizes the number of symbols needed to encode first the model and then the training images. It favors simple models as those that can be encoded concisely, and it favors accurate models as those that allow the the training images to be encoded concisely once the model has been provided. The clustering quality measure to be minimized is defined as $L(\mathcal{M}) + L(\mathcal{X} \mid \mathcal{M})$, where $L(\mathcal{M})$ is the number of bits needed to encode the model $\mathcal{M}$, and $L(\mathcal{X} \mid \mathcal{M})$ is the number of bits needed to encode the training images $\mathcal{X}$ when $\mathcal{M}$ is known.

To define $L(\mathcal{M})$ we specify a coding scheme for models that concisely enumerates each of a model's graphs along with its nodes, arcs, attribute vectors and position vectors. Then $L(\mathcal{M})$ is simply the number of bits needed to encode $\mathcal{M}$ according to this scheme.

To define $L(\mathcal{X} \mid \mathcal{M})$ we draw on the fact that given any probability distribution $P(x)$, there exists a coding scheme, the most efficient possible, that achieves essentially $L(x) = -\log_2 P(x)$. Recall that the match quality measure is based on an estimate of the probability that a match represents a true occurrence of the modeled object in the image. We use this probability to estimate $P(X \mid \bar{G}_i)$, the probability that the appearance represented by image X may occur according to the appearance distribution represented by model graph $\bar{G}_i$:

$$P(X \mid \bar{G}_i) = \max_{\langle E, T \rangle} P(H \mid E, T) \ .$$

This probability can be computed for any given image graph X and model graph G_i, using the matching procedure (Sect. 3.5) to maximize $P(H \mid E, T)$ over matches $\langle E, T \rangle$. $P(X \mid \bar{G}_i)$ is then used to estimate the length of an encoding of X given $\bar{G}_i$:

$$L(X \mid \bar{G}_i) = \min_{\langle E, T \rangle} \left(-\log_2 P(X \mid \bar{G}_i) + L_u(X, E) \right) \ .$$

The $L_u(X, E)$ term is the length of an encoding of unmatched features of X, which we define using a simple coding scheme comparable to that used for model graphs. Finally, we define $L(\mathcal{X} \mid \mathcal{M})$ by assuming that for any $X \in \mathcal{X}_i \subseteq \mathcal{X}$, the best match between X and any $\bar{G}_j \in \mathcal{M}$ will be that between X and $\bar{G}_i$ (the model graph obtained by generalizing the group containing X). Then the length of the encoding of each $X \in X$ in terms of the set of model graphs $\mathcal{M}$ is the sum of the lengths of the encodings of each in terms of its respective model graph:

$$L(\mathcal{X} \mid \mathcal{M}) = \sum_i \sum_{X \in \mathcal{X}_i} L(X \mid \bar{G}_i) \ .$$

As each training image is acquired it is assigned to an existing cluster or used to form a new one. Choices among clustering alternatives are made to minimize the resulting $L(\mathcal{M}) + L(\mathcal{X} \mid \mathcal{M})$. When evaluating an alternative, each cluster's subset of training images $\mathcal{X}_i$ is first generalized to form a model graph $\bar{G}_i$ as described below.

Generalizing Training Images to Form a Model Graph. Within each cluster, training images are merged to form a single model graph that represents a generalization of those images. An initial model graph is formed from the first training image's graph. That model graph is then matched with each subsequent training image's graph and revised after each match according to the match result. A model feature j that matches an image feature k receives an additional attribute vector $\mathbf{a}_k$ and position $\mathbf{b}_k$ for its series $\bar{A}_j$ and $\bar{B}_j$. Some unmatched image features are used to extend the model graph, while model features that remain largely unmatched are eventually pruned. After several training images have been processed in this way the model graph nears an equilibrium, containing the most consistent features with representative populations of sample attribute vectors and positions for each.

4 Experimental Results

The method has been implemented in a system that recognizes 3–D objects in 2–D intensity images using a basic repertoire of features. The lowest-level features are straight, circular and elliptical edge segments. Additional features, representing perceptually-significant groupings, are junctions, pairs and triples of junctions, pairs of parallel segments, and convex regions. Although this feature repertoire has proven adequate for recognizing a wide variety of objects, it can also be readily extended to extend the range of objects handled by the system.

Figs. 3 through 6 present one example of model learning and recognition. Other examples of objects the system has learned to recognize are shown in Fig. 7.

The learning procedure was applied to 112 training images of a bunny acquired at 5° intervals over camera elevations of 0° to 25° and azimuths of 0° to 30° (Fig. 3). The learning procedure clustered these training images to produce the groups shown in Fig. 4. (Although presenting training images to the system in different orders yielded different clusterings, those clusterings were all qualitatively similar in that each contained approximately the same number of groups and the same distribution of group sizes.) Each group of training images was generalized to produce a model graph describing the range of appearances contained in that group. One such model graph is depicted in Fig. 5; other model graphs describe other views of the bunny in similar detail. Note that individual model features differ widely in uncertainty, as indicated by the standard deviation ellipses shown in Fig. 5.

0° elevation, 90° azimuth   25° elevation, 0° azimuth

Fig. 3. Two of 112 training images used to learn an appearance model of the bunny.

Fig. 6 shows an image in which the bunny is successfully recognized by matching features of the image with those of one of the model graphs. In this case, the model graph that best matches the image is that derived from group D of the training images, which is as expected since group D encompasses that aspect of the bunny visible in the scene.

5   Summary

We have presented a general method for recognizing complex, real-world objects using appearance models acquired from training images.

Appearance in an image is represented by an attributed graph of discrete features and their relations, with a typical object described by many features. Since one object can vary greatly in appearance when viewed under different conditions, a model is represented by a probability distribution over such graphs. The range of this distribution is divided among characteristic views, allowing a simplified representation for each view as a model graph of independent features.

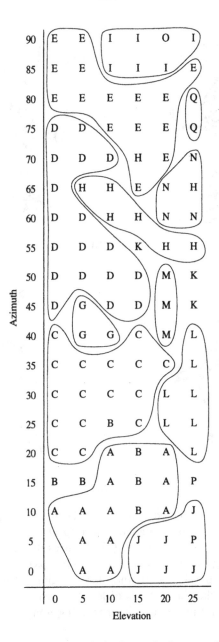

Fig. 4. Seventeen groups, designated A through Q, were formed from the 112 bunny training images. Contours delineate the approximate scope of the model views defined by some of the groups. Note, however, that because the model views are defined probabilistically, their boundaries are actually indefinite.

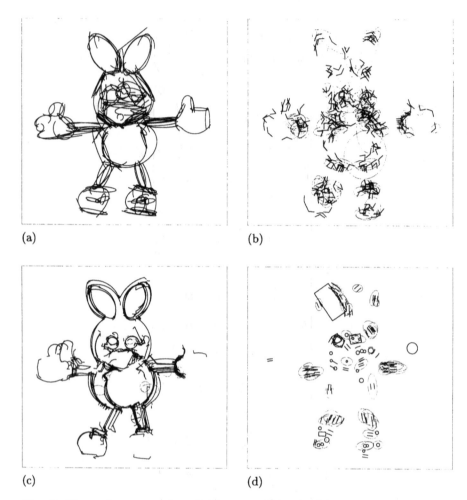

Fig. 5. Shown here are selected features of the model graph obtained by generalizing the training images assigned to group D (delineated in Fig. 4). Each feature is drawn at its mean location. (a) Features denoting edge segments. (b) Features denoting edge segment junctions. (c) Groups of edge segment junctions. (d) Groups denoting parallel pairs and closed regions of edge segments. In (b) through (d), ellipses showing 2 s.d.'s of feature location uncertainty are drawn for those features found in a majority of training images.

A model feature is described by probability distributions for probabilities of detection, various internal attribute values, and various image positions. All three distributions are estimated from samples supplied by training images.

A match quality measure provides a principled means of evaluating a match between a model and an image. It combines probabilities that are estimated using the distributions recorded by the model. The measure leads naturally to an efficient matching procedure called probabilistic alignment. In searching for

Fig. 6. Bunny test image. *Above:* Image. *Below:* Edge segment features of the image that were matched by those of the model. Additional features, such as junctions and regions, were also matched, but they are not shown here.

a solution, the procedure can employ constraints arising both from the topology of the model graph and from the probability distributions describing individual features.

The model learning procedure has two components. A conceptual clustering component identifies groups of training images that correspond to characteristic views by maximizing a global measure of clustering quality. That measure uses the minimum description length principle to combine a simplicity criterion favoring concise models, with a fit criterion based on the match quality measure. A generalizing component merges the images within each group to form a model graph representing a generalization of that group. It uses the matching procedure to determine correspondences among the group's images.

In principle the method can recognize any object by its appearance, given a sufficient range of training images, sufficient storage for model views, and an appropriate repertoire of features. In practice, however, highly flexible objects will require impractical numbers of training images and model views. For such objects, reducing the complexity of models, learning and recognition remains a topic for further study.

References

1. N. Ayache, O.D. Faugeras. HYPER: A new approach for the recognition and positioning of two-dimensional objects. *IEEE Trans. Patt. Anal. Mach. Intell.* PAMI-8:44–54, 1986.
2. J.S. Beis, D.G. Lowe. Learning indexing functions for 3-D model-based object recognition. In *Proc. Conf. Computer Vision and Patt. Recognit.*, 275–280, 1994.
3. G.J. Bierman. *Factorization Methods for Discrete Sequential Estimation.* Academic Press, 1977.
4. T.M. Breuel. Fast recognition using adaptive subdivisions of transformation space. In *Proc. Conf. Computer Vision and Patt. Recognit.*, 445–451, 1992.
5. J.B. Burns, E.M. Riseman. Matching complex images to multiple 3D objects using view description networks. In *Proc. Conf. Computer Vision and Patt. Recognit.*, 328–334, 1992.
6. O.I. Camps, L.G. Shapiro, R.M. Haralick. Object recognition using prediction and probabilistic matching. In *Proc. of the IEEE/RSJ Int. Conf. on Intell. Robots and Systems*, 1044–1052, July 1992.
7. T.A. Cass. Polynomial-time object recognition in the presence of clutter, occlusion, and uncertainty. In *Proc. European Conf. on Computer Vision*, 834–842, 1992.
8. C.H. Chen, P.G. Mulgaonkar. Automatic vision programming. *CVGIP: Image Understanding* 55:170–183, 1992.
9. J.H. Connell, M. Brady. Generating and generalizing models of visual objects. *Artificial Intell.* 31:159–183, 1987.
10. D.H. Fisher. Knowledge acquisition via incremental conceptual clustering. *Machine Learning* 2:139–172, 1987.
11. P. Gros. Matching and clustering: Two steps towards automatic object model generation in computer vision. In *Proc. AAAI Fall Symp.: Machine Learning in Computer Vision*, AAAI Press, 1993.
12. Y. Hel-Or, M. Werman. Pose estimation by fusing noisy data of different dimensions. *IEEE Trans. Patt. Anal. Mach. Intell.* 17:195–201, 1995.
13. D.G. Lowe. The viewpoint consistency constraint. *Int. J. Computer Vision* 1:57–72, 1987.
14. H. Murase, S.K. Nayar. Learning and recognition of 3-D objects from brightness images. In *Proc. AAAI Fall Symp.: Machine Learning in Computer Vision*, 25–29, AAAI Press, 1993.
15. A.R. Pope, D.G. Lowe. Learning object recognition models from images. In *Proc. Int. Conf. Computer Vision*, 296–301, 1993.

Fig. 7. Other examples of objects the system has learned to recognize. *Left*: One element drawn from each object's set of training images. *Right*: Recognition of the objects in test images.

16. A.R. Pope. *Learning to Recognize Objects in Images: Acquiring and Using Probabilistic Models of Appearance*. Ph.D. thesis, Univ. of British Columbia, 1995. WWW http://www.cs.ubc.ca/spider/pope/Thesis.html.

17. I. Rigoutsos, R. Hummel. Distributed Bayesian object recognition. In *Proc. Conf. Computer Vision and Patt. Recognit.*, 180–186, 1993.

18. J. Rissanen. A universal prior for integers and estimation by minimum description length. *Annals of Statistics* 11:416–431, 1983.

19. K.B. Sarachik, W.E.L. Grimson. Gaussian error models for object recognition. In *Proc. Conf. Computer Vision and Patt. Recognit.*, 400–406, 1993.

20. M. Seibert, A.M. Waxman. Adaptive 3-D object recognition from multiple views. *IEEE Trans. Patt. Anal. Mach. Intell.* 14:107–124, 1992.

21. M.A. Turk, A.P. Pentland. Face recognition using eigenfaces. In *Proc. Conf. Computer Vision and Patt. Recognit.*, 586–591, 1991.

22. S. van Huffel, J. Vandewalle. *The Total Least Squares Problem: Computational Aspects and Analysis*. SIAM, 1991.

23. W.M. Wells III. *Statistical Object Recognition*. Ph.D. thesis, MIT, 1992.

An Image Oriented CAD Approach [*]

Cordelia Schmid, Philippe Bobet, Bart Lamiroy and Roger Mohr

INRIA
655 avenue de l'Europe
38330 Montbonnot Saint Martin
France

Abstract. Matching abstract CAD models with images is a well studied problem. It includes the problems of identifying modelled objects for which 3D CAD data is available in images, and of locating them with respect to a given reference frame. Some authors have concluded that this problem has no real general solution as the representation levels are too different (see for instance the discussion in the workshop of CAD model-based vision [Bow91]).

We are developing an alternative approach which overcomes this problem by representing each CAD model by several images to which 3D CAD features are added. This representation allows to solve the standard vision problems to be solved much more easily such as "where is this CAD feature in the image?" or "what is the object pose?". In addition it supports fast and robust recognition using recently developed hashing techniques. Several annotated images must then be stored along with CAD data.

Experiments with different kinds of images illustrate the validity of the approach.

1 Introduction

1.1 Context of this work

CAD data includes information about shape, material, mechanical structure, etc. Such information allows the automation of design and manufacturing processes. For manufacturing processes, such as inspection or (visually guided) manipulation, direct matching of parts and image features using the shape information is required.

Several authors have addressed the CAD recognition problem. Most of them use 3D range data as e.g. in [BH86]. Others, such as in [Low86] try to match image primitives such as line segments to the contours of their CAD models either using the prediction/verification paradigm either relying on a Hough transform. More recent works in this domain mainly derive from these methods, like [CS94] introducing numerical optimization or [YC94] using aspect graphs.

Despite these mentioned successes there is a common opinion that in general CAD based vision is not very robust (see for instance the reported discussion in

---

[*] This work was performed in the Movi project which a joint project with Cnrs, Inpg, Inria, Ujf

[SB91], or the results given in [CJ90]). One of the major reasons for failure is the fact that the images do not directly reflect the 3D CAD information on the object : on the one side we have abstract 3D structure and on the other a sampled 2D intensity signal. A convincing example is given by Chen and Mulgaonkar [CM91]. After many experiments with quite simple objects, they were never able to predict from the CAD model what would be seen in the real images, even when they used sophisticated image generation programs: the predicted line segments were not the ones observed.

This has led some researchers to suggest representations closer to the images. One example are aspect graphs, which were first developed by Kœnderink [KD79], see [GCS91] for a complexity analysis. Aspect graphs consider 2D topological aspects of apparent line features. However, once more the model relies on the features that should theoretically be seen in the image and not on what is actually present in the sensed signal.

In order to avoid the unnecessary complexity of such graphs, Ikeuchi[IK88] contructs them from a finite set of synthetized views, but still with predicted features. Gros [Gro94][Gro95] addressed this problem in a new way : he considered also a finite set of views but with real feature appearing instead of working with features which are predicted to appear in the images. From there he built up models from existing image data. Such an approach bridges the gap between the model and perception and therefore makes matching easier. Following this idea, we introduce in this paper "aspect images", which will be referred hereafter as *model images*. Hence, we use 2D representations originated from the images to model each aspect of the object. A major difference with aspect graphs is that we only need to collect enough information on each aspect to allow matching and camera oriented pose estimation.

1.2 Proposed approach

Each object is represented by a number of *model images* which correspond to different views of it. The number of images necessary will be discussed later. Model images are taken in a clean environment, so that all of the features detected are due to the object. Figure 1 shows some model images of the "Dinosaur" object.

However, 3D CAD data is not present in such model images. Thus, we need a link between images and the CAD data. Such links are built by virtually adding the projection of the 3D CAD data to each model image. For instance points, lines etc. can be added. The 2D locations of these features are stored in a file attached to each model image. Other kinds of information might also be linked to the images, but we will concentrate on feature information in this paper. Figures 2 and 3 illustrate the principle. In figure 2 significant points of the 3D dinosaur are marked: end of the tail, claws, eyes, etc. In figure 3 a rotation axis, ellipses and points are added. On these examples, the 3D CAD features have been detected manually. However, if numerous features have to be attached to the model images, this can be done automatically by projecting the 3D model onto the image: once a few features have been located manually, the 3D–2D projection matrix can be estimated and all of the CAD data can then

Fig. 1. Some model images of the "Dinosaur".

be projected. Obtaining these 3D features can thus be computer assisted, and partially automated.

Fig. 2. CAD data defined for the dino object.

Having recovered the corresponding model and being given the matches between the retrieved model image and the image, we can compute the projective relation. This allows us to locate the CAD information precisely in the image to be recognized.

As different images present different aspects of the object, several images are needed. However, the number of images required is usually much smaller

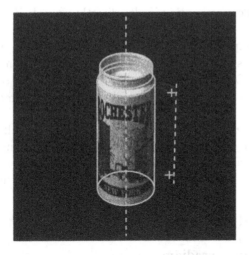

Fig. 3. Image with additional CAD data

theoretical number of topologically different aspects [GCS91] because the images capture only the most significant features: in order to be detected a feature must at least cover several pixels. For a discussion on aspect graphs for vision applications see [Bow91]. Therefore, even though each image is voluminous, the overall storage required is still reasonable with respect to current technology.

Furthermore, every detail does not need to be seen: only sufficient aspects are required to allow a discriminant matching. The position of the remaining 3D data in the image will be inferred (section 3). The set of model images must be sufficient to allow the location and matching of an arbitrary input image of the object. This is context dependent:
- if each object is unique, matching will be much easier than if several similar objects are observed with only a small variation of the viewing direction;
- matching will also be harder if the number of potential candidates is large.

1.3 Outline of the paper

Section 2 presents the robust matching paradigm, robustness is achieved by computing local invariant properties (allowing occlusion to be handled, and using a global consensus method to deal with spurious or missing features). When dealing with many objects this approach has to manipulate hundreds or even thousands of image models, but sequential search is avoided using a specially tuned indexing technique. Section 2 describes how all these steps are implemented on two different kinds of features.

From these matches we use recent results on the algebraic links between several images to derive the locations of the CAD features in the newly considered image (section 3). This would directly allow manufacturing applications such as visual servoing, for which only 2D feature locations are required (see for instance [EFR92]). Pose estimation is also directly possible if camera parameters are available.

A final set of experiments is described in section 4. We evaluate the number of image models that are required for good identification and illustrate the performance of CAD feature locations. The goal is to locate the features in new images with intermediate object pose and complex environments (background, occlusion).

For the remainder of this paper, the context will be the following: the objects can be viewed from any angle and a viewing distance varying by a factor of 2. The model images are stored in a database containing more than 1000 images (see appendix A for a desciption of the images contained in the database).

2 Robust and fast matching

2.1 The matching paradigm

We do not claim having a general image matching solution. However we describe a general matching scheme which contains two different matching tools for two kinds of images: images with nice line features (section 2.2) and images with texture (section 2.3). We feel that matching has to be done globally for robustness, in order to be able to filter out the various types of errors that can occur in the preprocessing steps. In fact the spirit is similar to the work done by Huttenlocher et al. [HKR93] and we agree with them that it would be even better if we could avoid complete segmentation.

When the camera moves, the images and the corresponding features change too. It is known that there is no general property invariant over all images of a general 3D scene ([BWR90],[CJ90], [MU92]). However it was experimentally shown by [BA90] that angles and length ratios are locally approximately constant with camera motion. This result has a strong theoretical foundation: in [BL93], Binford proved that these values are invariant up to a first degree approximation of the camera motion and zoom. As angles and length ratios are the basic invariants of the similarity group, this indicates that this group captures local properties of images rather well: with small camera motion, an image can be locally mapped onto the new view using just uniform scaling, translation and rotation: the transformations of the similarity group.

It must be emphasized that this approximation will only be valid locally. As expected, multiple model images are required to capture the global changes.

The two matching techniques described in this section rely on the following steps:

- *Feature extraction*: at this stage, we know from experience that many features are spurious, missing, or badly extracted; line bundles are used in the first implementation, interest points in the second one;
- *Computation of local characteristics*: invariants with respect to a considered group of image transforms are computed; for the reasons stated above, we will limit ourselves to the group generated by uniform scaling and rigid motion

(similarity group) in this paper; other invariant can easily be obtained for other groups for instance by limiting the kinds of camera motion permitted; unsurprisingly, the simpler the group (just translations for instance) the better the results;

- *Robust matching:* a Hough-like voting technique is used to allow all of the features to be taken into account and to filter out the noise: this captures a global consensus on the correct matching;
- *Hash based indexing:* to speed up the search for the right model, an indexing technique similar to the hashing developed in [GA94] is used (cf [LW88][SM92]).

2.2 Implementation for objects with line features

In this section we shall describe a matching and recognition technique for mainly polyhedral objects. Obviously this kind of objects does not span the entire scope of applications. We assume that the images used are suitable for line segmentation.

We follow the previously described paradigm using the following steps:

- *Feature Extraction:* Image contours are extracted in a standard way, followed by straight line approximation. Junctions of theses lines are then detected and these line segments and junctions are the basic features. The general process is described in [HSV90]. Figure 4 displays a segmented images for which the given objects were identified and located.

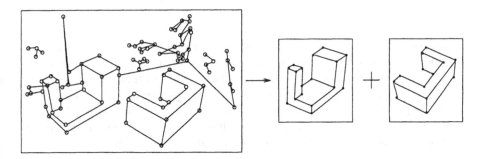

Fig. 4. Image in which the objects were correctly localized and identified.

- *Similarity invariant computation:* As here we have segments, local angles (i.e. angle between lines in junction) and length ratios are the suitable invariants we want to consider here. Notice that matched junctions completely define the similarity mapping between them: the junction point provided the translation, corresponding lines define the rotation, and the segment lengths define the scaling factor. Of course this is only meaningful for junctions with the same invariants.

– *Robust matching:* As each feature with corresponding invariants defines a unique 2D image transform, and as these transforms have to be similar for the whole object (the quasi-invariant hypothesis), we should observe a peak in the 4D space of all possible similarity transforms. So we extract all of the possible matches based on the invariants and project the corresponding transforms into a 4D Hough space (see [LG96] for details). Using this coherence filter allows the numerous false matches and spurious features to be rejected. Using the model images displayed in Figure 5 this process allows all of the objects in the noisy image of figure 4 to be identified.

Fig. 5. Collection of images used as models.

– *Indexing:* If k is the average number of features in an image (including each model image) and n the number of model images, a sequential search would require $O(k^2 n)$ accesses to the Hough space. To avoid this, the set of invariants is organized so that each invariant (α, r) (where α is the angle and r the length ratio) can be accessed rapidly by a tree search (see [LG96] for details). This allows a complexity of close to $O(k)$ accesses in the Hough space, as the tree is bounded if the invariants are accurate.

The two final steps are summarized in figure 6. Typical recognition time is about 1 second on Spar10 workstation for a database containing 100 objects.

2.3 Implementation for objects with texture

In the previous section a method based on line features was presented. However, many real world objects do not contain line features. Illustrative examples are displayed in figure 7 and figure 1. In general, for any textured object line feature detection is difficult. An alternative method can be based on the local characterization of keypoints. Keypoints are lower level image features than lines and

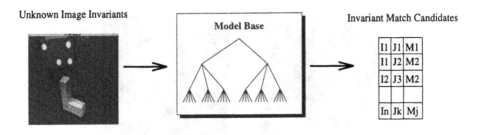

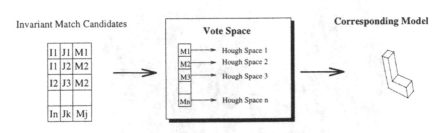

Fig. 6. Recognition and matching algorithm. We calculate the quasi-invariants, search for them in the model base, and vote in the corresponding Hough space for each plausible match found.

they are more reliable for non-polyhedral objects. Use of such feature already proved to be robust for computing the epipolar geometry [ZDFL94].

The method presented here is based on work on object recognition previously presented in [SM96, SM95]; for more details the reader is referred to these papers. The robustness of the method allows the CAD-model of an object to be retrieved in presence of occlusions and changes of background.

Following the same steps as the previous method we have:

- *Feature extraction:* In this approach, the choice of points for which the vectors of invariants are computed is very important as the positions of these points determine the quality of the local invariants. We have used the feature detector described in [HRvdH+92] for which good positioning is obtained even when detecting at different scale - an important factor here as we want to be robust to scale changes. In order to avoid the influence of the background keypoints close to the object boundary are removed (see the model images 1 to be convinced that it is easily performed).
- *Local invariant computation :* For object recognition purposes, detected keypoints have to be characterized. This characterization has to be discriminant and reliable. It also has to be invariant to some group of transformations (in our case, the group of similarity transformations).

The characteristics used are based on differential greyvalue invariants which have been studied theoretically by Kœnderink and co-workers [KvD87, Flo93, Rom94]. Owing to our stable implementation of these invariants, a reliable

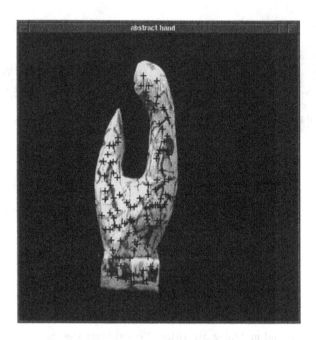

Fig. 7. Keypoints detected in one view of the "Abstract Hand".

characterization of the signal is obtained and third order derivatives can be used. A multi-scale approach [Wit83, Lin94] makes the characterization robust to scale changes up to a factor 2 (see [SM96]).

An efficient way to stabilize the calculation of derivatives is to use the derivatives of a smoothing function : the Gaussian. This choice also coincides with the definition of scale-space, which will be important for our multi-scale approach.

Given the derivatives of a function at a point up to Nth order, the Taylor expansion describes the function locally to this order. Thus, the image in a neighbourhood of a point can be described by the set of derivatives. This set is stacked into a vector. Such a vector has been used by Kœnderink [KvD87] who named it the "local jet". Given the local jet it is possible to combine the derivatives to obtain differential invariants under the group of rigid displacements in the image (translation and rotation). These can be stacked into a vector, denoted by $\vec{\mathcal{V}}$.

The first part (cf. equation 1) of this vector contains the complete and irreducible set of differential invariants up to 2nd order. The L_x and L_y signify the derivatives in x and y direction respectively.

$$\vec{\mathcal{V}}_i \, [0..4] = \begin{bmatrix} L \\ L_x L_x + L_y L_y \\ L_{xx} L_x L_x + 2 L_{xy} L_x L_y + L_{yy} L_y L_y \\ L_{xx} + L_{yy} \\ L_{xx} L_{xx} + 2 L_{xy} L_{yx} + L_{yy} L_{yy} \end{bmatrix} \tag{1}$$

where $L_{i_1 \ldots i_n}(\vec{x}, \sigma)$ is the convolution of intensity image I with the Gaussian derivatives $G_{i_1 \ldots i_n}(\vec{x}, \sigma)$:

$$L_{i_1 \ldots i_n}(\vec{x}, \sigma) = (G_{i_1 \ldots i_n} * I)(\vec{x}, \sigma)$$

It is possible to calculate the invariants for different sizes σ of the Gaussian. The second part of the vector contains a complete set of invariants of third order. Equation 2 presents this set in Einstein notation. Readers interested in Cartesian notation are referred to [Flo93] or previous work [SM95].

$$\vec{\mathcal{V}} \, [5..8] = \begin{bmatrix} \varepsilon_{ij}(L_{jkl} L_i L_k L_l - L_{jkk} L_i L_l L_l) \\ L_{iij} L_j L_k L_k - L_{ijk} L_i L_j L_k \\ -\varepsilon_{ij} L_{jkl} L_i L_k L_l \\ L_{ijk} L_i L_j L_k \end{bmatrix} \tag{2}$$

with ε_{ij} the 2D antisymmetric epsilon tensor defined by $\varepsilon_{12} = -\varepsilon_{21} = 1$ and $\varepsilon_{11} = \varepsilon_{22} = 0$.

At this stage we have, for each point, a vector $\vec{\mathcal{V}}$ of 9 invariants which makes object recognition or matching possible in the presence of any rigid displacement. To allow for scale changes, this vector has to be adapted to the corresponding scale. To be resistant to different scale changes, that is to similarity transformations, the vector of invariants has thus to be calculated at several scales. A methodology to obtain such a multi-scale representation of a signal was first proposed by [Wit83] and Kœnderink [Koe84]. Lindeberg [Lin94] has extended and summarized this approach. It should be noted that in a multi-scale context the calculation support must also be adjusted. This means that theoretically valid invariants to scale changes can not be used in practice.

Matching and Indexing: For database indexing, the vectors of the image to be recognized are used in a voting algorithm. The indexing and hashing algorithm is similar to the one described in 2.2. To take into account the uncertainties of the components of the invariant vector as well as their differences in magnitude, the Mahalanobis distance is used to compare invariant vectors.

However, in contrast to the method presented in 2.2 we do not use global coherence, but rather we add a second level of locality : the neighbourhood of each point is structured. The semi-local constraint requires that the neighbors of a point stay the same. As spurious or missing detection is possible, we only require 50% of the points to correspond. A geometrical constraint also requires that the angles between corresponding points are similar.

The robustness of the method is illustrated in [SM96]. Paintings, aerial views and toy objects can be recognized correctly under any image rotation and any scale change up to a factor of two. Using an efficient multi-dimensional hashing algorithm we can retrieve images from a database containing close to 1000 images in less than 5 seconds. It has also been shown that given small parts of images the corresponding entire image can be correctly recognized. In this work we show that the 3D position of the objects can be recovered correctly by setting up a representation of each object using several 3D views. These views are added to the above described database. Some images of the database are displayed in appendix A.

An example To set up a representation of an object, a collection of different model images is used. Some of them are shown in figure 8.

Fig. 8. Several views of the "Abstract Hand".

The model image set contains 18 images, with views spaced by 20 degrees, allowing to go around the object. Using such a basis the object in figure 7 can be correctly retrieved. As a matter of fact the two matched model images are the two closest regarding the 3D rotation. The input image and the two corresponding model images are shown in figure 9; the input image displayed in the middle. The matched corners used for indexing are displayed too. Figure 10 shows that for the same rotation angle plus a scale change of 1.2; at this scale correct recognition is still possible without using the multi-scale approach.

Figure 11 shows that even with a densely textured background, correct retrieval is still possible. As before, the two closest views are detected correctly. Note that the number of keypoints has significantly increased due to the structure of the background. The model contains about 100 points and the forest background more than 500.

Further experimental results including the retrieval of CAD-data are given in section 4.

Fig. 9. In the middle the view to be recognized. On the left and right the views retrieved from the database. Images in the database are spaces at 20 degrees. Keypoints used for voting are displayed.

Fig. 10. A viewpoint change plus a scale change of 1.2 still allows the corresponding views to be correctly recognized.

3 Recovering the CAD data

Once an object has been recognized in an image, the position of the CAD data has to be recovered. The idea is to use the matches $\mathcal{M}$ obtained during the recognition process to recover the CAD data and pose of the model. The solution presented relies in recent results on projective geometry: the so-called "trilinearity constraint" ([Sha94]).

3.1 Definition of the problem

Let us formally describe the image to be retrieved by $\mathcal{I}$, and the closest image in the model database by $\mathcal{I}_i$. The matched features of $\mathcal{I}_i$ are a subset of the features stored with this image. We also know the positions of all CAD features in $\mathcal{I}_i$.

Fig. 11. The object in front of a dense background. On the right all detected keypoints are displayed. Correct recognition still occurs.

The problem is to compute the positions of these features in the test image. As no depth is associated with the matched point, such a recovery has no solution when the input image is matched with only one model image; only the estimated similarity between the images could be used, but as it is only locally valid, this solution would be a bad approximation.

3.2 The trilinearity constraint

Results of projective geometry have shown that a one to one pixel relation exists for three images [Sha94, FM95, MZ92, FLM92]. In the following, this relation will be noted as $\mathcal{R}$. The equation 3 presents one form of this relation (four such forms exist). It expresses the constraint between the coordinates of three corresponding image points p, p' and p'' in three different images. A given set of matches between these images allows to compute the parameters $\alpha_{[1..18]}$. Then, given a match between two images, the corresponding point in the third image can be directly computed.

$$
\begin{cases}
\alpha_1 + \alpha_2 x + \alpha_3 x'' + \alpha_4 y + \alpha_5 y' + \\
\alpha_6 x x'' + \alpha_7 y y' + \alpha_8 x y' + \alpha_9 x'' y + \\
\alpha_{10} x'' y' + \alpha_{11} x'' y y' + \alpha_{12} x x'' y' \qquad = 0 \\
\alpha_{13} + \alpha_{14} x + \alpha_{15} y + \alpha_{16} y' + \alpha_3 y'' + \\
\alpha_{17} y y' + \alpha_9 y y'' + \alpha_{10} y' y'' + \alpha_{18} x y' + \\
\alpha_6 x y'' + \alpha_{12} x y' y'' + \alpha_{11} y y' y'' \qquad = 0
\end{cases}
\tag{3}
$$

There are several different methods available to compute these relations. The method used here is based on a geometric formulation due to [BZM94] and [Har92]. This method relies on an implicit projective reconstruction of the model.

3.3 Recovering the CAD data with $\mathcal{R}$

The CAD-data can be retrieved by finding a new model image for which we can compute matches with the input image. Let $\mathcal{I}_{i+1}$ be the next matched model image. Let $\mathcal{M}_{i\leftrightarrow i+1}$ be the set of matches between images $\mathcal{I}_i$ and $\mathcal{I}_{i+1}$. Using these two sets $\mathcal{M}$ and $\mathcal{M}_{i\leftrightarrow i+1}$, we obtain a set of triple matches between images $\mathcal{I}$, $\mathcal{I}_i$ and $\mathcal{I}_{i+1}$. This set describes completely the geometric relations existing between the three images. We can then compute the relation $\mathcal{R}$ between the three images.

For images $\mathcal{I}_i$ and $\mathcal{I}_{i+1}$, we also know the CAD feature position. Using $\mathcal{R}$ the corresponding position in $\mathcal{I}$ is directly obtained from equation 4. This equation shows that the computation of the CAD position (x, y) is solved by a linear system.

$$\begin{cases} (\alpha_1 + \alpha_3 x'' + \alpha_5 y' + \alpha_{10} x'' y') + \\ (\alpha_2 + \alpha_6 x'' + \alpha_8 y' + \alpha_{12} x'' y')x + (\alpha_4 + \alpha_9 x'' + \alpha_7 y' + \alpha_{11} x'' y')y & = 0 \\ (\alpha_{13} + \alpha_{16} y' + \alpha_3 y'' + \alpha_{10} y' y'') + \\ (\alpha_{14} + \alpha_{18} y' + \alpha_6 y'' + \alpha_{12} y' y'')x + (\alpha_{15} + \alpha_{17} y' + \alpha_9 y'' + \alpha_{11} y' y'')y & = 0 \end{cases}$$

$$(4)$$

The figure 12 sums up this step. Once we have obtained the position of the CAD model points in the unknown image, the computation of the pose of the object can be done using standard methods.

3.4 Numerical computation of the relation $\mathcal{R}$ and remarks

The quality of the position of the retrieved CAD data relies on the quality of the trilinear relation $\mathcal{R}$. The computation of $\mathcal{R}$ has been heavily studied in the literature. It is well known that this computation is very sensitive to false matches. Due to scene clutter, the recognition process is prone to false matches. It is therefore important to reject these, so a least-median squares based method is used to compute relation $\mathcal{R}$. This allows up to 50% of false matches to be rejected.

The use of this method has also the nice side-effect of providing a global consistency constraint on the matching data. The results presented in section 4 are good. This is due to the fact that the retrieval process uses local, semi-local and global constraints. Local and semi-local constraints are used during the matching process (section 2.3).

3.5 Computation cost consideration

Retrieving the CAD data in the unknown image can be seen as a complex process involving a lot of data. However, almost of these data and process can be driven off-line. For instance, the matches $\mathcal{M}_{i\leftrightarrow i+1}$ between model images contained in the data base can be computed off-line during the storage process of the base. The relation $\mathcal{R}$, between the CAD model and the unknown image $\mathcal{I}$,

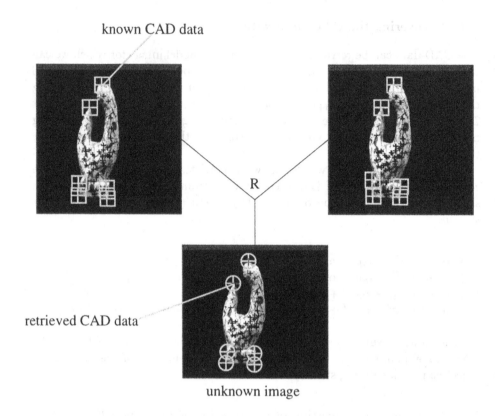

known CAD data

R

retrieved CAD data

unknown image

Fig. 12. Given two model images from the data base and for which CAD data are known and an input image, the trilinear relation $\mathcal{R}$ allows to project these data on the input image.

depends on $\mathcal{I}$ and thus cannot be pre-computed. However, this relation relies on matches between the three images $\mathcal{I}$, $\mathcal{I}_i$ and $\mathcal{I}_{i+1}$. The correspondences between $\mathcal{I}_i$ and $\mathcal{I}_{i+1}$ depend only on the data base and can thus be stored during the computation of the model. Moreover, the relation $\mathcal{R}$ depends on epipolar geometry between the three images. Since epipolar geometry links only two images, a subset (corresponding to $\mathcal{I}_i$ and $\mathcal{I}_{i+1}$) of this geometry can also be computed off line. Almost all computation of the retrieval process can thus be driven off line letting very few and simple computations to be done on line.

4   An example

In this section we show that the proposed method which integrates recognition and 3D location works well. The object used for the experiment described here is a toy dinosaur already displayed in Figure 1. We recall that the model images of this object are part of a database containing more than 1000 images (see appendix A). As no real 3D data was available, we want want to locate some of the 3D feature from the image: the position of the dinosaur's head, tail, the ends of its fingers etc.

4.1 The Dino CAD base

To built up our set of model images, different views of the dinosaur in front of a uniform background were taken. This allowed us to easily eliminate border points which would not be useful in the presence of a background change. The required number of views depends on the complexity of the object. Experiments have shown that 18 images are sufficient to build the dinosaur's model data base. Figure 13 shows the number of votes for a view close to views 1 and 2. 36 views are considered for this experiment. The votes are significant for images 0, 1, 2 and 3. The number of views can thus be reduced by a factor 2 without decreasing the quality of the matching process. 18 images are then needed to represent the dinosaur's model. For less complex objects, such as the "Abstract Hand", 9 views proved to be sufficient to set up a model base.

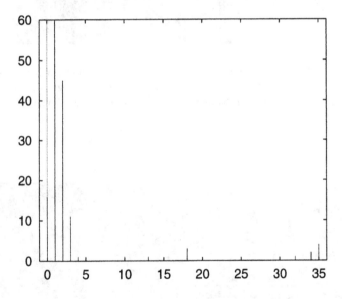

Fig. 13. Votes in Hough space for 36 model images.

Figure 1 shows some of the model images. To all these images, we have added the 2D locations of the different 3D features we want to find (see Figure 2).

4.2 Experiments on matching

A wide variety of test images were taken in order to test recognition and retrieve the so called CAD data. We first tested robustness to 3D position: given an arbitrary viewing angle, the corresponding closest views of the correct object are recognized, for an example see the left image of figure 14. The method is robust to the presence of any background as well as to the presence of occlusions, see the middle and right images of figure 14.

The local characteristics, presented in section 2.3, are invariant to image rotation and scale change. Our approach is therefore not only robust to 3D position, but also robust to scale changes, rotations about the optical axis and translations. Examples are presented in figures 15 and figure 16 for different backgrounds.

Fig. 14. Robustness of the method to a 3D viewpoint change, a background change and occlusion. The object and its corresponding views are recognized correctly in case of any 3D position of the object, in the presence of a changed background and in the presence of occlusion.

Fig. 15. The object and its nearest corresponding views are correctly recognized in case of any 3D position, scale change and translation. Note the different backgrounds.

Figure 17 shows the points used for indexing. On the left and on the right the retrieved images (the corresponding closest views), and in the middle the object to be recognized. Cross indicated the matched points. Noted that two points in

Fig. 16. The object and its nearest corresponding views are recognized correctly in case of any 3D position, image rotation, scale change and translation.

the forest are incorrectly matched. These points can be easily rejected in the following, as the projective transformation has to be global consistent (section 3.4). The matched points are used below to calculate the 3D position.

Fig. 17. Matched images

4.3 Locating the CAD data

Given the matches between the image to retrieve and the closest image stored, it is possible to compute the relation $\mathcal{R}$ between the model and this image. Figure 2 shows the manually selected CAD features in a model image. The figure 18 presents the projection of the CAD data on the test image. For this example the mean distance between the points is 0.23 pixels. The robust method has rejected 2 false matches.

In order to illustrate the quality of the position recovery, figure 19 shows a blow up of the input image with the the retrieved position of the CAD data point "left-eye".

Fig. 18. Retrieved CAD data for the dino object. They correspond accurately to the CAD model defined in figure 2.

Fig. 19. A blow up of the eye. The position of its CAD data has been precisely retrieved.

5 Discussion and conclusion

This paper presents and validates the use of a set of *model images* linked to CAD data in order to provided an image oriented CAD data base. This approach allows robust matching and recovers CAD data from new images. The robustness of the approach relies on invariant based matching: 3D features admit quasi-invariants based on the similarity group[BL93]; so we compute similarity invariants from image features, and use a global voting technique to robustly select the model which corresponds best to the given image data and simultaneously estimate an approximate 2D transformation that maps the model onto the new object image. Using local invariants allows the method to be robust to occlusion; even if half the object is hidden, the remaining features still have the same invariants. Using a global matching technique allows bad segmentation and noise to be handled, provided enough evidence can be accumulated from the features extracted. The approach is illustrated with two kinds of features: line segments for polyhedral scenes using local angles and length ratios as invariants; and points of interest for gray level images, for which a set of complete local differential invariants is used for indexing. Further experiments on robustness against large occlusions and indexing into large image databases are reported in [SM96].

Our experiments illustrate not only the robustness of the method, but also the fact that the model images need not to be too numerous. Typically 18 images are enough to encompass the object. Despite high theoretical complexity of aspect graphs, such low numbers are not surprising: all the tiny details do not need to be represented in the database; it is sufficient to observe enough significant features to allow reliable matching.

Using matches of feature points, the positions of known 3D features can be computed using the trilinear relations linking the image coordinates of the 3 images. This can be performed done when the 3D CAD features themselves were not detected.

It has to be pointed out that this method can only be used if features with invariants can be extracted. This might fail for the current grey level invariant based implementation if the scene light was changed drastically, for instance for aerial images between morning and noon. Therefore there is still a challenge to find quantities invariant not only under geometric transformations, but also under illumination changes. Our greylevel invariants, based on derivatives, can be easily extended to allow for affine changes in intensity; however, as far as we know nothing has been done for more general lighting changes, except in the case of color images [LB92].

Feature extraction is a delicate process, particularly for large features like the contour lines used in 2.2. Therefore it is probably more robust to use local features such as point of interest. Huttenlocher [HKR93] and Rucklidge [Ruc95] have proved that matching based on sets of noisy points is possible, but here we have preferred to add higher level information linked to the structure (lines) or the grey levels, as we deal with extremely complex shapes and backgrounds in our experiments.

Our present approach is still far from dealing with generic shapes, but how to perform this is an open research problem [DBB$^+$93],[ZY95]. The only realistic example we know of is limited to a specific world: cars in road scenes [KDN93], [KN95]. With the approach proposed here, we would have to store all possible models of everything and we could not hope to be generic. However we can envision an extension towards genericity through a learning stage. Such learning would estimate which features or groups of feature are representative of a given view of a class of objects. Such an approach was already explored in [Gro95] in which different images of a same model are collected into a single abstract model, but there is still a long way to go towards generic shape.

References

[BA90]   J. Ben-Arie. The probabilistic peaking effect of viewed angles and distances with application to 3-D object recognition. IEEE *Transactions on Pattern Analysis and Machine Intelligence*, 12(8):760–774, August 1990.

[BH86]   R.C. Bolles and R. Horaud. 3DPO : A three-dimensional Part Orientation system. *The International Journal of Robotics Research*, 5(3):3–26, 1986.

[BL93]   T.O. Binford and T.S. Levitt. Quasi-invariants: Theory and exploitation. In *Proceedings of* DARPA *Image Understanding Workshop*, pages 819–829, 1993.

[Bow91]   K. Bowyer. Why aspect graphs are not (yet) practical for computer vision. In *Proceedings of the* IEEE *workshop on Direction on automated* CAD-*based Vision, Maui, Hawaii, USA*, pages 97–104, 1991.

[BWR90]   J.B. Burns, R. Weiss, and E.M. Riseman. View variation of point set and line segment features. In *Proceedings of* DARPA *Image Understanding Workshop, Pittsburgh, Pennsylvania, USA*, pages 650–659, 1990.

[BZM94]   P. Beardsley, A. Zisserman, and D. Murray. Sequential update of projective and affine structure from motion. Technical Report 2012/94, University of Oxford, Oxford, United Kingdom, August 1994.

[CJ90]   D.J. Clemens and D.W. Jacobs. Model-group indexing for recognition. In *Proceedings of* DARPA *Image Understanding Workshop, Pittsburgh, Pennsylvania, USA*, pages 604–613, September 1990.

[CM91]   C.H. Chen and P.G. Mulgaonkar. CAD-based feature-utility measures for automatic vision programming. In *Direction in Automated CAD-Based Vision*, pages 106–114. IEEE Computer Society Press, 1991.

[CS94]   JL. Chen and G.C. Stockman. Matching curved 3D object models to 2D images. In A.C. Kak and K. Ikeuchi, editors, *Proceedings of the Second CAD-Based Vision Workshop*, pages 210–218, Los Alamitos, California, February 1994. IEEE Computer Society Press.

[DBB$^+$93] S.J. Dickinson, R. Bergevin, I. Biederman, J.-O. Eklund, R. Munck-Fairwood, and A. Pentland. The use of geons for generic 3D object recognition. In *Proceedings of the 12th International Joint Conference on Artificial Intelligence, Chambery, France*, pages 1693–1699, 1993.

[EFR92]   B. Espiau, F.Chaumette, and P. Rives. A new approach to visual servoing in robotics. IEEE *Trans. on Robotics and Automation*, 8(3):313–326, 1992.

[FLM92]   O.D. Faugeras, Q.T. Luong, and S.J. Maybank. Camera self-calibration: Theory and experiments. In G. Sandini, editor, *Proceedings of the 2nd*

European Conference on Computer Vision, Santa Margherita Ligure, Italy, pages 321–334. Springer-Verlag, May 1992.

[Flo93]   L. Florack. *The Syntactical Structure of Scalar Images*. PhD thesis, Universiteit Utrecht, m11 1993.

[FM95]   O. Faugeras and B. Mourrain. On the geometry and algebra of the point and line correspondences between n images. In *Proceedings of the 5th International Conference on Computer Vision, Cambridge, Massachusetts, USA*, pages 951–956, June 1995.

[GA94]   A. Guéziec and N. Ayache. Smoothing and matching of 3-D space curves. *International Journal of Computer Vision*, 12(1):79–104, 1994.

[GCS91]   Z. Gigus, J. Canny, and R. Seidel. Efficiently computing and representing aspect graphs of polyhedral objects. IEEE *Transactions on Pattern Analysis and Machine Intelligence*, 13(6):542–551, 1991.

[Gro94]   P. Gros. Using quasi-invariants for automatic model building and object recognition: An overview. In *Proceedings of the NSF-ARPA Workshop on Object Representations in Computer Vision, New York, USA*, December 1994.

[Gro95]   P. Gros. Matching and clustering: Two steps towards object modelling in computer vision. *The International Journal of Robotics Research*, 14(6):633–642, December 1995.

[Har92]   R. Hartley. Invariants of points seen in multiple images. Technical report, G.E. CRD, Schenectady, 1992.

[HKR93]   D.P. Huttenlocher, G.A. Klanderman, and W.J. Rucklidge. Comparing images using the Hausdorff distance. IEEE *Transactions on Pattern Analysis and Machine Intelligence*, 15(9):850–863, September 1993.

[HRvdH+92]   F. Heitger, L. Rosenthaler, R. von der Heydt, E. Peterhans, and O. Kuebler. Simulation of neural contour mechanism: from simple to end-stopped cells. *Vision Research*, 32(5):963–981, 1992.

[HSV90]   R. Horaud, T. Skordas, and F. Veillon. Finding geometric and relational structures in an image. In *Proceedings of the 1st European Conference on Computer Vision, Antibes, France*, Lecture Notes in Computer Science, pages 374–384. Springer-Verlag, April 1990.

[IK88]   K. Ikeuchi and T. Kanade. Applying sensor models to automatic generation of object recognition programs. In *Proceedings of the 2nd International Conference on Computer Vision, Tampa, Florida, USA*, pages 228–237, 1988.

[KD79]   J. Koenderink and A.V. Doorn. The internal representation of solid shape with respect to vision. *Biological Cybernetics*, 32:211–216, 1979.

[KDN93]   D. Koller, K. Daniilidis, and H.-H. Nagel. Model-based object tracking in monocular image sequences of road traffic scenes. *International Journal of Computer Vision*, 10:257–281, 1993.

[KN95]   H. Kollnig and H.H. Nagel. 3D pose estimation by fitting image gradients directly to polyhedral models. In *Proceedings of the 5th International Conference on Computer Vision, Cambridge, Massachusetts, USA*, pages 569–574, 1995.

[Koe84]   J.J. Koenderink. What does the occluding contour tell us about solid shape? *Perception*, 13:321–330, 1984.

[KvD87]   J. J. Konderink and A. J. van Doorn. Representation of local geometry in the visual system. *Biological Cybernetics*, 55:367–375, 1987.

[LB92]   S.W. Lee and R. Bajcsy. Detection of specularity using color and multiple views. In G. Sandini, editor, *Proceedings of the 2nd European Conference on Computer Vision, Santa Margherita Ligure, Italy,* pages 99–114. Springer-Verlag, May 1992.

[LG96]   B. Lamiroy and P. Gros. Rapid object indexing and recognition using enhanced geometric hashing. In *Proceedings of the 4th European Conference on Computer Vision, Cambridge, England,* volume 1, pages 59–70, April 1996. Postscript version available at ftp://ftp.imag.fr/pub/MOVI/publications/Lamiroy_eccv96.ps.gz.

[Lin94]   T. Lindeberg. *Scale-Space Theory in Computer Vision.* Kluwer Academic Publishers, 1994.

[Low86]   D. Lowe. Three-dimensional object recognition from single two-dimensional images. *Artificial Intelligence,* ??:355–395, 1986.

[LW88]   Y. Lamdan and H.J. Wolfson. Geometric hashing: a general and efficient model-based recognition scheme. In *Proceedings of the 2nd International Conference on Computer Vision, Tampa, Florida, USA,* pages 238–249, 1988.

[MU92]   Y. Moses and S. Ullman. Limitations of non model–based recognition. In *Proceedings of the 2nd European Conference on Computer Vision, Santa Margherita Ligure, Italy,* pages 820–828, May 1992.

[MZ92]   J. Mundy and A. Zisserman. Projective geometry for machine vision. In J. Mundy and A. Zisserman, editors, *Geometric Invariance in Computer Vision,* chapter 23, pages 463–519. MIT Press, 1992.

[Rom94]   Bart M. ter Haar Romeny, editor. *Geometry-Driven Diffusion in Computer Vision.* Kluwer Academic, computational imaging and vision edition, 1994.

[Ruc95]   W.J. Rucklidge. Locating objects using the Hausdorff distance. In *Proceedings of the 5th International Conference on Computer Vision, Cambridge, Massachusetts, USA,* pages 457–464, 1995.

[SB91]   L. Shapiro and K. Bowyer, editors. IEEE *Workshop on Directions in Automated CAD-Based Vision,* Maui, Hawai, 1991. IEEE Computer Society Press.

[Sha94]   A. Shashua. Trilinearity in visual recognition by alignment. In Jan-Olof Eklundh, editor, *Proceedings of the 3rd European Conference on Computer Vision, Stockholm, Sweden,* pages 479–484. Springer Verlag, May 1994.

[SM92]   F. Stein and G. Medioni. Structural indexing: Efficient 3-D object recognition. IEEE *Transactions on Pattern Analysis and Machine Intelligence,* 14(2):125–145, 1992.

[SM95]   C. Schmid and R. Mohr. Matching by local invariants. Technical report, INRIA, August 1995.

[SM96]   C. Schmid and R. Mohr. Combining greyvalue invariants with local constraints for object recognition. In *Proceedings of the Conference on Computer Vision and Pattern Recognition, San Francisco, California, USA,* m4 1996. Accepted for CVPR 96. ftp://ftp.imag.fr/pub/MOVI/publications/Schmid_cvpr96.ps.gz.

[Wit83]   A. P. Witkin. Scale-space filtering. In *Proceedings of the 8th International Joint Conference on Artificial Intelligence, Karlsruhe, Germany,* pages 1019–1023, 1983.

[YC94]   J. Ho Yi and D.M. Chelberg. Rapid object recognition from a large model database. In A.C. Kak and K.Ikeuchi, editors, *Proceedings of the Second CAD-Based Vision Workshop*, pages 28–35, Los Alamitos, California, February 1994. IEEE Computer Society Press.

[ZDFL94]   Z. Zhang, R. Deriche, O. Faugeras, and Q.T. Luong. A robust technique for matching two uncalibrated images through the recovery of the unknown epipolar geometry. Rapport de recherche 2273, INRIA, May 1994.

[ZY95]   S.C. Zhu and A.L. Yuille. FORMS: a flexible object recognition and modelling system. In *Proceedings of the 5th International Conference on Computer Vision, Cambridge, Massachusetts, USA*, pages 465–472, 1995.

A The database

The database contains paintings, aerial images, toy objects of the Columbia database and 3D CAD test objects. Examples of the paintings are shown in row one of figure 20. Some of the aerial images are displayed in row two (courtesy of Istar). Sample objects of the Columbia database are given in row three and in row four we display 3D objects for which CAD information has been added.

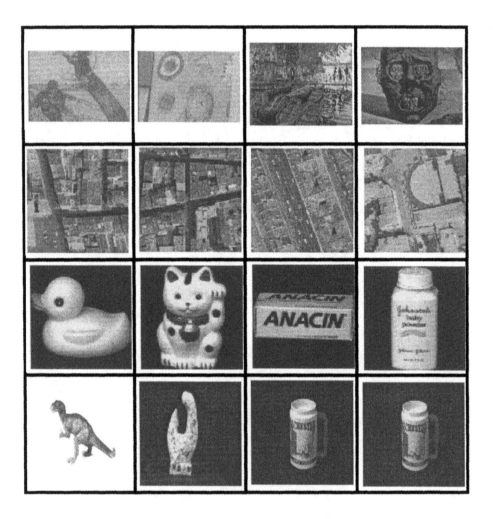

Fig. 20. Some images of the database. The database contains more than 1000 images.

An Experimental Comparison of Appearance and Geometric Model Based Recognition

J. Mundy[1], A. Liu[1], N. Pillow[2], A. Zisserman[2], S. Abdallah[2], S. Utcke[2], S. Nayar[3] and C. Rothwell[4]

[1] General Electric Corporate Research and Development, Schenectady, NY, USA
[2] Robotics Research Group, University of Oxford, Oxford, UK
[3] Dept. of Computer Science, Columbia University, NY, USA
[4] INRIA, Sophia Antipolis, France

Abstract. This paper describes an experimental investigation of the recognition performance of two approaches to the representation of objects for recognition. The first representation, generally known as appearance modelling, describes an object by a set of images. The image set is acquired for a range of views and illumination conditions which are expected to be encountered in subsequent recognition. This image database provides a description of the object. Recognition is carried out by constructing an eigenvector space to compute efficiently the distance between a new image and any image in the database. The second representation is a geometric description based on the projected boundary of an object. General object classes such as planar objects, surfaces of revolution and repeated structures support the construction of invariant descriptions and invariant index functions for recognition.

In this paper we present an investigation of the relative performance of the two approaches. Two objects, a planar object and a rotationally symmetric object are modelled using both approaches. In the experiments, each object is intentionally occluded by an unmodelled distractor for a range of viewpoints. The resulting images are submitted to two separate recognition systems. Appearance-based recognition is carried out by SLAM and recognition of invariant geometric classes by Lewis/Morse.

1 Introduction

Over the last few years, there has been increasing interest in object recognition based on a set of images of an object. This representation of an object is called an *appearance* model. Over the same period, a number of recognition systems have been implemented which employ a geometric description of an object. These geometric descriptions are based on general object classes, such as planar structures or surfaces of revolution (SORs). Each approach has strengths and limitations which can complement each other to form a more competent overall recognition system.

In this paper, we present some initial results of experiments to characterize the performance of implemented recognition systems for each approach. It is emphasized that these results are preliminary, but do represent an attempt to directly compare the two recognition processes under identical conditions, i.e. using the same set of test images and model libraries. We chose to focus on the performance of each system with respect to camera viewpoint and in the presence of occlusion. The performance of recognition under varying illumination is certainly also of great interest but was not considered here.

This sort of comparison of recognition methodologies is badly needed to advance our understanding of object recognition, but it is usually difficult to obtain such data due to the state of implementation of most research software. The results of this investigation highlight a number of strengths and weaknesses of each approach and consequently the type of shape representation that is appropriate for model based object recognition.

The appearance (SLAM) and geometry based systems (Lewis/Morse) are reviewed in sections 2 and 3 respectively. The two systems are compared on the same test images in section 4.

2 Appearance Models for Recognition — SLAM

The representation of an object by its image appearance is an empirical model which is defined for the range of viewing conditions under which the object is to be recognized [11, 14]. The assumption is that if the intensity pattern in a new image is *near* a stored image of some object then the image contains the object. The appearance model makes no commitment to constraints which might be embodied in a class such as object shape or surface texture. Therefore a recognition system based on appearance can acquire descriptions of any type of object. The only requirement is that a sufficient number of views of the object have to be acquired to provide an image close in appearance to any image in which the object is to be recognized. The major assumptions are that the object can be segmented from the scene and is not (significantly) occluded.

An appearance model can be generalized by interpolating between the acquired views. A new view of an object can be generated by assuming that an object induces a manifold in the space of image pixel measurements. For example, suppose that a series of images of an object are collected for a sample of two dimensional translations on the ground plane. It is assumed that views of the object at other translation positions can be closely approximated by interpolating the image intensities of images in the neighbourhood of a given position.

A full appearance model requires six degrees of freedom to account for object pose variation and two parameters to account for illumination direction. Then a sample of images is collected for an object to sample this eight-dimensional manifold in the high dimensional space of image pixels. In the current experiments, an object is represented by a 128×128 pixel array so the appearance manifold is embedded in a space of 16384 dimensions.

The efficiency of computing distances in such a high dimensional space can be vastly improved by the use of principal components analysis [3] (PCA). The use of PCA and appearance models has been used quite effectively for face recognition [1, 2, 6, 15]. PCA captures the variation in image data by projecting the image onto a low dimensional subspace of eigenvectors which are constructed from the covariance matrix of the image intensities. Since image intensities are typically correlated, it is possible accurately to represent a full 16384 vector in 10- or 20-dimensional eigenvector space.

The eigenvectors are computed by accumulating the covariance matrix, C_{ij} of image vectors over the database of stored object appearances. That is,

$$C_{ij} = \sum_k (I_i(k) - \mu_i)(I_j(k) - \mu_j)$$

where $I_i(k)$ is the intensity of pixel i for image k in the database; μ_i is the mean value of pixel i over the set of stored images. The eigenvectors of C_{ij} define a linear transformation of the image data onto a vector subspace. An eigenvector can be dropped if its contribution to the overall variance of the image data is small. For images with smooth patches of relatively small image variation, the variance over the database can be captured with only 10 or 20 of the original 16384 dimensions.

The SLAM software used in these experiments carries out the following specific steps.

1. Image normalization. A bounding box is constructed around the object by thresholding the object from the background. A standard image resolution, e.g. 128×128 is used to sample the bounding box. The intensity is normalized by subtracting the average pixel intensity, and then converted to a unit vector in the 16384 dimensional space.
2. Each resulting image vector is added to the image database with a label corresponding to the object class.
3. A suitable vector subspace is constructed which accounts for the variance in the database [10]. In the current experiments, 10 eigenvectors define the subspace.
4. The manifolds for each object class are approximated with a spline. In the current experiments, the appearance variation is limited to one dimension of object rotation, so the manifold is a spline curve.
5. The fitted spline curve or surface is sampled to provide an adequate coverage of the required view conditions. For example, for a rotation interval of 5°, the manifold can be sampled to 1° to provide detailed coverage.
6. To recognize an object in a new image instance, the image is segmented and normalized (as described above) and then projected onto the eigenspace. The closest stored manifold sample point in the database is located by a binary search strategy; the object is then assigned the class label of that closest manifold.

3 Geometry-Based Recognition — Lewis/Morse

There are two recognition systems that are used. The first, Lewis, is for planar object recognition. This system uses indices based on plane projective invariants.

The second, Morse, is for 3D curved object recognition. In the experiments here, we are using the Lewis and Morse systems in tandem to provide an overall recognition for both planar and SOR object classes in a given image.

3.1 Lewis

Here we summarize the main features of the Lewis planar object recognition system. The system is built upon plane projective invariants of the object outline. These invariants have the same value in any perspective image of the object. Recognition proceeds by measuring plane projective invariants in the target image. The invariants are used to construct index vectors to select models from the library. If the index value coincides with that associated with a model, a recognition hypothesis is generated. Recognition hypotheses corresponding to the same object are merged to form joint hypotheses, provided they are geometrically compatible. The (joint) hypotheses are then verified. The stages of recognition are shown in Fig.1. In more detail:

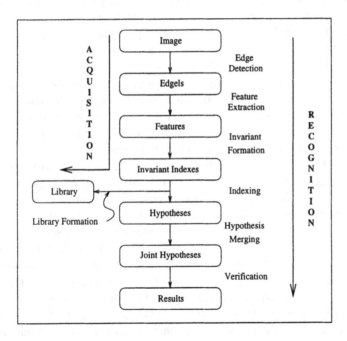

Fig. 1. The recognition system has a single grey scale image as input and the outputs are verified hypotheses with associated confidence values. Many of the processes are shared by the acquisition and the recognition paths.

Projective Invariants Used. There are three different algebraic invariant constructions used in the system: five lines; a conic and two lines; and a conic pair. These are applicable to image curves that are 'algebraic' (lines, conics). More details of these invariants are given in [9].

In all cases there is tolerance to partial occlusion, i.e., the invariants can still be formed if part of the outline is occluded. This is a result of using *semi-local* invariant descriptions — i.e. not global ones such as moments of the entire boundary — and *redundancy*: there are a number of different descriptors for each object so that there is not an excessive requirement for any single object region to be visible. In the algebraic case, lines and conics can still be extracted if part of the curve is occluded.

Feature Extraction and Invariant Formation. The goal of the segmentation is the extraction of geometric primitives suitable for constructing invariants. In the algebraic case, this involves straight lines and conics, and for non-algebraic curves, concavities delineated by bitangents.

Once sets of grouped features, f, have been produced, the algebraic and canonical invariants are computed. Each set of grouped features, or concavity curve, generally produces a number of invariant values which are collected into a vector $M(f)$. The invariant vector formed by the above process represents a point in the multidimensional invariant space. The space is quantized to enable hashing. Each object feature group is represented by a collection of points that define a region in the invariant space, the size of which depends upon the measured variance in the invariant value.

Indexing to Generate Recognition Hypotheses. The invariant values computed from the target image are used to index against invariant values in the library. If the value is in the library a preliminary recognition hypothesis is generated for the corresponding object. Each type of invariant (e.g., five lines, conic pair) separately generate hypotheses.

Hypothesis Merging. Many collections of primitives may come from the same model instance: for example, an object consisting of a square plate with a circular hole in it admits four collections, each consisting of a conic and two connected lines. Each collection has an invariant which may generate a recognition hypothesis. Such a set of recognition hypotheses is *compatible* if a single model instance could explain all of them simultaneously. Prior to verification, compatible hypotheses are combined into *joint hypotheses*.

Verification. There are two steps involved in verification, both of which can reject a (joint) recognition hypothesis. The first is to attempt to compute a common projective transformation between the model features and the putative corresponding features in the target image. The second is to use this transformation to project the entire model onto the target image, and then *measure* image support.

Back-projection and subsequent searching involves the entire model boundary, not just the features used to form the invariant. Projected model edgels must lie close to image edgels with similar orientation (within 5 pixels and 15^0). If more than a certain proportion of the projected model data is supported (the threshold used is 50%), there is sufficient support for the model, and the recognition hypothesis is confirmed.

Model Acquisition and Library Formation. A model can be acquired directly from a single image. No special orientations or knowledge of the camera calibration are required. Acquisition is simple and semi-automatic (for instance, curves do not have to be matched entirely by hand between images), using the same software for segmentation and invariant computation as used during recognition.

A model consists of the following: a name; a set of edges from an acquisition view of the object (used in the back-projection stage of verification); the lines, conics and concavities fitted to the edges; the expected invariant values and to which algebraic features and curve portions they correspond (the mean and variance of the invariant values being computed from a variety of 'standard' viewpoints of the object); and, finally, topological connectivity and geometric relations between feature groups used in the construction of joint invariants.

The library is partitioned into different sub-libraries, one for each type of invariant (e.g. one for the five-line invariant, another for the conic pair). Each sub-library then has a list of each of the invariant values tagged with an object name, and is structured as a hash table.

A detailed description of this system appears in [17].

3.2 Morse

This is a recognition system for 3D objects. The system is organized around a number of geometrically defined object *classes*. The classes include surfaces of revolution, canal surfaces (pipes) and polyhedra. These three classes cover a large number of manufactured objects. A class provides two functions. First, it provides a grouping relationship in the image. The geometric class defined in 3D *induces* relationships in the image which must hold between points on the image outline (the perspective projection of the object). The resulting image constraints enable both identification and grouping of image features belonging to objects of that class. Second, the class also supports the computation of 3D invariant descriptions including symmetry axes, canonical coordinate frames and projective signatures. Both grouping and invariant formation are viewpoint invariant, and proceed with no information on object pose.

Recognition consists of two major processes. The first is establishing an object as belonging to one of the classes. This is achieved by grouping or feature organization directed by the constraints provided by a particular geometric class. The second stage is identification which is achieved by using invariant indices derived for specific instances of the class. This differs from Lewis where recognition is targeted directly at particular objects, not first at a class of objects.

For example, if a scene contains several SORs, the grouper first identifies curve pairings that could have arisen from an SOR (they satisfy a particular transformation, see section 3.2), and then groups those arising from the same SOR. Such identification and grouping is possible because the 2D image curve which results from imaging the 3D class is tightly constrained. In turn these constraints can be used during grouping to test the validity of the class assumption. Subsequently, the organized SOR boundaries can be used to provide invariant indices constructed from portions of the boundary (see section 3.2).

Indexing the model library via the invariants generates recognition hypotheses for particular *models* of that class. For example, if an SOR image outline had been grouped the SOR invariants might index particular models, e.g. vase #2 or bottle #4, and recognition hypotheses for these models would then be verified by projecting the specific model boundary onto the image.

In the test examples of this paper the 3D object is of the SOR class. So in the following we describe in more detail the grouping and invariant index formation for this class. Full details of the Morse system are given in [18].

SOR Grouping. The imaged outline of a surface of revolution can be separated into two 'sides' by the projected symmetry axis. The two sides are tightly constrained: they are related by a particular four-degree-of-freedom plane projective transformation — a planar harmonic homology [8]. This relationship is exact. The transformation is represented by a non-singular 3×3 matrix T, where $T^2 = I$. Two pairs of point correspondences determine T. This transformation is fundamental to grouping outlines of SORs: the line of fixed points of T is the imaged symmetry axis; T provides point to point correspondence between the sides of the outline; this disambiguates the matching of bitangents used to form the invariants; and finally, T can be used to *repair* missing outline portions, filling in gaps by transforming over points from the other side of the outline.

Based on these properties, grouping for this class can be carried out by associating curves which are projectively equivalent, and then testing if the transformation between two projectively related curves is a planar harmonic homology. If it is not then the associated curves can be ruled out as members of this class. This is simply tested by checking if $T^2 = I$.

For each SOR in an image, the outcome of the grouping is an identified axis, pairs of bitangents (one half of each pair on either side of the axis), and repaired outline contours around the object; as illustrated in Fig.12.

SOR Invariants. It has been shown previously [7] that intersections of corresponding pairs of bitangents on an SOR's imaged outline are projections of points on the axis of the object. Constructing the set of such bitangent intersection points and computing their cross-ratio provided a means of object identification. However that required finding four distinct bitangent pairs which limits the model SORs to ones with a suitably rich geometry.

The invariant used here is again a cross-ratio of distinguished points on the SOR axis, but these four points are determined from a single bitangent pair (e.g. from a single concavity) in an image. This invariant is a *quasi-invariant* — it is invariant to an excellent approximation under perspective imaging. Its construction requires the image aspect ratio to be correct. The construction of the invariant is illustrated schematically in Fig.2.

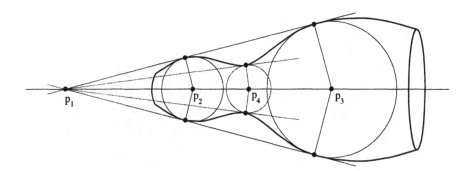

Fig. 2. The construction used to generate the SOR quasi-invariant. The invariant is the cross-ratio of four distinguished axis points. The first, p_1, is the intersection of the concavity's bitangent lines; p_2 and p_3 are the centres of circles tangent to the outline at each of the bitangent points; by a similar circle construction, p_4 comes from tangent lines cast from p_1 to the interior of the concavity.

As in Lewis, the invariants are used to index into a model-base to generate hypotheses, which are then merged and finally verified using back-projection.

4 Experiments

Four objects are used for the experiments, three SORs and one planar shape (a floppy disk). The four objects are shown in Fig. 3.

Acquisition Images A set of acquisition images of each object is required for SLAM in order to sample the pose space. For the SORs, the slant of the symmetry axis was varied in 5° steps, keeping the SOR at the horizontal centre of the image. The 15 acquisition images for SOR1 are shown in Fig.4. Similar sets are used for SOR2 and SOR3. The planar object was imaged in an oblique view and the orientation about the normal to the object's plane is varied in 10° steps to acquire its appearance model. The 36 acquisition images are shown in Fig.5. The illumination in all cases was from a diffuse source and maintained constant over the image collection.

Test Images There are three sets of test images:

1. **SOR1 + distractor:** There are 9 images corresponding to three degrees of occlusion, varying from no occlusion, slight occlusion to heavy occlusion; and for each degree of occlusion three viewpoints are imaged. An unmodelled distractor is included to allow partial occlusion of the SOR. The images are shown in Fig. 6.
2. **Disk + distractor:** There are 9 images corresponding to three degrees of occlusion, varying from no occlusion, slight occlusion to heavy occlusion; and for each degree of occlusion three viewpoints are imaged. An unmodelled

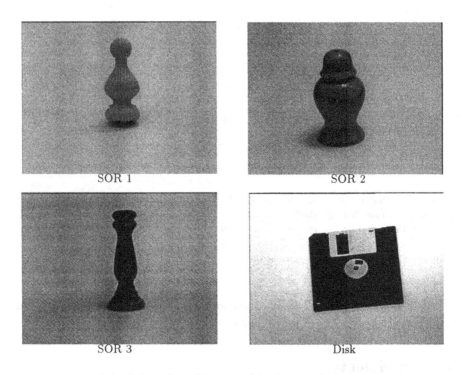

Fig. 3. The four objects used in the experiments.

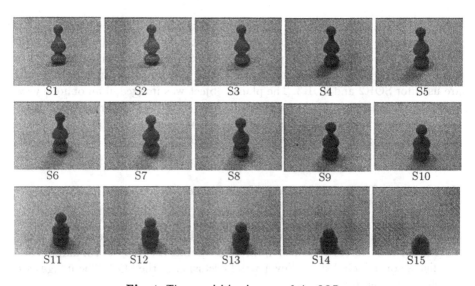

Fig. 4. The acquisition images of the SOR.

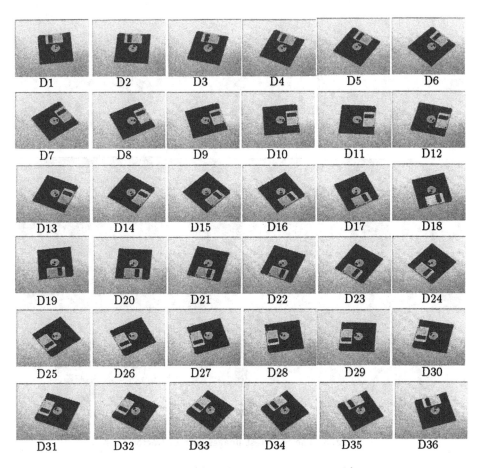

Fig. 5. The acquisition images for the planar object.

distractor is included to allow partial occlusion of the disk.. The images are shown in Fig. 7.

3. **SOR1 + disk**: There are 6 images corresponding to two viewpoints for the SOR, with the disk varying in pose, causing different occlusions for both the SOR and disk. The images are shown in Fig.8.

4.1 Appearance-Based System

Each set of acquisition images is normalized as described in section 2. In the experiments below recognition is tested against two different model libraries. Each model library is constructed by including all the acquisition images (i.e. in this case, one set for each object that is modelled), and constructing the PCA eigenspace.

The two model libraries are: first, only the three SORs of Fig. 3; and second, all four objects of Fig 3 (i.e. now including the disk) together with 20 other

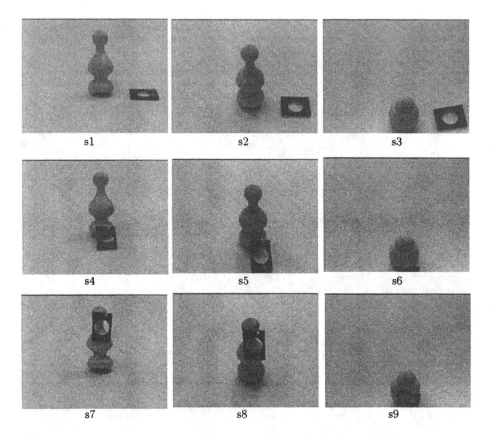

Fig. 6. The SOR test images with distractor.

objects shown in Fig.9. The first model library allows SLAM and Morse to be compared on the same set of objects.

The manifold representing each object can be displayed as a three dimensional subspace by projecting the acquisition images onto the first three principal components. In the case of these experiments, the manifolds are curves since only one parameter is varied in generating the appearance model. A typical example is shown in Fig.16.

The actual eigenspace used for recognition has ten dimensions, i.e. recognition proceeds by normalizing a new image and then projecting it onto the sub-space defined by ten eigenvectors. The closest stored point, within a tolerance threshold, is retrieved and associated with the stored label to classify the object.

4.2 Geometry-Based System

In the Lewis and Morse systems only one image is needed in principle to acquire a model. However, to reduce the effects of measurement error and to determine

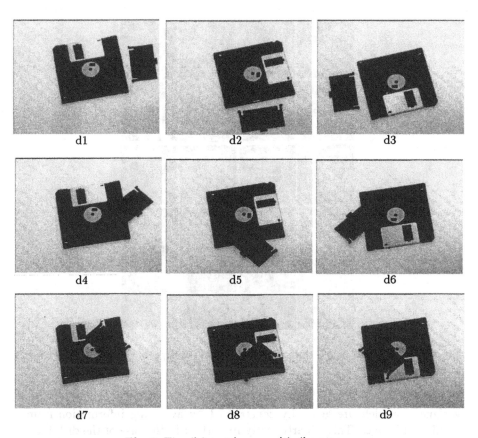

Fig. 7. The disk test images with distractor.

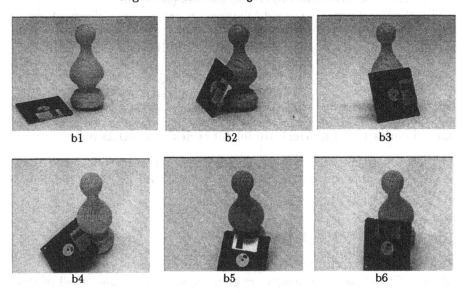

Fig. 8. Test images containing both SOR1 and the disk.

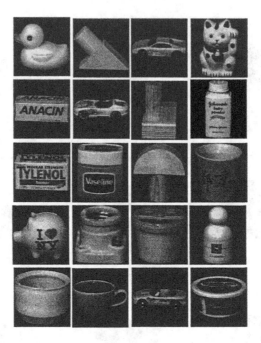

Fig. 9. The objects which, with those of Fig.3, form the larger of the model-bases used by SLAM.

variances, models are generally constructed by averaging information from a number of images. This is particularly important in the case of the disk for the Lewis system, because of non-rigidity of the disk (see below).

For the Morse system two model libraries are used. First, a library consisting of only the three SORs of Fig.3. Second, a library consisting of these three SORs, together with 21 others. The complete set of 24 SORs is shown in Fig.10.

For the Lewis system there are 8 planar objects in the model library; these are the disk of Fig.3, together with the 7 objects shown in Fig.11.

In all the following experiments the same parameters are used for all images.

4.3 Results I — Identical Model Libraries — 3 SORs only

This test is not applicable to Lewis, since it cannot represent SORs. Only SLAM and Morse are applied here, each having only the 3 SORs (SOR1, SOR2, SOR3) in their model library.

Test images: SOR1 + distractor (Fig. 6)

SLAM: SOR1 is correctly recognized in all nine images.

Morse: Five of the nine images are grouped successfully, and in all of these cases the SOR1 is correctly recognized — i.e. where grouping succeeds there are no false positives or negatives. Grouping fails for two reasons: first, in

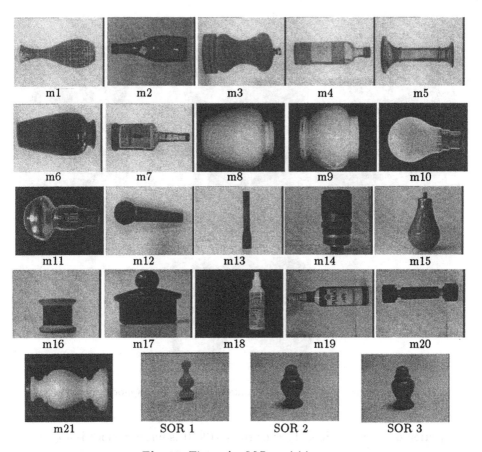

Fig. 10. The entire SOR model-base.

images s3, s6 and s9, the SOR is essentially viewed from above and the 'sides' of the SOR, necessary to drive the grouping and measure the characteristic geometry, are not visible; secondly, in s8 the bitangent points are occluded, and grouping can not begin. The recognized SORs are shown in Fig.12.

Test images: SOR1 + disk (Fig.8)

SLAM: SOR1 is recognized in two of the six images, b1 and b5. In the other four images an incorrect SOR (SOR2 or SOR3) is recognized. The results are listed in table 1.

Morse: Three of the six, b1, b5 and b6, are recognized. In the three cases where recognition fails this is due to a grouping failure because too much of the outline is occluded for grouping to begin. There are no false positives. The recognized SORs are shown in Fig.13.

The occlusion of one object by another causes two problems for SLAM: first, it makes segmentation more difficult; second, it perturbs the position of the

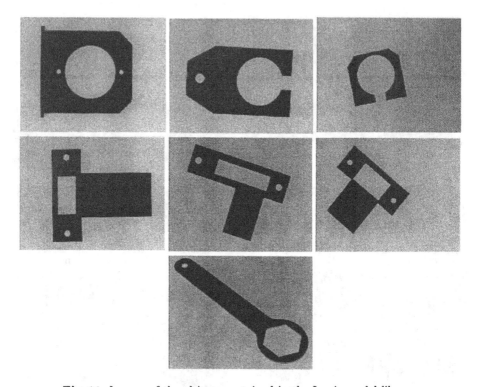

Fig. 11. Images of the objects contained in the Lewis model library.

projection of the image into the eigenspace [12]. It is unclear which is applicable here.

Summary The conclusion from this test is that the geometric system, Morse, is fail safe — i.e. there are no false positives, but grouping is its weakness at present. It correctly recognizes SOR1 in 8 of the 15 test images which contain it (7 false negatives, no false positives). The appearance system, SLAM, can have a problem distinguishing between the 3 SORs. It correctly recognizes SOR1 in 11 of the 15 images. However, there are 4 false positives where the wrong SOR is recognized (in total: 4 false negatives, 4 false positives).

4.4 Results II — Variation with Model Libraries

Here the number of objects in each model library is increased to investigate how this affects performance. For SLAM, 24 objects are used (the three test SORs, the disk, and twenty others). For Morse, 24 SORs are used (i.e. the three test SORs and 21 others), and for Lewis 8 planar objects are included (the disk and 7 others).

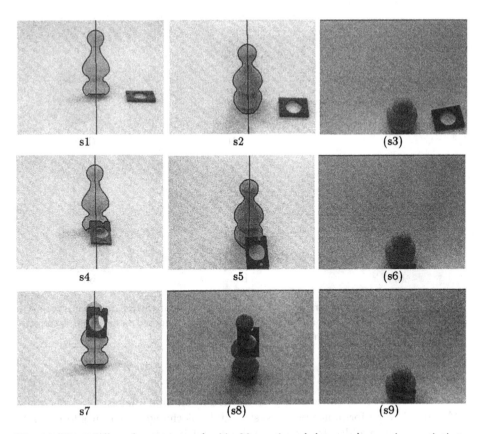

Fig. 12. The SOR test images recognized by Morse. An axis is super-imposed on each that is correctly identified; those with labels in brackets failed.

Image	SLAM-Classified Object	Nearest Distance $/10^7$
b1	SOR1	1.13
b2	SOR2	1.02
b3	SOR3	1.20
b4	SOR2	0.73
b5	SOR1	1.68
b6	SOR2	1.31

Table 1. SLAM recognition results for images in Fig.8 (SOR1 and disk) against a model library of SOR1, SOR2 and SOR3.

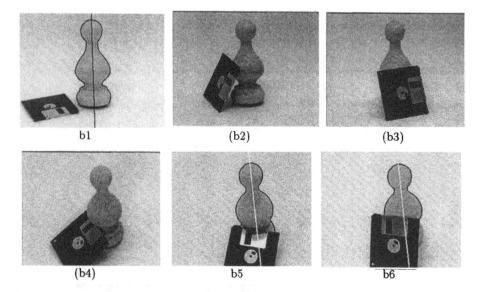

Fig. 13. Morse recognition results for the images of Fig.8 (SOR1 and disk) against a model library of SOR1, SOR2 and SOR3. An axis is super-imposed on each that is correctly identified; those with labels in brackets failed.

Test images: SOR1 + distractor (Fig. 6)

SLAM: SOR1 is again correctly recognized in all nine images.

Morse: The performance of the system did not change when an additional 21 models were added to the library, i.e. SOR1 is again correctly recognized in five of the nine images as shown in Fig.12. The recognition was successful in all cases where the SOR could be grouped. Although recognition hypotheses were generated for other SORs in a few cases, these were always eliminated by the verification stage so there were no false positives.

Lewis: None of the objects in the Lewis model-base appear in these images, so no models should be recognized. This is indeed the case, i.e. there are no false positives.

Test images: Disk + distractor (Fig.7)

SLAM: The disk was correctly recognized in all cases.

Morse: Neither of the objects in these images is in the Morse library (SORs only), and no SORs were recognized in these images i.e. there were no false positives.

Lewis: The disk was correctly recognized in all nine images. The recognized disks are shown in Fig.14.

Test images: SOR1 + disk (Fig.8)

SLAM: In all cases the nearest manifold was SOR2, i.e. recognition failed on all images.

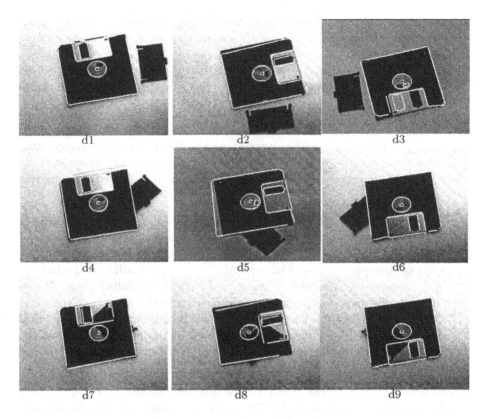

Fig. 14. The disk + distractor images all successfully recognized by Lewis. The images show the model outline back-projected onto the image.

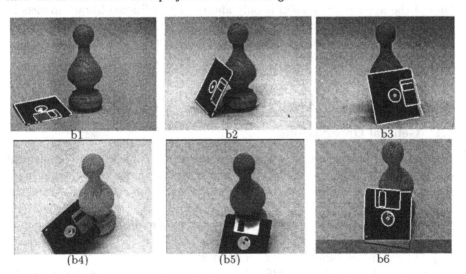

Fig. 15. The disks recognized by Lewis in the disk + SOR images. Where successful, the back-projected outline is displayed; those that failed have labels in brackets.

Morse: Again, the performance does not change on adding additional models. Three of the six, b1, b4 and b5, are recognized, and there are no false positives. Results are shown in Fig.13.

Lewis: The disk was correctly recognized in four of the six images, b1, b2, b3 and b6, and there are no false positives. The recognized disks are shown in Fig.15.

The two Lewis failures, b4 and b5, were caused by a relative translation of the disk itself and the disk case: The invariant used for these two images is computed from two lines and one conic; the conic is the central circular component of the disk, and this can move relative to the case. There are two consequences of this non-rigidity:

1. The invariants deviate from a fixed value. However, the value and variance of this invariant is obtained by averaging measurements from a number of acquisition images in Fig.5 (where there is relative motion between the disk and case between images). Consequently, the invariant will have a relatively high variance and measured values should lie in the predicted range. This is the case: disk *hypotheses* are generated for b4 and b5.

2. The back-projection of the model outline will be displaced from its veridical image position. The back projection is used for verification, and it is the verification stage which prevents recognition in these two cases.

Summary Again, both the geometric systems are fail-safe. This is due to the employment of a verification stage. The appearance system does not have a verification test, and does produce false positives. The performance of the geometric systems is not affected by adding additional objects to the library. The performance of the appearance system is only affected by additional library models in the case of the SOR1 and disk test images. However, an SOR is still recognized in these images, even though this is not always the correct SOR. A possible explanation for this is shown in Fig.16.

5 Summary and Proposal for Further Comparisons

5.1 General Observations

It is clear that both approaches have strengths and drawbacks that complement one another. Appearance models have the great advantage of not requiring a formal description of the constraints peculiar to an object. In the geometric approach this description is required to facilitate pose invariant recognition. Currently, there is not a theory for the description of even simple curved shapes, except for a limited set of geometric classes such as SORs and canal surfaces (pipes) [16], and certainly not for arbitrary shapes.

On the other hand, it is difficult to see how an appearance model can be generalized to incorporate objects which should be considered to be in the same geometric class. Incidental variations in appearance, such as surface albedo or

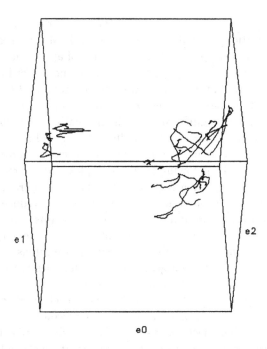

Fig. 16. Each object generates a one-dimensional manifold parametrized by the rotation applied during image acquisition. Here the manifolds (curves) are plotted in a sub-space defined by the first three eigenvectors. The one dimensional manifolds for the three SORs (left) are well separated from the one dimensional manifolds of the other objects (to the right). This may explain why an SOR is correctly distinguished from other objects but can be confused with other SORs

texture of otherwise identical objects must be treated as completely separate object instances.

A significant advantage of the geometric approach is that geometric class models also provide constraints which support figure-ground segmentation. For example, in the case of rotational symmetry, the relation between corresponding points on the imaged boundary provides a grouping mechanism which can extract SOR features in the presence of complex textured backgrounds and significant occlusion.

On the other hand if grouping is not successful, then the geometric approach cannot proceed. This is the problem with the current implementation state of Morse: a very powerful grouping constraint is available for SORs, but at present it is only applied to curves if they contain a bitangent. Consequently, if the bitangent points are occluded no grouping occurs. However, first, it is clear why the system has failed; and second, there is considerable potential for improvement by employing the constraint more fully (i.e. considering other curves for grouping). The appearance approach is inferior in both these respects.

For appearance models, some alleviation of the segmentation and occlusion problems can be obtained by partial appearance matching — finding subparts of objects and checking the consistencies of the geometrical arrangements of the subparts for recognition, see [13] and also Schmid & Mohr (these proceedings). Nevertheless, SLAM, as a specific implementation of appearance modelling, currently has the deficiency that it can only recognize one object in an image (even though the image may contain several modelled objects) and this can be attributed to the difficulty of figure-ground segmentation in appearance systems.

A related problem to segmentation is the normalization required by appearance systems. In the presence of mutual illumination, and mutual shadows, this normalization is difficult to achieve. For example, it can be shown that effects of mutual illumination and shadowing lead to complex and unpredictable patterns of intensity in real scenes [4]. For surfaces with a Lambertian reflectance map, the dimension of the illumination manifold (for each view) is reduced to three, even in the presence of mutual illumination (see papers in this collection). However, varying shadow structures are still a significant obstacle. The only reliable invariants to illumination and mutual object placement are intensity discontinuities. Therefore, it can be expected that geometric boundary descriptions will be much more invariant than normalized intensity patterns.

Finally there is the issue of statistical variation. Geometric models impose hard constraints on image geometry. This is an advantage — in that it facilitates powerful grouping mechanisms and allows strict verification, but also a disadvantage in that variations from these constraints can result in recognition failing. This was exemplified by the relative motion of the disk and its case. This non-rigidity caused the Lewis system to fail on two examples, whilst SLAM tolerated the variation because the disk was still the nearest manifold. However, the lack of a verification stage in SLAM does result in false positives.

5.2   Future Investigation

This comparison of representations for recognition has raised many significant issues for further investigation. It is now clear to us that this sort of study is essential for rapid progress in object representation. The areas where various representations complement each other is a fertile direction for new research.

At the most abstract level, we have seen that appearance modelling is largely empirical while the geometric invariance model originates from a theoretical understanding of image formation and perspective projection. The contest is then based on the completeness of appearance model data acquisition vs the applicability of a geometric representation of actual shapes in the world. It is impossible to acquire a fully complete appearance model for all variations which could occur. On the other hand, currently there are suitable geometric representations for only a small number of classes, and also a 'model' should be more than geometry alone.

Clearly, a resolution is to combine the two representations. Chris Taylor [6] has made some moves in this direction by using a template to correct for geometry (faces) and then use eigenfunctions after geometric and intensity normalization. In some applications, starting with local features and following with geometry, may be a useful alternative.

In general, well-founded segmentation and grouping mechanisms should be used to derive object descriptions. Then statistical classifiers can account for the aspects of object appearance which cannot be modelled theoretically. Currently, the geometric invariant recognition systems use statistics only in defining tolerance on invariant values. Additional features can be added, such as surface markings, which can only be described by an appearance model, using geometric segmentation and grouping to isolate specific object surfaces. Such regions are typically defined for a specific object, rather than a class, and therefore are best used during verification.

Some of the significant questions which were raised by this initial study are:

1. What is the effect of a larger model-base? It is expected that the effective separation of objects in eigenspace (for SLAM) and in invariant space (for Lewis/Morse) will be reduced as the number of library objects is increased. A preliminary answer to this question has been given in this paper.
2. What is the effect of more degrees of freedom on the appearance model manifold? In the current experiments, we only varied one rotational degree of freedom. It might be expected that SLAM's recognition tolerance will be reduced as the dimension of the manifold increases. The invariant representation is not affected by object pose or internal camera parameters which would all have to be included in an appearance model.
3. How severe is the figure-ground isolation problem? The current experimental setup does not provide very challenging background or object textures. Will the geometric grouping constraints be sufficiently powerful to isolate an object with surface markings and texture from a textured and cluttered background?
4. How will the computational complexity of recognition scale with complexity of the scene? Can geometric grouping be made efficient in a textured and cluttered scene?
5. How will a statistically optimum set of eigenvectors compare to the principal components currently used in SLAM? It is not necessarily the case that the eigenvectors which rapidly converge to a good approximation of the image intensity also provide maximum separation of the object manifolds [5].

Acknowledgements

Lewis was originally developed by Charlie Rothwell, and was subsequently implemented in C++ primarily by Charlie Rothwell, with contributions from Bill Hoffman and Chien-ming Huang. Morse, has been implemented using the same basic C++ libraries as Lewis primarily by Nic Pillow, with contributions from Jane Liu and Sven Utcke. The SLAM software was developed by Sameer Nene, Shree Nayar, and Hiroshi Murase at Columbia University. We are grateful to Sameer A. Nene for his technical assistance in using the SLAM software. Financial support was provided by several agencies: ESPRIT BRA Project 'IMPACT'; the UK EPSRC; and General Electric.

References

1. Beymer D., 'Face Recognition Under Varying Pose', *Proc CVPR*, 756–761, 1994.
2. Craw I. and Cameron P., 'Parametrizing images for recognition and reconstruction', *Proc BMVC*, 367–370, 1991.
3. Duda R.O. and Hart P.E., *Pattern Classification and Scene Analysis,* Wiley, 1973.
4. Forsyth D. and Zisserman A., 'Reflections on shading', *PAMI*, 13, 7, 671–679, 1991.
5. Belhumeur N., Hespanha J. and Kriegman D., 'Eigenfaces vs. Fisherfaces: Recognition Using Class Specific Linear Projection', *Proc. ECCV*, 45–58, 1996.
6. Lanitis A., Taylor C.J. and Cootes T.F., 'A Unified Approach to Coding and Interpreting Face Images', *Proc ICCV*, 368–373, 1995.
7. Liu J.S., Mundy J.L., Forsyth D.A., Zisserman A. and Rothwell C.A., 'Efficient Recognition of Rotationally Symmetric Surfaces and Straight Homogeneous Generalized Cylinders', *Proc. CVPR*, 1993.
8. Mukherjee D.P., Zisserman A. and Brady J.M., 'Shape from symmetry—detecting and exploiting symmetry in affine images', *Phil. Trans. R. Soc. Lond. A*, 351, 77–106, 1995.
9. Mundy J.L. and Zisserman A. *Geometric Invariance in Computer Vision*, MIT Press, 1992.
10. Murakami H. and Kumar V., 'Efficient calculation of primary images from a set of images,' *IEEE Transactions on Pattern Analysis and Machine Intelligence*, 4, 511–515, 1982.
11. Murase H. and Nayar S.K., 'Visual Learning and Recognition of 3-D Objects from Appearance'. *IJCV*, 14, 1, 1995.
12. Murase H. and Nayar S., 'Illumination Planning for Object Recognition Using Parametric Eigenspaces' *IEEE Trans. PAMI*, 16, 12, 1219–1227, 1995.
13. Murase H. and Nayar S., 'Image Spotting of 3D Objects Using the Parametric Eigenspace Representation', *Proc. of 9th Scandinavian Conference on Image Analysis*, 325–332, June 1995.
14. Nene S.A., Nayar S. and Murase H., 'SLAM: Software Library for Appearance Matching,' *Proc. of ARPA Image Understanding Workshop, Monterey*, November 1994.
15. Pentland A., Moghaddam B. and Starner T., 'View-based and modular eigenspaces for face recognition', *Proc CVPR*, 84–91, 1994.
16. Pillow N., Utcke S. and Zisserman A., 'Viewpoint-Invariant Representation of Generalized Cylinders Using the Symmetry Set'. *Image and Vision Computing*, 13, 5, 1995.
17. Rothwell C.A. *Object Recognition through Invariant Indexing.*, OUP, 1995.
18. Zisserman A., Forsyth D., Mundy J., Rothwell C., Liu J. and Pillow N., '3D Object Recognition using Invariance', *AI Journal*, 78, 239–288, 1995.

3D Representations
and Applications

Virtualized Reality: Being Mobile in a Visual Scene *

Takeo Kanade, P. J. Narayanan, and Peter W. Rander

Robotics Institute
Carnegie Mellon University
Pittsburgh, PA 15213, U. S. A.

Abstract. The visual medium evolved from early paintings to the realistic paintings of the classical era to photographs. The medium of moving imagery started with motion pictures. Television and video recording advanced it to show action "live" or capture and playback later. In all of the above media, the view of the scene is determined at the transcription time, independent of the viewer.

We have been developing a new visual medium called *virtualized reality*. It delays the selection of the viewing angle till view time, using techniques from computer vision and computer graphics. The visual event is captured using many cameras that cover the action from all sides. The 3D structure of the event, aligned with the pixels of the image, is computed for a few selected directions using a stereo technique. Triangulation and texture mapping enable the placement of a "soft-camera" to reconstruct the event from any new viewpoint. With a stereo-viewing system, virtualized reality allows a viewer to move freely in the scene, independent of the transcription angles used to record the scene.

Virtualized reality has significant advantages over virtual reality. The virtual reality world is typically constructed using simplistic, artificially-created CAD models. Virtualized reality starts with the real world scene and virtualizes it. It is a fully 3D medium as it knows the 3D structure of every point in the image.

The applications of virtualized reality are many. Training can become safer and more effective by enabling the trainee to move about freely in a virtualized environment. A whole new entertainment programming can open by allowing the viewer to watch a basketball game while standing on the court or while running with a particular player. In this paper, we describe the hardware and software setup in our "studio" to make virtualized reality movies. Examples are provided to demonstrate the effectiveness of the system.

---

* A version of this paper appeared in the International Conference on Artificial Reality and Tele-Existence/Conference on Virtual Reality Software and Technology, Tokyo, Nov 1995. This paper is reprinted in these proceedings with the kind permission of ACM-SIGCHI

1   Introduction

We have a few visual media available today: paintings, photographs, moving pictures, television and video recordings. They share one aspect: the view of the scene is decided by a "director" while recording or transcribing the event, independent of the viewer.

We describe a new visual medium called *virtualized reality*. It delays the selection of the viewing angle till view time. To generate data for such a medium, we record the events using many cameras, positioned so as to cover the event from all sides. The time-varying 3D structure of the event, described in terms of the depth of each point and aligned with the pixels of the image, is computed for a few of the camera angles - called the *transcription angles* - using a stereo method. We call this combination of depth and aligned intensity images the *scene description*. The collection of a number of scene descriptions, each from a different transcription angle is called the virtualized world. Once the real world has been virtualized, graphics techniques can render the event from any viewpoint. The scene description from the transcription angle closest to the viewer's position can be chosen dynamically for rendering by tracking the position and orientation of the viewer. The viewer, wearing a stereo-viewing system, can freely move about in the world and observe it from a viewpoint chosen dynamically at view time.

Virtualized reality improves traditional virtual reality. Virtual reality allows viewers to move in a virtual world but lacks fine detail as their worlds are usually artificially created using simplistic CAD models. Virtualized reality, in contrast, starts with a real world and virtualizes it.

There are many applications of virtualized reality. Training can become safer and more effective by enabling the trainee to move about freely in a virtualized environment. A surgery, recorded in a virtualized reality studio, could be revisited by medical students repeatedly, viewing it from positions of their choice. Telerobotics maneuvers can be rehearsed in a virtualized environment that feels every bit as real as the real world. True telepresence could be achieved by performing transcription and view generation in real time. And an entirely new generation of entertainment media can be developed: basketball enthusiasts and broadway aficionados could be given the feeling of watching the event from their preferred seat, or from a seat that changes with the action.

Stereo or image-matching methods, which are the key components in virtualized reality, are well-studied. Precise reconstruction of the whole scene using a large number of cameras is, however, relatively new. Kanade [5] proposed the use of multi-camera stereo using supercomputers for creating 3D models to enrich the virtual world. Rioux, Godin and Blais [11] outlined a procedure to communicate complete 3D information about an object using depth and reflectance. Fuchs and Neuman [2] presented a proposal to achieve telepresence for medical applications. Some initial experiments were conducted at CMU using the video-rate stereo machine [7], by the team of UNC, UPenn and CMU [1], and at Tsukuba by Satoh and Ohta [12]. Laveau and Faugeras [8] attempt "view transfer" with uncalibrated cameras using epipolar constraints alone.

We presented some early results from virtualized reality in an earlier paper [6]. This paper presents it in greater detail, in three stages of creating a virtualized real scene - scene transcription, structure extraction and view generation. Examples from our virtualizing studio are interspersed with the discussion to elucidate the concepts.

2 Scene Transcription

The central idea of this research is that we can virtualize real-world scenes by capturing scene descriptions - the 3D structure of the scene aligned with its image - from a number of transcription angles. The scene can be synthesized from any viewpoint using one or more scene descriptions. The facility to acquire the scene descriptions is called the virtualizing studio. Any such studio should cover the action from all angles. Stereo techniques used to extract the scene structure require images corresponding to precisely the same time instant from every camera to be fed to them in order to accurately recover 3D scene structure. We potentially need to virtualize every frame in video streams containing fast moving events to satisfactorily reproduce the motion. Therefore, the studio should have the capability to record and digitize every frame of each video stream synchronously. We elaborate on the physical studio, the recording setup and the digitizing setup in this section.

2.1 Virtualizing Studio Setup

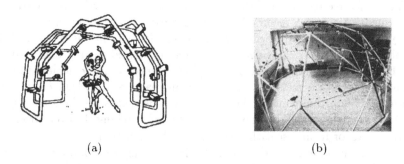

(a) (b)

Fig. 1. The virtualizing studio. (a) Conceptual (b) The dome

Figure 1(a) shows the studio we have in mind. Cameras are placed all around the dome, providing views from angles surrounding the scene. Figure 1(b) show the studio we have built using a hemispherical dome, 5 meters in diameter, constructed from nodes of two types and rods of two lengths. We started the studio with 10 cameras to transcribe the scene, arranged in two clusters, each providing a scene description. We have since upgraded the system to a 51 camera system with coverage from a hemispherical space. The cameras are mounted on special L-shaped aluminum brackets that can be clamped on anywhere on the rods.

2.2 Synchronous Multi-camera Recording

To synchronously acquire a set of video streams, a single control signal can be supplied to the cameras to simultaneously acquire images and to the digitizing equipment to simultaneously capture the images. In order to implement this approach directly in digital recording hardware, the system would need to handle the real-time video streams from many cameras. For a single monochrome camera providing 30 images per second, 512×512 pixels per image with 8 bits per pixel, the system would need to handle 7.5 MBytes of image data per second. A sustained bandwidth to store the captured data onto a secondary storage device is beyond the capabilities of typical image capture and digital storage systems, even with the best loss-less compression technology available today. For example, our current system - a Sun Sparc 20 workstation with a K2T V300 digitizer - can capture and store only about 750 KBytes per second. Specialized hardware could improve the throughput but at a substantially higher cost. Replicating such a setup to capture many video channels simultaneously is prohibitively expensive.

We developed an off-line system to synchronously record frames from multiple cameras. The cameras are first synchronized to a common sync signal. The output of each camera is time stamped with a common Vertical Interval Time Code (VITC) and recorded on tape using a separate VCR. The tapes are digitized individually off-line using a frame grabber and software that interprets the VITC time code embedded in each field. We can capture all frames of a tape by playing the tape as many times as the speed of the digitizing hardware necessitates. The time code also allows us to correlate the frames across cameras, which is crucial when transcribing moving events. Interested readers can refer to a separate report [9] for more details on the synchronous multi-camera recording and digitizing setup. Figure 2 shows a still frame as seen by five cameras of the virtualizing studio digitized using the above setup.

3 Structure Extraction

We use the multi-baseline stereo (MBS) technique [10] to extract the 3D structure from the multi-camera images collected in our virtualized reality studio. Stereo algorithms compute estimates of scene depth from correspondences among images of the scene. The choice of the MBS algorithm was motivated primarily by two factors. First, MBS recovers dense depth maps - that is, a depth estimate corresponding to every pixel in the intensity images - which is needed for image reconstruction. Second, MBS takes advantage of the large number of cameras that we are using for scene transcription to increase precision and reduce errors in depth estimation.

Fig. 2. Five captured images to be used to compute one scene description

3.1 Fundamentals of Multi-Baseline Stereo

To understand the MBS algorithm, consider a multi-camera imaging system in which the imaging planes of the cameras all lie in the same physical plane and in which the cameras have the same focal length F. For any two of the cameras, the disparity d (the difference in the positions of corresponding points in the two images) and the distance z to the scene point are related by

$$d = BF\frac{1}{z} \tag{1}$$

where B is the baseline, or distance between the two camera centers. The simplicity of this relation makes clear one very important fact: the precision of the estimated distance increases as the baseline between the cameras increases. In theory, the cameras can be placed as far apart as possible. Practical experience using stereo systems reveals, however, that increasing the baseline also increases the likelihood of mismatching points among the images. There is a trade-off between the desires for correct correspondence among images (using narrow baselines) and for precise estimates of scene depth (using wide baselines).

The multi-baseline stereo technique attempts to eliminate this trade-off by simultaneously computing correspondences among pairs of images from multiple cameras with multiple baselines. In order to relate correspondences from multiple image pairs, we rewrite the previous equation as

$$\frac{d}{BF} = \frac{1}{z} = \zeta \tag{2}$$

which indicates that for any point in the image, the inverse depth ζ is constant since there is only one depth z for that point. If the search for correspondences is computed with respect to ζ, it should consistently yield a good match at the correct value of ζ independently of the baseline B. With multiple (more than 2) cameras, correspondences can now be related across camera pairs, since the searching index is independent of the baselines. The resulting search combines the correct correspondence of narrower baselines with the higher precision of wider baselines, and has been proven to yield a unique match of high precision.

One way to find correspondences between a pair of images is to compare a small window of pixels from one image to corresponding windows in the other image. The correct position of the window in the second image is constrained by the camera geometry to lie along the epipolar line of the position in the first image. The matching process involves shifting the window along this line as a function of ζ, computing the match error - using normalized correlation or sum of squared differences (SSD) - over the window at each position, and finding the minimum error. The estimate of inverse depth, $\hat{\zeta}$, is the ζ at this minimum.

3.2 Depth Map Editing

Window-based correspondence searches suffer from a well-known problem: inaccurate depth recovery along depth discontinuities and in regions of low image texture. The recovered depth maps tend to "fatten" or "shrink" objects along depth discontinuities. This phenomena occurs because windows centered near the images of these discontinuities will contain portions of objects at two different depths. When one of these windows is matched to different images, one of two situations will occur. Either the foreground object will occlude the background object so that depth estimates for the background points will incorrectly match to the portion of the foreground in the window, or both the foreground and background regions will remain visible, leading to two likely candidate correspondences. In regions with little texture - that is, of fairly constant intensity - window-based correspondence searches yield highly uncertain estimates of depth. Consider, for example, a stereo image pair with constant intensity in each image. With no intensity variation, any window matches all points equally well, making any depth estimates meaningless.

To address this inaccuracy in depth recovery, we could reduce the window size used during matching, potentially matching individual pixels. This approach reduces the number of pixels effected by depth discontinuities. By doing so, however, we also reduce the amount of image texture contained within the window, increasing the uncertainty of the recovered depth estimate. Conversely, we could increase the size of the window to give more image texture for matching. This action increases the image texture contained in the window, but also increases the area effected by the discontinuities. Optimizing the window size requires trading off the effects of the depth discontinuities with those of the low-texture regions.

In order to work around this trade-off, we have incorporated an interactive depth map editor into our process of structure extraction. Rather than send the MBS-computed depth maps directly on to the next processing stage, we instead manually edit the depth map to correct the errors that occur during automatic processing. While a good window size still helps by reducing the number of errors to be corrected, it is less important in this approach because the user can correct the problems in the depth maps. We are currently exploring modifications to the stereo algorithm in an effort to reduce or eliminate this need for human intervention.

3.3 General Camera Configurations

Fig. 3. Results of multi-baseline stereo algorithm

For general camera positions, we perform both intrinsic and extrinsic camera calibration to obtain epipolar line constraints, using an approach from [14]. Using the recovered calibration, any point in the 3D coordinate system of the reference camera can be mapped to a point in the 3D coordinate system of any of the other cameras. To find correspondences, we again match a reference region to another image as a function of inverse depth. To find the position in the second image corresponding to this inverse depth, we convert the reference point and inverse depth into a 3D coordinate, apply the camera-to-camera mapping, and project the converted 3D point into the other image. As with the parallel-camera configuration, the full search is conducted by matching each reference image point to the other images for each possible ζ. We then add the match error curves from a set of image pairs and search for the minimum of the combined error function. Figure 3 shows the depth map recovered by applying this approach to the input images shown in Figure 2. The depth map has 74 levels for a depth range of 2 meters to 5 meters.

4 View Generation

We described how to "virtualize" an event in terms of a number of scene descriptions in the previous sections. The medium of virtualized reality needs to synthesize the scene from arbitrary viewpoints using these scene descriptions. To render the scene from other viewpoints using graphics workstations, we translate the scene description into an object type, such as a polygonal mesh. We texture map an intensity image onto the rendered polygons, generating visually realistic images of the scene. Graphics workstations have specialized hardware to render them quickly. A Silicon Graphics Onyx/RE2 can render close to 1 million texture mapped triangles per second.

We describe how new views are generated from a single scene description first. The generated view will be lower in quality as the viewpoint gets far from the transcription angle. We discuss how we can use multiple scene descriptions to get realistic rendering from all angles.

4.1 Using a Single Scene Description

A scene description consists of a depth map providing a dense three dimensional structure of the scene aligned with the intensity map of the scene. The point (i, j) in the depth map gives the distance of the intensity image pixel (i, j) from the camera. We convert the depth map into a triangle mesh and the intensity map to texture to render new views on a graphics workstation. There are two aspects of performing this translation realistically: object definition and occlusion handling.

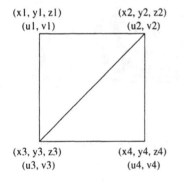

Triangle 1:
 Vertex 1: (x1, y1, z1), texture coordinate: (u1 / m, v1 / n)
 Vertex 2: (x2, y2, z2), texture coordinate: (u2 / m, v2 / n)
 Vertex 3: (x3, y3, z3), texture coordinate: (u3 / m, v3 / n)

Triangle 2:
 Vertex 1: (x2, y2, z2), texture coordinate: (u2 / m, v2 / n)
 Vertex 2: (x3, y3, z3), texture coordinate: (u3 / m, v3 / n)
 Vertex 3: (x4, y4, z4), texture coordinate: (u4 / m, v4 / n)

Fig. 4. Triangle mesh and texture coordinate definition

Object Definition Graphics rendering machines synthesize images of a scene from an arbitrary point of view given a polygonal representation of the scene. Texture mapping pastes an intensity image onto these rendered polygons, generating visually realistic images of the scene from arbitrary view points. We currently generate a triangle mesh from the depth map by converting every 2×2 section of the depth map into two triangles. Figure 4 illustrates how the mesh is defined. The (x, y, z) coordinates of each point in the image are computed from the image coordinates and the depth, using the intrinsic parameters of the imaging system. Each vertex of the triangle also has a texture coordinate from the corresponding intensity image. This simple method results in $2(m-1)(n-1)$ triangles for a depth map of size $m \times n$. The number of triangles for the depth map shown in Figure 3 is approximately 200,000. Though this is a large number of triangles, the regularity makes it possible to render them efficiently on graphics workstations.

We reduce the number of triangles in our scene definition by adapting an algorithm developed by Garland and Heckbert that simplifies a general dense elevation/depth map into planar patches [3]. The algorithm computes a triangulation using the smallest number of vertices given a measure for the maximum deviation from the original depth map. The procedure starts with two triangles defined by the outer four vertices. It repeatedly grows the triangle mesh by adding the vertex of maximum deviation and the corresponding triangle edges till the maximum deviation condition is reached. Using this technique, we have reduced mesh size by factors of 20 to 25 on typical scenes without affecting the visual quality of the output.

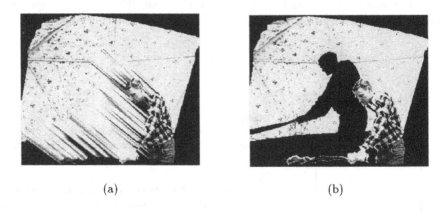

(a)   (b)

Fig. 5. (a) View without discontinuity compensation (b) With compensation

Occlusion Handling The simple rendering technique described above treats the entire depth map as one large surface, connecting pixels across depth discontinuities at object boundaries. This introduces an artificial surface bridging the discontinuity, with the few pixels of texture stretched over the surface. When generating views for angles far from the transcription angle, these surfaces become large and visually unrealistic; in Figure 5(a), for instance, the person and the wall appear to be connected. We therefore delete these artificial surfaces by not rendering the triangles that overlap discontinuities, resulting in "holes" as seen in Figure 5(b). We fill these holes using other scene descriptions as explained in Section 4.2.

Multi-frame Sequences The discussion to this point has focussed on virtualizing a single, static scene. It is also possible to virtualize moving scenes by virtualizing each frame separately. The resulting virtualized reality movie can

Fig. 6. Seven frames of a basketball sequence. Original images are shown on top. The event synthesized from a moving viewpoint is shown below that. The viewer motion starts left and above the original transcription angle and moves to the right

be played with the viewer standing still anywhere in the world by rendering each frame from the viewer's position. The scene can also be observed by a viewer whose movement through the world is independent of the motion in the scene. Figure 6 shows seven frames of a basketball sequence from the reference transcription point and from a synthetically-created moving viewpoint.

4.2 Merging Multiple Scene Descriptions

There are two reasons for combining the scene descriptions from multiple transcription angles while generating new views. First, as discussed in Section 4.1, depth discontinuities appear as holes in views far from the transcription angle when using a single scene description. We should "fill" these holes using a scene description from another transcription angle for which the portion of the scene is not occluded. Second, the intensity image used for texturing gets compressed or stretched when the viewing angle is far from the transcription angle, resulting in poor quality of the synthesized image. If the viewer strays far from the starting position, we should choose the most direct transcription angle for each viewing angle to minimize this degradation.

One merging strategy is to combine the scene descriptions from all transcription angles ahead of time to generate a model of the scene that contains all the necessary detail. Several methods are available to register and model objects from multiple range images [4, 13, 15]. Such a consolidated model attempts to give one grand description of the entire world. We only require the best partial description of the world visible from a particular viewing angle at any time. Such a partial description is likely to be more accurate due to its limited scope; inaccuracies in the recovery of the portion not seen will not affect it. It is likely to be simpler than a consolidated model of the scene, lending easily to real time view generation. The partial description we use consists of a reference scene description from the transcription angle closest to the viewing angle plus one or two supporting ones. The reference description is used for rendering most of the view and the supporting ones are used for filling the gaps.

Fig. 7. The baseball scene from 3 different viewpoints using one supporting scene description to fill holes. (a) Same view as Figure 5. (b) From far left. (c) From below and left

We do not combine the triangle meshes generated using the reference and supporting scene descriptions into one triangle mesh. We render most of the view using the reference scene description in the first pass. While doing so, the pixels belonging to the holes – corresponding to triangles at depth discontinuities that we opt not to render – are identified and marked. The view is rendered from the supporting scene descriptions in subsequent passes, limiting the rendering to these hole pixels. Figure 7(a) shows the results of filling the holes of Figure 5(b) using one supporting view. Notice that the background pattern and the right shoulder of the person has been filled properly. Figure 7(b) (c) show the same baseball scene from viewpoints very different from the original transcription angle. The "holes" left in the image corresponds to the portion of the scene occluded from both the reference and supporting transcription angles.

5 Conclusions

We introduced and elaborated on the concept of virtualized reality in this paper. It combines techniques from computer vision and computer graphics to virtualize a real world event and to let a viewer move about freely in the virtualized world.

284

We also demonstrated the efficacy of virtualized reality using scenes virtualized in our studio to make such movies.

Fig. 8. Depth-Key image merging technique enables mutual occlusion of virtual and real scenes

A promising new technology with applications in Virtualized Reality is a new image keying technique called Depth-Key [7]. Image keying is a method of merging images by switching among images based on some information (or key) attached to each image pixel. Chroma-key, for example, is a standard video keying method used in TV indsutry to select part of real images - e.g. a weather reporter in front of a blue screen - by using chromaticity as the selection key. This approach works well when the real scene always lies in front of the virtual scene, but does not allow the virtual scene to occlude the real one. In contrast, Depth-Key uses pixel-by-pixel depth information as the key, allowing mutual occlusion of the real and virtual scenes. For example, in Figure 8, the person actually reaches around the virtual object, generating mutual occlusion of the real and virtual scenes.

It is today possible to virtualize an event such as a surgery and let trainees move about it in a realistic recreation of the surgery in a manner they prefer. We plan to combine Depth-Key with Virtualized Reality, enabling the merging of the user's environment with the virtualized world. Multiple users could co-exist in a common virtualized world and see each other in addition to this world. We also plan to push the training and entertainment applications of virtualized reality in the future.

Acknowledgments: We would like to thank Atsushi Yoshida and Kazuo Oda for their discussions about and graphics of the Depth-Key system.

References

1. H. Fuchs, G. Bishop, K. Arthur, L. McMillan, R. Bajcsy, S. Lee, H. Farid, and T. Kanade. Virtual Space Teleconferencing using a Sea of Cameras. In *Proceedings of the First International Symposium on Medical Robotics and Computer Assisted Surgery*, pages 161–167, 1994.
2. H. Fuchs and U. Neuman. A Vision Telepresence for Medical Consultation and other Applications. In *In Sixth International Symposium of Robotics Research*, pages 555–571, 1993.
3. M. Garland and P. S. Heckbert. Fast Polygonal Approximation of Terrains and Height Field. Technical Report Computer Science Tech Report CMU-CS-95-181, Carnegie Mellon University, 1995.
4. H. Hoppe, T. DeRose, T. Duchamp, M. Halstead, H. Jin, J. McDonald, J. Schweitzer, and W. Stuetzle. Piecewise Smooth Surface Reconstruction. In *SIGGRAPH94*, pages 295 – 302, 1994.
5. T. Kanade. User Viewpoint: Putting the Reality into Virtual Reality. *MasPar News*, 2(2), Nov 1991.
6. T. Kanade, P. J. Narayanan, and P. W. Rander. Virtualized Reality: Concept and Early Results. In *In IEEE Workshop on the Representation of Visual Scenes*, 1995.
7. T. Kanade, A. Yoshida, K. Oda, H. Kano, , and M. Tanaka. A Stereo Machine for Video-rate Dense Depth Mapping and its New Applications. In *Proceedings of Computer Vision and Pattern Recognition*, 1996.
8. S. Laveau and O. Faugeras. 3-D Scene Representation as a Collection of Images and Fundamental Matrices. Technical report, INRIA, 1994.
9. P. J. Narayanan, P. W. Rander, and T. Kanade. Synchronizing and Capturing Every Frame from Multiple Cameras. Technical Report CMU-RI-TR-95-25, Carnegie Mellon University Robotics Institute, 1995.
10. M. Okutomi and T. Kanade. A multiple-baseline stereo. *IEEE Transactions on Pattern Analysis and Machine Intelligence*, 15(4):353 – 363, 1993.
11. M. Rioux, G. Godin, and F. Blais. Datagraphy: The Final Frontier in Communications. In *In International Conference on Three Dimensional Media Technology*, 1992.
12. K. Satoh and Y. Ohta. Passive Depth Acquisition for 3D Image Displays. *IEICE Transactions on Information and Systems*, E77-D(9), 1994.
13. M. Soucy and D. Laurendeau. Multi-Resolution Surface Modelling from Multiple Range Views. In *Proceedings of Computer Vision and Pattern Recognition*, pages 348 – 353, 1992.
14. R. Tsai. A versatile camera calibration technique for high-accuracy 3D machine vision metrology using off-the-shelf tv cameras and lenses. *IEEE Journal of Robotics and Automation*, 3(4):323 – 344, 1987.
15. G. Turk and M. Levoy. Zippered Polygon Meshes from Range Images. In *SIGGRAPH*, pages 311 – 318, 1994.

Generic Shape Learning and Recognition[1]

Alexandre R.J. François[2] and Gérard Medioni

Institute for Robotics and Intelligent Systems
University of Southern California
Los Angeles, California 90089-0273
{afrancoi,medioni}@iris.usc.edu

Abstract. We address the problem of generic shape recognition, in which exact models are not available. We propose an original approach, in which learning and recognition are intimately linked, as recognition is based on previous observation.

The input to our system is in the form of segmented descriptions of objects in terms of parts. In 2-D, the shape is incrementally decomposed into parts suggested by curvature sign changes, and for each part an axial description is derived from both local and global information. The parts are organized into a connection hierarchy. For 3-D objects, we intend to use segmented tridimensional descriptions, the parts being modeled by generalized cylinders. In this case, the connection graph is not necessarily a hierarchy, but can still be used with our algorithms.

The part description obtained at this point is still too detailed and fine grained in order to easily categorize and compare shapes. Hence, we use a simplified description of parts, capturing part local geometry and connection with the superpart information. The local geometry parameters are qualitative and symbolic, and are quasi-invariants under projection and viewpoint change. Both types of parameters take discrete values derived from the available fine description. The connection parameters are normalized to be scale-independent. These simplified part descriptions are organized into a connection hierarchy as provided by the original decomposition. The parameters are chosen to ensure that the information carried in these descriptions is sufficient to perform shape recognition.

Actual shape descriptions are stored in a data-base, from which they must be efficiently and specifically retrieved when a new shape is proposed for recognition. We define a hierarchical indexing system based on the structure of the descriptions and the local description of parts. This mechanism allows for dynamic updating of the data-base with a minimum computing cost.

When a shape is submitted for recognition, the data-base is searched for the closest known shapes. A partial match, based on the connection structure and the aggregation of dissimilarities between parts, is computed in-

---

1. This research was supported in part by the Advanced Research Projects Agency of the Department of Defense and was monitored by the Air Force Office of Scientific Research under Contract No. F49620-90-C-0078 and/or Grant No F49620-93-1-0620. The United States Government is authorized to reproduce and distribute reprints for governmental purposes notwithstanding any copyright notation hereon.

2. Sponsored for this research by NOESIS S.A., Immeuble Ariane, Domaine Technologique de Saclay, 4 rue René Razel - Saclay, F-91892 Orsay Cedex, FRANCE.

crementally level by level between the shape and the possible candidates. The combination of the incremental process with the hierarchical indexing makes the number of shapes processed at each step decrease rapidly, therefore dramatically reducing the average complexity of the retrieval. The selected retrieved shape(s) are used to give a classification for the submitted shape.

This approach to recognition is influenced by the Case-Based Reasoning (CBR) paradigm, which embeds all the characteristics required to meet our goals, such as the ability to process noisy, incomplete and new data. It also provides an interesting framework for higher-level intelligent processing (e.g. justified interpretation, automatic learning).

We describe our implementation for 2-D shapes recognition and present results. The current implementation should also work on 3-D descriptions as described above, with minor changes. We also intend to use this system as the core of a higher-level vision-based reasoning system.

1 Introduction

Generic shape recognition is a major problem in image understanding. In order to perform shape recognition without *a priori* exact geometric or semantic knowledge about the observed objects, we need to produce context independent high-level shape descriptions. Many researchers have discussed this problem (see [16] for an overview). The conclusion is that a "good" shape description should produce segmented parts organized in a hierarchy. Recent progress in the quality of resulting descriptions in 2-D and 3-D makes it relevant to consider such recognition (see e.g. [17][20]).

Our design of a recognition engine is influenced by the Case-Based Reasoning (CBR) paradigm successfully applied in Artificial Intelligence [11][18]. In this context, the system uses a visual memory, which is an organized set of previously described and identified shapes, to propose a documented classification of a given shape. Therefore shape descriptions are dynamically and incrementally stored in a database (learning process), from which they must be efficiently and specifically retrieved when a new shape is proposed for recognition. This implies an efficient *indexing*, combined with an appropriate *retrieval* algorithm. The originality of this approach is that recognition is based on previous observation, which means that the problem addressed is really shape *recognition* as opposed to shape identification (model matching, pattern recognition, etc.).

Here, we describe a recognition system featuring symbolic hierarchical description of shapes, hierarchical indexing of the database and incremental partial-matching retrieval. The principles presented are illustrated by experimental results obtained with an implementation based on the hierarchical 2-D shape description method described in [17]. We start with a brief overview of the work in shape recognition, and an overview of the CBR paradigm. In section 3, we present the symbolic hierarchical description model we are using. Section 4 describes the dynamic database organization process, and section 5 the retrieval process with a complexity study. Tests performed on real and simulated databases are presented in section 6. Finally, we summarize our contribution and discuss possible extensions for the system.

We do not address the actual processing of the images to obtain shape descriptions (figure-ground problem, segmentation, feature extraction, etc.), and assume that the input to the system consists of adequate descriptions of shapes, segmented into parts, as produced in [17] or [20].

2 Previous Work

2.1 Recognition in Image Understanding

Most of the studies in object recognition are concerned with recognizing one object, for which an exact geometric model already exists, by finding its position and orientation (see [6]). In a sense, this should be referred to as pose estimation and calibration. We are interested in the issues associated with a very large number of entries, and the absence of exact models (genericity).

A number of systems developed use low-level primitives for recognition. These primitives are chosen to exhibit invariance properties to viewpoint and occlusion. Representative examples of such work are Bolles and Cain's local-feature-focus method [3], and Grimson and Lozano-Pérez's interpretation tree [7][8]. The major drawback of these methods is that recognition relies on low-level features, which usually occur in large numbers in a scene and in relatively low number in objects, and have a low discrimination power, because of the large search space involved.

These methods usually ignore the complexity associated with the number of models. In order to be able to handle large model data-bases, indexing techniques were introduced, and even became the recognition engine foundation. A few examples of "indexing-based" recognition systems are [10],[12],[4],[19] and [9]. However, we do not believe that low-level primitives allow to perform high-level, human-like recognition.

Recently, Murase and Nayar proposed in [13] an original approach using actual object images instead of models. As a corollary, learning is an inherent feature of their system. Also, they do not explicitly address shape, which simplifies the process of recognition. While this system is useful for a number of application scenarios, several aspects do not meet the goals we stated: it does not allow occlusion, and higher-level interpretation cannot be conducted from the recognition (and pose estimation) result. We also note that the system's visual memory can only be updated off-line, involving the recomputation of the universal eigenspace, which is extremely costly in terms of image storage since all the learning images have to be available at the time of an update.

Three main bodies of work present ideas that give partial answers to the problems identified above:

- Nevatia and Binford [14] developed a system that can recognize curved objects from range images (3-D from $2^{1/2}$-D). They present a technique for generating structured, symbolic descriptions of objects by segmenting them into simpler subparts. The description is based on generalized cones. An object is represented by a connection graph in which the parts are described with a small number of symbolic parameters. The recognition is performed at a symbolic level, but lacks an efficient indexing method taking advantage of the description model and therefore cannot handle a large number of models. Another important aspect of this work is that the "models" used for recognition are object descriptions generated from actual object observations.

- Biederman's recognition-by-components (RBC) theory [2] shows that this part-oriented, symbolic approach of object recognition is a mechanism used in human image understanding. According to this work, only a relatively small set of part primitives is enough to describe all possible complex objects. The information describing the object is a description of the parts, and the description of their relations. Moreover, the recognition of a complex object can be performed considering its few main parts, and eventually refined if necessary.

- Ettinger uses in the recognition paradigm he defines in [5] several notions that complement some aspects of the ideas found in Nevatia and Binford [14]:
    - Decomposition of object models into subparts hierarchies.
    - Non-exact matching recognition.
    - Hierarchical indexing of the data-base.

    However, the model object representation uses low-level features and therefore the recognition cannot be performed (and explained) at a symbolic level.

Given these works, and keeping in mind that we want to perform re-cognition, explicitly based on previous observation, we derive four principles to achieve our goal:

- *Hierarchical graph description*: the shapes should be described in terms of elementary parts and their connections, and should be organized hierarchically in function of the scale or size (at least a coarseness order has to be defined on the parts to orient the description graph).

• *Symbolic part description*: the description of parts used for recognition should be *symbolic*, and clearly separated from the low-level feature extraction process. It should also be context independent (the context parameters are separated from the generic description): scale normalization, orientation independence, etc.

• *Hierarchical database organization*: in order to handle large databases, the visual memory has to be organized, and a hierarchical indexing is expected to be an efficient solution in the RBC model context.

• *Partial matching*: The recognition paradigm is built on a partial-match hypotheses generation, followed by interpretation and validation.

Recently, Provan *et al.* [15] described learning of 3-D object models using Bayesian networks. They use volumic segmented object description, the parts being modeled by generalized cylinders. Although their approach is motivated by similar goals, the general framework in which they work is different from ours, as they use a *probabilistic* method to perform learning of object *models*.

2.2 The Case-Based Reasoning paradigm

Given the principles above, we have found that they can be accommodated within the general framework of CBR defined in Artificial Intelligence. Clearly, CBR is not expected to be a turnkey solution to the difficult problem enunciated above; rather, it provides guidelines which need to be significantly refined to be turned into a working system.

CBR is a general paradigm for reasoning from experience [11][18]. Its developments in Artificial Intelligence include the representation of episodic knowledge, memory organization, indexing, case modification and learning. Since it is also a psychological theory of human cognition, it may be a natural reasoning model for performing intelligent tasks.

The underlying principle in CBR is that the reasoning is based on actual previous experiences (or *cases*) rather than on artificially theorized knowledge [1], such as for instance Rule-Based Reasoning or Neural Networks. This mainly avoids the difficult to control loss of information that occurs in the production of a set of rules or training of a neural network. The low-level processes of the recognition, i.e. case description and retrieval of the closest known cases, are only based on case data and meta-knowledge. The synthesis, without which there is no reasoning, occurs at a higher level, using more meta-knowledge, and specific knowledge if needed. In any case, this integration is clearly separated from the early steps of the reasoning process (retrieval), and the introduction of knowledge is precisely confined and controlled. Furthermore, interpreting the retrieval result means analyzing the similarities and inferring a relations between the problem to be solved or the situation to be interpreted, and the closest known problems or situations. Therefore a solution is always documented.

All these principles allow us to seriously consider *generic* shape recognition. Since the system relies on previous experience, it is *adaptive* and *expandable*: learning is an inherent feature of the system.

Since it is not our point to present CBR in this paper, we will not develop in detail its specific aspects. We will just point out that in the context of shape recognition (which is an interpretation problem), a case contains two sets of data. The *shape description* includes the symbolic description (as described below) which will be used for the retrieval of similar shapes, and information about the context in which this description has been produced (e.g. normalization parameters, original image, low-level data, reference to the original image(s), etc...). The *identification* data can simply be the class to which the object belongs (as in our current implementation, in which the qualitative reasoning following the retrieval is mainly left to the operator), or it can be a more complex and organized object which will be used for the interpretation of retrieved cases, and for the interpretation of complex scenes for example. In the remainder of this paper, the word *case* will sometimes be used to refer to the symbolic shape description used for recognition (which is the part of the case which is considered for the recognition steps studied) and consequently, the term *case-base* will refer to the shape database. We want to emphasize that any shape description used in the system is supposed to be the description of an actual observed shape, and that the system never relies on some artificial object model.

3 Symbolic Hierarchical Generic Shape Description

In this section, we describe a general representation model for generic objects, and discuss how to build a case for the recognition system. We illustrate all the principles described with 2-D shapes originally processed as described in [17]. The extension of these principles to 3-D objects description is discussed in section 3.3.

Shape recognition is an interpretation problem. The solution is based on the interpretation of similarities and differences between a proposed shape and a set of known shapes. Therefore, the description model used to manipulate the shapes has a fundamental influence on the quality of the recognition. The first step to a good representation is a good low-level processing of the original data. We assume that such a description is available (see example in Figure 1).

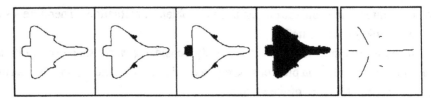

Fig. 1. Decomposition of an "f106" shape (from [17])

3.1 Hierarchical Shape Decomposition

Principle

We characterize complex shapes using a small number of elementary parts, consistently with Biederman's geons theory [2]. We apply this process to 2-D shapes and it can easily be adapted for 3-D objects. We briefly outline below the process to generate these descriptions from real images, more details can be found in [17].

Planar shapes are first described by a B-spline approximation of their contour. The contour is initially segmented at curvature sign changes into potential local parts, that are described by their Smooth Local Symmetry axis. The local description is complemented by the computation of parallel symmetries. Given this local information and global relationships, the shape is hierarchically decomposed into parts, by first removing the small and well defined parts and then by analyzing the remaining shape. This results is a natural axial description of the shape together with a hierarchical decomposition into its parts (see Figure 1).

Once the parts are identified and described, they are organized into a directed connection graph (see Figure 2). The edges are oriented from the larger to the smaller part, according to the size parameter chosen. For the 2-D shape description used, the size of the part is defined by the length of the axis. The shape description used for recognition is composed of the description of its parts translated into a symbolic description model (presented below), organized into a connection graph. This allows qualitative reasoning by finding and analyzing correspondences between the graph describing an input shape and a graph describing a possible model.

Properties of the Description Graph.

The description graph in the case of 2-D shapes is a tree. In our implementation, the root contains no information used for the recognition. The root is the only node of level 0. For a description hierarchy of depth d, the levels are numbered from 0 (root) to d-1 (see Figure 2).

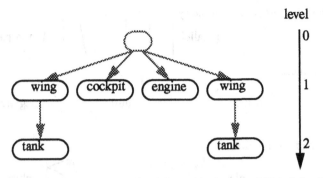

Fig. 2. Directed connection graph obtained from the decomposition of the "f106" shape shown in Figure 1 (hierarchy of depth 3).

General properties of objects and the construction process for the connection graph allow us to infer structural and semantic information of such a description graph:

- A complex object is made of a limited number of main parts which, given our orientation of the description graph, will appear on the first level of the description hierarchy. Hence the first level is a coarse description of the shape (only the main parts), which is enough to infer an initial classification for the shape [2]. Each additional level refines the description.

- The depth of the description is limited to very few levels by the resolution of observation. As a corollary, the total number of parts is also limited by the resolution of the image.

3.2 Symbolic Description of Parts

We first generate symbolic parameter values derived from the low-level description of the shape and of the segmented parts.

Intrinsic Part Description: Local Geometry

Parts are characterized by intrinsic geometrical properties similar to the geons in [2]. For our 2-D parts description, we consider the following set of parameters (see Figure 3): the type of boundaries symmetry (parallel or non-parallel boundaries), the type of curvature (positive, negative or null) and the type of termination (mono-angular or pluri-angular). These symbolic parameters are obtained from the low-level description. Their qualitative nature makes them relatively insensitive to noise (the larger the part, the less sensitive the parameter values). The combinations of the possible values for these parameters define a limited set of elementary 2-D patterns that can describe any segmented part. A small number of special types of parts

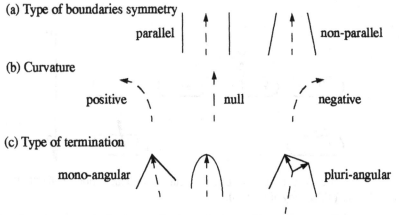

Fig. 3. Local geometrical characterization of parts: (a) symmetry; (b) curvature; (c) termination

may also be defined to describe particular cases (e.g. blobs). It is important to emphasize the fact that these local characteristics of the part are determined by a process taking into account both local and global features of the shape [17].

Geometrical Organization of Parts

A shape is also described by relations between connected parts. The description process used produces a connection hierarchy, which means that a part is only related to its superpart. The parameters we consider to describe planar shape parts relations are (see Figure 4): the size of the part (normalized length of the axis), the type of the connection with the superpart and the angle between the superpart axis and the part axis (if the axis are curved, we consider the angle between the tangents at the projected intersection point). The type of connection parameter takes symbolic values and is rather noise tolerant. The other two parameters are more subject to noise alteration, so their quantization for the recognition is delayed until the part comparison process (see section 5.2). Their values are directly computed from the low level description. The length of the axis is normalized by the length of the longest part axis in the shape in order to produce a scale independent description. There exist many other possibilities that could for instance make this normalization insensitive to occlusion on the largest part, and more experimentation is needed to fine tune them. The normalization parameters are stored as part of the shape description, so that they can be used during the interpretation of the retrieved shapes.

The description obtained contains the information necessary to rebuild a scale-independent symbolic version of the original shape in which the geometrical rela-

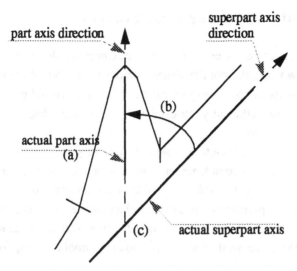

Fig. 4. Relative parameters for part description: (a) size of the part (normalized length of axis); (b) relative angle between the part axis and its superpart axis; (c) type of connection.

tions are kept. An example of such a symbolic shape description is shown in Figure 2.

3.3 Extension to 3-D Objects Description

Very few modifications would be necessary to adapt the system for 3-D object recognition. The low-level description process used is the one described in Zerroug and Nevatia [20].

The object decomposition and the part description is directly adaptable from Biederman's geons [2] and Zerroug and Nevatia's [20] 3-D object segmented description. The result is a symbolic description of the 3-D object as a set of 3-D part symbolic descriptions organized in a connection graph. Compared to a 2-D shape part description, a 3-D object part description requires several additional parameters to describe spatial relations. Moreover, the importance of some parameters may differ. For instance, the type of connection between parts is a fundamental information for 3-D objects. The description graph can be oriented using the size of the parts (as we orient the 2-D shape description graph). In general it is not a tree, but a mono-rooted oriented acyclic graph (parts can have more than one connected part in a 3-D object). This structure is compatible with all the processes described later in the paper.

4 Dynamic Database Organization

If a good description of cases is the first requirement for shape recognition, the retrieval of known shapes from the visual memory is the core of the recognition engine. Once a shape is translated into a case, it has to be stored in a data-base from which it can later be efficiently retrieved when a described shape is proposed to the system for recognition.

Since we want the system to be able to handle very large databases, an efficient retrieval relies on two complementary aspects: an adequate organization of the case-base (identified as the *indexing problem*) and an adapted retrieval algorithm. In this section, we present the indexing mechanism we have developed and implemented. The complementary retrieval algorithm is described in the next section. In these two sections, the word "shape" refers to the symbolic description used for recognition.

4.1 Hierarchical Indexing of Shape Descriptions

In the recognition process, the search for similar cases is data-driven, which implies an efficient indexing based on shape descriptions for retrieval. The most natural and efficient data structure for indexing objects for retrieval is a hierarchy. Organizing the shapes in a hierarchy requires the definition of a partial order on the shapes, which we will infer from the description graph properties outlined in 3.1. Since the retrieval is data-driven, the hierarchical indexing should be based on the structure of the description hierarchies. This structural "indexing key" must be complemented by a simplified description of the parts (*i.e.* a subset of the description parameters).

Definition: We call *I-Structure* (for Indexing Structure) the simplified description graph obtained from the original symbolic description graph by reducing the number of parameters for the description of parts. The I-Structure is isomorphic to the original description graph and since the parameters used to describe the parts are a subset of the parameters used in the symbolic description used for recognition, the I-Structure information is strictly contained in the description of the shape.

Choice of Part Description Parameters for Indexing

Consistently with our part-oriented approach of recognition, the description of parts used for the I-Structure includes the local geometrical information. This choice is supported by the fact that the local geometry of the parts is less subject to variations due to noise, and is described with symbolic parameters. Since we would like the recognition system to be noise tolerant, it seems logical not to consider the param-

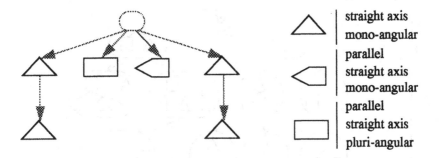

	straight axis
	mono-angular
	parallel
	straight axis
	mono-angular
	parallel
	straight axis
	pluri-angular

Fig. 5. I-Structure derived from the "f106" shape symbolic description

eters describing the relative position of parts. This also allows not to assume the objects to be rigid, but rather to be made of articulated rigid parts.

Finally, we use as I-Structures for our index the partial descriptions of shapes that is obtained by considering for the description of parts only the local geometry description parameters (see Figure 5).

Hierarchical Indexing of Hierarchical Structures

We propose a hierarchical indexing method for hierarchical structures. To each node of the index is associated one (and one only) I-Structure. The cases referenced by the node are those which description matches *exactly* the node's I-Structure. We organize the nodes into a *specialization hierarchy*, by defining a partial order on the I-Structures:

An I-Structure S' is a direct specialization of an I-Structure S iff

- $depth(S') = depth(S)+1$
- S and S' are identical down to depth $depth(S)$

We present in Figure 6 an example of hierarchical indexing of hierarchical structures built from three types of parts, based on the structure itself and on the type of the nodes at a given level.

Given the partial order defined above, the hierarchical index inherits properties from the description hierarchies:

- Since the first level of a description hierarchy is a coarse description of the shape, the first level of the index represents a coarse filter (it allows initial discrimination hypotheses based on the few main parts of the shapes).
- The deeper levels allow to focus and refine the early recognition hypotheses.

Therefore, the branching factor of the root is expected to be high because of the diversity expected among observed shapes, even in terms of their main parts. On the contrary, the branching factor at deeper levels should be much lower due to the re-

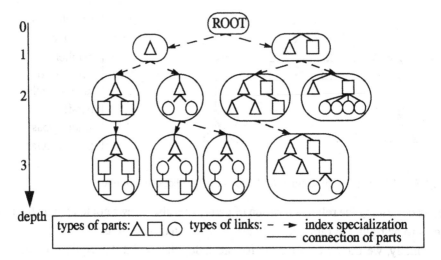

Fig. 6. Example of hierarchical index, discriminating shapes on the structure of their description and on local characterization of parts.

duction of possible details of the shapes once they are coarsely characterized (and to the reduction of observable types as the parts become smaller in the image).

We want to point out that this indexing technique easily applies to acyclic oriented mono-rooted graphs which would occur for description of 3-D shapes (see section 3.3). Given the importance of the type of connection between two parts in the 3-D object description, this parameter should also be used in the I-Structure for 3-D object indexing.

4.2 Dynamic Evolution of the Database

One important feature of the indexing mechanism presented above is that a shape can be added to the data-base at anytime without any recomputation of the existing structure. If the corresponding nodes already exists, the new shape is added to the list of shapes pointed by the node. If the deepest node with a compatible structure does not have exactly the same structure (which means that the description hierarchy of the current node is deeper than the I-Structure attached to this node), then the needed nodes are created.

This dynamic updating principle allows to start either with an important list of shapes in the data-base or with a minimum data-base which is incrementally updated with new shapes (involving supervised learning).

4.3 Discussion

Since no specific knowledge about the described shapes is used in the description and indexing processes, the shapes in the index are not grouped semantically but *geometrically*. For example, a front and side views of a car will not be related directly through the index hierarchy. From the recognition point of view, this is coherent with a symbolic generic shape processing. Practically, this also means that the indexing is a pre-processing for the retrieval, consistent with the data-driven approach, which is a good point for retrieval efficiency, as we shall prove in section 5.4.

5 Shape Retrieval

In this section, we present a retrieval mechanism based on a partial matching of shape descriptions which takes advantage of the index. After justifying our preliminary choices, we define a method for the evaluation of both digital and symbolic similarities between parts, then describe the core of the retrieval process.

5.1 Exact Matching vs. Partial Matching

In most of the previously considered approaches, the recognition is based on an exact matching search of the model memory.

The indexing method described above allows to conduct such a search efficiently. Indeed, if one uses shape models instead of actual shape descriptions, each model is expected to have its own I-Structure which correspond to *one* node in the index. Since exact matching by itself doesn't meet our goals, we don't want to discuss here in detail the complexity of this search process, but one can see that the worst case cost, which is linear in the number of shapes in the database without using the index, is independent of the total number of shapes and becomes linear against the branching factor and the depth of the index hierarchy which are determined by the properties of the set of models. The hierarchical indexing of hierarchical structures principle presented in the previous section can be used very efficiently in a traditional model-based approach, without however making it more relevant for generic shape recognition. An exact matching search presents many draw-backs incompatible with generic recognition, especially in our "recognition from previous actual observations" approach. It ends in failure if the same exact shape is not found in the data-base. If the description process is not stable enough or if the data is incomplete or noisy, recognition cannot be performed, and no interpretation of the input shape is proposed. This is the main reason why we consider a partial matching retrieval approach.

The partial matching retrieval produces hypotheses for the recognition. It is based on a metric to evaluate a similarity (or a dissimilarity) between shape descriptions, complemented by qualitative information about the comparison to allow "high-level" (symbolic) solution generation and explanation.

5.2 Dissimilarity Between Shapes

Our evaluation of the dissimilarity between two shapes is based on the definition of "transition costs" that represent the cost of the assumption that two different "symbolic" objects (parameter values, parts, shape descriptions) have been obtained from the same real object, because of noise in the original data, variation of the observation conditions, etc. The processing of such costs between parts and their aggregation in costs between shapes is described in the following subsections, along with the definition of a symbolic structure that facilitates the digital computation and allows to perform as well a symbolic, qualitative comparison of shapes.

Transition Cost Between Two Parts

At the lower level, transition costs are defined for each parameter. For symbolic parameters, a cost is attributed to each possible couple of its range of values. This transition cost is therefore a discrete function expressing the cost of the assumption that two different symbolic values for a parameter have been determined from compatible original data (as a result of noise in the image for example). For numerical parameters, we define discrete cost functions expressing the cost of the assumption that the difference observed between two values was caused by noise or context variations. In our implementation, we made sure the costs we defined were distances.

The aggregation function used to compute the dissimilarity (transition cost) between two parts is a weighted sum of the parameters transition costs. Weights can be used to give more importance to some of the parameters. In our current implementation, all the parameters are given the same weight.

Pairing the Parts: Correspondence Hierarchy

In order to easily compute the transition costs, we introduce a comparison structure which is a *correspondence hierarchy* instantiated between two shape descriptions (see Figure 7). Each node of this graph points to two matching parts of the considered shapes, and its position in the correspondence hierarchy is similar to the position of the parts it links in their respective description hierarchies. A special type of

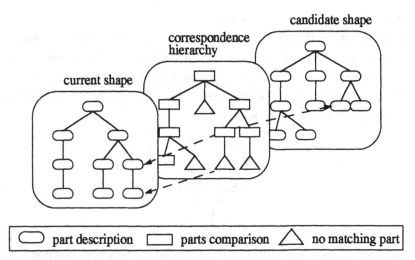

Fig. 7. Example of comparison structure instantiated between two hierarchical shape descriptions. Only a few links are shown to preserve readability.

node is defined for non paired parts, and the transition cost associated to such a node is maximum (i.e. automatically set to 1).

The strategy we used in our implementation is the following:

• First try to find a pairing part in the known shape subparts list for each of the parts of the proposed shape subparts list, by selecting the closest (according to the defined transition cost) among the remaining subparts. If the cost between the two paired parts is equal to 1 (which can occur if parts are available for pairing but are completely different), then a missing part node is generated for the proposed shape part.

• As soon as one of the subparts list is completely paired, missing part nodes are generated for the remaining subparts in the other list.

In this approach, the matching search is based on the proposed shape data, which is consistent with our data-directed retrieval paradigm. The algorithm only ensures that associated parts have the smallest transition cost. A "security" threshold prevents the system from pairing two completely different parts.

The correspondence hierarchy is an adequate structure for the comparison process. It allows to store both digital and symbolic comparison data for later interpretation and reasoning about the retrieved shapes. It is also a key component for the retrieval algorithm. For efficiency, in our implementation the index nodes actually point to pre-instantiated comparison structures for each shape description in the database.

Transition Cost Between Two Shapes

Let d be the depth of the correspondence structure, and k_l the number of parts comparison nodes on level l ($0 < l < d$). Each node in the correspondence hierarchy (characterized by its number i ($1 < i < k_l$ on level l) is attributed the cost $c_{l,i}$ between the two parts it pairs. The cost between the two shape descriptions associated to this comparison structure is defined as:

$$C = \frac{\sum_{l=1}^{d-1} \left(A(l) \sum_{i=1}^{k_l} c_{l,i} \right)}{\sum_{l=1}^{d-1} k_l} \tag{5.1}$$

The factor $A(l)$ is a decreasing function of the level which expresses that the presence of a detail is a strong argument in favor of the match, whereas the absence of a detail is not a reliable information. We propose to use the following function:

$$A(l) = \left(\frac{d-l}{d-1} \right)^q$$

defined for $0 < l < d$, and where q is an integer (in our implementation, $q=2$).

5.3 An Incremental Partial-Matching Retrieval Algorithm

Principle

The most computationally expensive task in the retrieval process is to find the correspondence between two shape descriptions. This is a subgraph isomorphism problem, which is NP-complete. We have developed an algorithm which uses the index and the correspondence structures to avoid computing this expensive correspondence for *all* shapes.

The principle of our algorithm is to build the comparison structures level by level, keeping at each stage only the structures which point to a *compatible* case for the current level. This is made possible by the hierarchical indexing of cases. At the beginning of the retrieval process, all *compatible* nodes of the first level of the index are selected, and the first level of the comparison structure is built for all the cases pointed by these nodes and the nodes in their sub-trees. The order defined for the descriptions and the indexing ensures that this first level correspondence is a coarse comparison. Hence an early diagnosis, highly discriminative, can be based on the closest shapes for this level of detail. To build the next level (which makes sense only for cases which actually have parts of a deeper level, *i.e.* indexed by deeper index nodes), the *partially compatible* subnodes of the previously selected nodes are selected and the process is reiterated, increasing the precision of the comparison.

In the average case, the algorithm ends after the last level of the proposed shape has been processed, or after no candidate is left in the database for further investigation. In most cases, at least one shape will be retrieved for interpretation. The only case in which no shape can be retrieved is when no compatible first level node can be found, which means that the system has to learn more shapes.

Selection of the (Partially) Compatible Nodes

A critical notion in this algorithm is the (partial) *compatibility* between a shape description (or rather the subdescription called I-Structure) and the I-Structure attached to an index node. This determines the nodes that are selected on each level and therefore have both semantic and efficiency implications.

As suggested by the previously outlined semantic properties of the index (see section 4.1), this compatibility has to be considered separately for the first level (level 1) of the index and for the deeper levels:

• The selection of the first-level nodes is a key point of the use of the index. Since the large parts local geometry is less subject to noise alteration, the only factors we have to consider are the occlusion problem, and the ability to find close shapes even if the proposed shape is unusual (in the sense of different from the shapes in the database). Occlusion can result either in the absence of a part in the shape or in its replacement by a "degenerate" part. We have to keep in mind that the chosen strategy only affects the first level nodes selection, which means that the occlusion considered here occurs on main parts of the shape. Even a human cannot recognize a complex object whose main parts are occluded: consider for example a plane shape with occluded wings, it is easy to see a pencil, a rocket, etc. Therefore strong assumptions about the quality of this level of description will not degrade dramatically the expected performance of the system in terms of quality of the recognition. Hence a first possible selection strategy consists of keeping on the first level the one and only node exactly compatible with the input shape, assuming that no occlusion occurs on the main parts. It should not allow the system to handle extreme cases, but should make the retrieval very efficient since the number of selected nodes on the first level is equal to 1 (see complexity analysis below). Occlusion of main parts can be handled by considering a node partially compatible with the current shape if the node's I-Structure first level contains a percentage of matching parts (from the local geometry point of view) with the shape first level. This percentage has to be adjusted to allow the system to process shapes with occlusion and to find at least one compatible node for new shapes. A 100% requirement, which means that a node is compatible if all the parts on level 1 of its I-Structure match one part in the shape I-Structure level 1, does not allow to process bad cases of occlusion (the occluded part is replaced with a degenerate part geometrically different) but allows to deal with total occlusion of parts and with nice partial occlusion cases since the node's I-Structure is allowed to have more parts than the input shape on the considered level. A low percentage results in the selection of many nodes and compromises the efficiency of the retrieval. In general, this parameter depends on the properties of the shapes in the database and on the variability expected for the input shapes. For example, this approach would not be efficient (in term of quality of the recognition) with a database containing occluded shape descriptions. We present tests of these strategies in section 6.2.

• For the selection of deeper nodes, we consider that the next comparison level between the current shape and a shape in the data-base is worth computing if they present similarities down to this level of the description. In our current implementation, we require that at least one entire branch (down to the current depth) can be matched between the two I-Structures (parts matching is based on local geometry). This may be considered as a rather loose constraint, but gives satisfactory results (see section 6.2). It is also easy to implement recursively and is not computationally expensive in the average case compared to a distance computation (see complexity analysis below).

5.4 Complexity Study

In this section, we perform a theoretical study of the complexity of the retrieval algorithm, as described above. Experimental validation is presented in section 6.2.

For this complexity study, we consider a case-base in which all the shapes descriptions have the same depth, so that the all cases are in indexed by the leaves of the index tree, on the same level. The cases are supposed to be uniformly distributed in the leaves of the index tree. Given these assumptions, the parameters of a retrieval process are:

• n: number of shapes in the database.

• d: depth of the shapes in the database, and under our assumption of depth uniformity of the shapes, depth of the index tree (d levels from 0 to $d-1$).

• k: number of parts on each level of a description hierarchy. This number is assumed to be independent of the level (see section 3). It is also independent of the number of shapes in the database.

The parameters d and k are characteristic of the shape descriptions, and are independent of the number of shapes in the database. Therefore they will be treated as constants in the remainder of this section.

Since we are mostly interested in the performance of the retrieval system with large data-bases (several hundreds to several thousands shapes), the complexity is evaluated as the number of operations computed between parts (distances and compatibility tests) as a function of n.

Searching Without Using the Index

As a baseline to evaluate the improvement in complexity provided by using the index, we first compute the number of distances processed to build the correspondence hierarchies for all shapes. The number of distances computed between the parts of two descriptions to build a level l in the correspondence hierarchy is a function $K(k)$, independent of l, and with a complexity of $O(k^2)$.

With these assumptions, the average number of distances processed is:

$$n \sum_{l=1}^{d-1} K(k) = n(d-1)K(k) \qquad (5.2)$$

and the retrieval is in $O(n)$.

Taking Advantage of the Hierarchical Indexing

Distances Computed: As explained above (see section 4), semantics of the hierarchical index make us separate the processing on the first level from the processing on deeper levels. Let b_l be the branching factor on level l of the index hierarchy (i.e. the average number of subnodes of a given node from level $l-1$) and c_l the number of subnodes kept at level l (compatibility factor: $c_l < b_l$).

- *First level:* The number of nodes on the first level (b_1) is not independent of the number of shapes in the database. We note $b_1 = f(n)$. Obviously, f is an increasing function of n and is always smaller than or equal to n. For small values of n, $f(n)$ will be close to n. The asymptotic limit of f is imposed by the descriptive power of the I-Structure nodes parameters (number of parts observable at first level times the number of possible combinations of the local geometry description parameters). If these description parameters are adequate for a good coarse discrimination of the shapes, $f(n)$ is expected to have a linear domain for values of n above the efficiency threshold:

$$f(n) = B \cdot n, \text{ where } B \text{ is a constant and } B<<1 \qquad (5.3)$$

We will consider c_1 constant, and provide experimental validation in section 6.2.

- *Deeper levels:* In accordance with the expected (and observed) structure of the index, we assume that the branching factor and the compatibility factor are constant for levels deeper than level 1: $\forall l \geq 2,\ b_l = \beta$, $\forall l \geq 2,\ c_l = \gamma$.

The average number of shapes considered for level 1 is nc_1/b_1. The average number of shapes considered at level $l > 1$ is $n(c_1/b_1)(\gamma/\beta)^{l-1}$ and the average total number of distances processed for the retrieval is:

$$n\frac{c_1}{b_1} K(k)\left[1 + \sum_{l=2}^{d-1} \left(\frac{\gamma}{\beta}\right)^{l-1}\right] \tag{5.4}$$

This formula shows that the gain in terms of distances computed provided by the indexing is mainly determined by the compatibility factor over branching factor ratio on the first level of the index hierarchy. With $c_1 = 1$ and $f(n) = B \cdot n$, (5.4) becomes:

$$B^{-1} K(k)\left[1 + \sum_{l=2}^{d-1} \left(\frac{\gamma}{\beta}\right)^{l-1}\right] \tag{5.5}$$

Therefore, the number of distances between parts computed when the index is used tends to be *independent of the number of shapes in the database*.

In the cases where c_1 is not equal to 1, it is bounded by a constant. The result is that when B is not strictly constant, its deviation from a constant value is amplified in the factor nc_1/b_1 (see simulations in section 6.2). In this case, the number of distances computed remains linear against n, but as shown in the simulations, the slope of the linear relation is very small and the improvement in complexity remains important.

Influence of Compatibility Tests: Formula (5.4) shows that the cost in terms of distance computations is optimum for a high branching factor on the first level, and a comparatively low and decreasing branching factor for the deeper levels, but using the index introduces another complexity parameter which has to be taken into account: the number of compatibility tests processed between parts. A high branching factor presents indeed the drawback of requiring more compatibility tests between the parts of the current shape description and the parts of the I-Structure nodes. This number does not depend on the number of objects in the database, but on the parameters of the index tree instead. We expect the cost added by compatibility test computations to be negligible against the overall gain in number of costly operations (distance computations) that have to be computed for the retrieval.

In order to validate our assumptions and computations, we performed simulations using different algorithms. The simulation protocol and results are presented in section 6.2.

5.5 Summary

The combination of the indexing method and the retrieval algorithm presented here, used with an adequate shape description model, provide several major properties that make the system described meet our original goals of generic shape recognition:

- *Efficient learning*: the dynamic organization of the database gives the system a very powerful and flexible way of manipulating knowledge, especially with the open-world and adaptability assumptions, which are required for genericity.
- *Efficient recognition*: the memory organization allows the retrieval to be nearly independent of the number of shapes in the database.

6 Experimental Results

In this section, we present examples of symbolic hierarchical shape description, database dynamic creation and updating, and recognition performed on both actual and simulated shapes with different algorithm implementations. We implemented the system in CLOS, which prevented us to consider computing accurate time statistics (garbage collecting introduces random and not negligible noise in the statistics...).

6.1 2-D Shape Hierarchical Descriptions

We illustrate the description process by following the different steps leading to the description of an "F106" shape, obtained from real data. The shape decomposition is shown in Figure 1.

Description Parameters

We describe here the details of the implementation of the description model and of the transition costs processing.

We present in Table 1. the different possible values for the symbolic parameters together with the corresponding transition cost matrix. The transition costs are not used for indexing, and therefore the definition of these functions does not influence the index efficiency. Of course they are critical for the quality of the retrieval.

The cost for the length of axes from two parts is a function of the ratio r equals length of the smallest axis over length of the longest axis; the cost for the angle parameter is a function of the axis angle difference α. Table 2. shows the costs associated with the intervals of the possible values for these variables. Note that the

parameter	possible values	transition cost matrix
boundaries symmetry	0: parallel 1: non parallel	$\begin{bmatrix} 0 & 2 \\ 2 & 0 \end{bmatrix}$
curvature	0: positive 1: straight axis 2: negative	$\begin{bmatrix} 0 & 1 & 2 \\ 1 & 0 & 1 \\ 2 & 1 & 0 \end{bmatrix}$
termination	0: mono-angular 1: pluri-angular	$\begin{bmatrix} 0 & 2 \\ 2 & 0 \end{bmatrix}$
part-superpart connection	0: end to close-end 1: end to close-body 2: end to mid-body 3: end to far-body 4: end to far-end	$\begin{bmatrix} 0 & 1 & 2 & 3 & 4 \\ 1 & 0 & 1 & 2 & 3 \\ 2 & 1 & 0 & 1 & 2 \\ 3 & 2 & 1 & 0 & 1 \\ 4 & 3 & 2 & 1 & 0 \end{bmatrix}$

Table 1. Values and corresponding transition cost matrix for the symbolic parameters.

definition of the transition costs for angles determines the degree of variation in the part angles allowed in the recognition process.

parameter	associated variable	variable intervals	associated cost
size of part (length of axis)	$r = \dfrac{min(l_1, l_2)}{max(l_1, l_2)}$	$r < 0.1$ $0.1 \leq r < 0.2$ $0.2 \leq r < 0.4$ $0.4 \leq r < 0.6$ $0.6 \leq r$	0 1 2 3 4
angle between part axes	$\alpha = (\alpha_1 - \alpha_2) \bmod \pi$	$\alpha < 0.2$ $0.2 \leq \alpha < 0.6$ $0.6 \leq \alpha < 1.3$ $1.3 \leq \alpha < 1.7$ $1.7 \leq \alpha$	0 1 2 3 4

Table 2. Discrete cost function for numerical parameters.

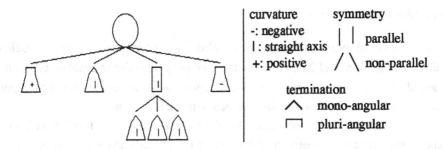

Fig. 10.   I-Structure associated with the "F16" shape shown in Figure 8.

Resulting Shape Description

A scheme of the description and the corresponding I-Structure for an "f106" shape have already been presented in Figure 2 and Figure 5 respectively. We have developed two graphical tools to display partial information of shape descriptions. One draws a schematic skeleton, i.e. the part axes and respecting the geometrical structure of the shape, and the other draws the I-Structure hierarchy, showing the connection graph and the local geometry of the parts. Screen hard copies of these displays for an "F16" shape (Figure 8) are presented in Figure 9 and Figure 10 respectively. It is should be clear that the "skeleton" display is not an exact skeleton of the original shape, but a symbolic representation of the shape geometry.

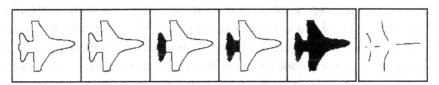

Fig. 8.   Decomposition of an "f16" shape (from [17])

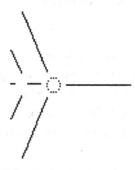

Fig. 9.   Symbolic skeleton for the "F16" shape shown in Figure 8

6.2 Shape Recognition

Two issues have to be considered when evaluating a recognition system: the quality of the recognition, and its computational efficiency. In order to evaluate the incremental retrieval method proposed in this paper, we implemented four recognition engines and compared the results obtained with each of them:

•   **Engine (1)**'s retrieval algorithm does not use the index. It builds all correspondence hierarchies between the submitted shape and the shapes in the databases, and sorts them by decreasing cost. This engine is expected to retrieve the closest shape in the data-base, according to the distance only which means that it may not be the best candidate, in any situation (noise, occlusion, etc.), with the worst computational efficiency.

•   **Engine (2)**'s algorithm uses the index on an exact compatibility selection basis: the deepest node exactly compatible with the submitted shape is found and the correspondence hierarchies are built (down to this node level) between the submitted shape and the shapes indexed by this node and its subnodes. This engine is expected to retrieve an input shape already stored in the database with the best computational cost. If the same exact shape is not in the database, the retrieved shape is not necessarily the closest, and the average complexity depends on the index statistics.

•   **Engine (3)** implements the variant of the incremental retrieval algorithm described in section 5.3 in which the one and only exactly compatible node on the first level of the index is selected. This engine is expected to retrieve the best candidate shape with a very good efficiency, but should have problems in the retrieval of badly occluded shapes (*i.e.* when the geometry of the main parts is degenerated).

•   **Engine (4)** implements the incremental algorithm described in section 5.3, with a selection ratio for first level index nodes of 75%. This engine is expected to retrieve the best candidate from the case-base with a good computational efficiency.

Quality of the Retrieval

We use a database of planes (such as the ones described above) and several other shapes processed in [17] (beans, human, seven-shape, etc.) to evaluate the quality of the recognition. The database contains a total of 14 shapes. There are 25 nodes in the index (8 on the first level). The evaluation of the quality is mainly performed with engines (3) and (4), to show the importance of the partial-match aspect of the retrieval.

We first propose to the system an "F4" shape already described and included in the data base (see Figure 11). Engines (3) and (4) retrieve the stored version of the shape as the closest shape. In this case we process the retrieval with engines (1) and

(a)

```
#<CASE recognized as a F4>
"plain"
The shape-description is:
#<2D-DESC of depth 2>
 #<2D-PART: 1 2 1 2 2.0 1.0>
 #<2D-PART: 1 1 0 2 0.0 0.8>
  #<2D-PART: 1 1 0 0 -0.3 0.1>
  #<2D-PART: 1 1 0 0 0.3 0.1>
  #<2D-PART: 1 1 1 2 3.14 0.3>
  #<2D-PART: 1 1 0 4 -1.1 0.3>
  #<2D-PART: 1 1 0 0 -0.1 0.1>
  #<2D-PART: 1 1 0 0 0.1 0.1>
  #<2D-PART: 1 1 0 4 0.0 0.1>
  #<2D-PART: 1 1 0 4 1.1 0.3>
 #<2D-PART: 1 0 1 2 -2.0 1.0>
```

(b)

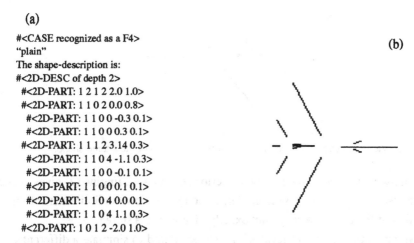

Fig. 11. Case built for an "F4" shape (a) and corresponding symbolic skeleton (b).

(a)   (b)

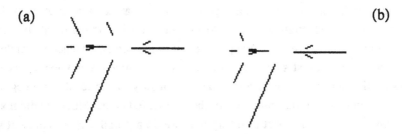

Fig. 12. Occluded shapes: (a) "good" conditions (the local geometry of the occluded part is not changed); (b) "bad" conditions (the part is totally occluded).

(2) as well, and obtain the same result. The only difference between the different algorithms for a shape already stored is the efficiency of the retrieval (see efficiency study below).

Then we perform the recognition for two altered descriptions of the same stored "F4" shape to simulate different types of occlusion: in the first case, the geometry of the occluded part is kept, in the other case, the part is completely occluded. The symbolic skeletons of these altered shapes are presented in Figure 12 (a) and (b) respectively. Of course, a strictly exact matching retrieval would end in failure for these cases. Engine (4) retrieves the original "F4" stored shape in both cases, pointing out the differences in the correspondence hierarchies. Engine (3) is able to retrieve the original "F4" shape only for the "good" occlusion case (the correspondence hierarchy is of course the same as the one built with engine (4)). No shape can be retrieved for the bad case. Note that if the database was much larger, one or several closest cases could be found and the analysis of the correspondence

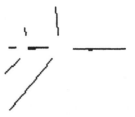

Fig. 13. View-point variation simulation: symbolic skeleton of the altered shape.

hierarchy would determine if they are serious candidates. Engine (2) gives also the same result, but the database is not large enough to show the limitations of this algorithm in the case of a shape not exactly identical to a stored shape.

We also process the retrieval with a shape altered to simulate a different viewpoint. The resulting symbolic skeleton is presented in Figure 13 a. Small parts are missing from self-occlusion, and the axis angles and part sizes are altered. However the main parts (first level of the description hierarchy) are kept, with the same local geometry. Therefore, engines (3) and (4) are able to retrieve the original "F4" shape description and the differences are pointed out in the correspondence hierarchy.

The retrieved closest shape(s) in the case of a completely unknown shape depends on the quality of the database, and the quality of the resulting recognition highly depends on the interpretation of these retrieved cases. Since we did not develop this higher-level aspect of the system, we do not find useful to present such tests. It is however obvious that when an algorithm cannot be used for the retrieval of shapes that are variations of stored shapes, it cannot be used for the retrieval of shapes that have no such direct relation with the stored shapes.

The results presented here show the interest of performing partial matching retrieval (engines (3) and (4)), which allows to deal with noisy and occluded shapes. Engine (4) is able to retrieve a good candidate in any case. Engine (3) can do the same in most cases. The next natural step in our tests is the comparison of their computing performances.

Computing Performances

The Efficiency of the recognition system in terms of computing performance highly depends on the use of the index. To perform tests on large data-bases in the short time we had, we build random shapes as follows:

• *Part generation:* Since the goal of these tests is the evaluation of the indexing performances, the only parameters used are those describing the local geometry. They all take symbolic values. For each parameter, a value is randomly chosen among the possible values (a probability is associated with each value). The other parameters are given arbitrary values.

- Shape generation

    - The depth of the description is randomly chosen among 3, 4 and 5 with same probability.

    - The number of parts for level 1 is randomly chosen among 2, 3, 4 and 5 with equal probabilities. Then the corresponding number of parts is generated with a probability of 0.8 for a parallel symmetry (0.2 for non-parallel), 0.2 for a positive curvature, 0.6 for a straight axis (0.2 for negative curvature) and 0.3 for a mono-angular termination (0.7 for pluri-angular).

    - For the next level, the number of parts is chosen among 2, 3 and 4 with equal probabilities, and the probabilities for the part generation are 0.2 for a parallel symmetry (0.8 for non-parallel), 0.2 for a positive curvature, 0.6 for a straight axis (0.2 for negative curvature) and 0.8 for a mono-angular termination (0.2 for pluri-angular).

    - The distribution of the part connections between two adjacent levels (part and subpart levels) is done randomly: for each part on the part level, a number of subparts is drawn among 1, 2 and 3 with probabilities of 0.6, 0.3 and 0.1 respectively. The corresponding number of parts is taken from the subpart level and the process is reiterated while available subparts remain.

Then cases are built from the random shape descriptions, and collected into casebases. This generation process allows to build databases which properties are similar to the properties expected for databases of real objects.

Since we are mostly interested in the influence of the number of shapes in the database (n), we build five databases for each value of n in 10, 20, 30, 40, 50, 60, 70, 80, 90, 100, 125, 150, 175, 200, 250, 300, 400, 500, 600, 700, 800, 900, 1000, 1200, 1400 and 1600 shapes. We also build one database for each value of n in 1800, 2400, 4000 and 5000 shapes, being aware that the random generation algorithm is such that the quality of the databases produced deteriorates for large values of n (the random generation cannot provide the variety expected in natural shapes), going against the algorithms using the index. We verify that the average number of nodes on the first level (b_1) is an increasing function n of with a large linear domain, which supports the assumptions made in the theoretical complexity study (see section 5.4).

For our recognition tests, we ran engines (2), (3) and (4) on each shape in each database (*i.e.* for n=10, 10 recognition instances were processed with each database, for n=1000, 1000 recognition instances were processed with each database). Engine (1) was used the same way on databases whose number of shapes does not exceed 300 because of the time complexity of the process. The retrieval statistics were collected, and the data presented are average values for each value of n.

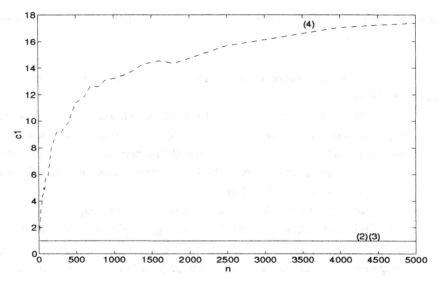

Fig. 14. Average values of c_1 against n for engines (2), (3), (4).

In order to produce numerical data to evaluate the theoretical complexity computed in section 5.4, we are first interested in the relative behavior of the variable terms in formula (5.2) and in formula (5.4). We therefore have to compare n and nc_1/b_1. A plot of the average values of $c1$ against n for engines (2), (3) and (4) (engine (1) does not use the index) is presented in Figure 14 Of course for engines (2) and (3), it is a constant equal to 1. For engine (4) it is an increasing function of n that is bounded by a constant. We have also studied the term nc_1/b_1 against n for engines (2) (same as for engine (3)) and (4). The curve for engine (2) is in fact a plot of $y = n/b_1 = B^{-1}$ since $c_1 = 1$. It is not exactly a constant, which shows that $b1$ is not strictly a linear function of n, but its increase rate is negligible against 1, which supports our conclusion that the cost of the retrieval in terms of distances computed should tend to be independent of the number of shapes in the database, or linear against n with a factor much smaller than 1, depending on the first level index nodes selection strategy.

We present in Figure 15 the average number of distances between parts computed for the retrieval with engines (1), (2), (3) and (4) As expected, when compared with the results obtained for engine (1), this number is almost a constant for engine (2), and increases very slowly for engine (3). The number of distances processed for the retrieval with engine (4) is not constant, but the complexity is still dramatically reduced when compared to a retrieval not using the index.

The average total number of part operations (number of distances plus number of compatibility tests) is shown in Figure 15. These curves show that the compati-

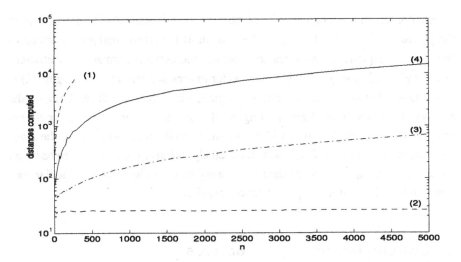

Fig. 15. Average number of distances computed between parts with engines (1), (2), (3) and (4).

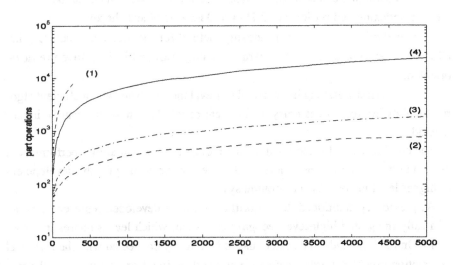

Fig. 16. Average total number of operations computed between parts (distances+compatibility tests) with engines (1), (2), (3) and (4).

bility tests performed do not modify significantly the behavior of the engines (the number of operations is multiplied by a constant approximately equal to 2 in our tests). Moreover, the computing cost of a compatibility test is much lower that the cost of a distance computation.

Finally, we find that engine (2)'s complexity is independent of n, but as seen in the previous subsection, this engine does not allow to perform real generic recognition. After the quality tests conclusions, we are interested in comparing the efficiency of engines (3) and (4). Engine (3)'s performances are very close to engine (2)'s, which makes it a very efficient retrieval engine, with a very low dependence on the number of shapes in the database. Engine (4)'s performances are not as good, but this indexing-based algorithm still represents a major gain compared to the algorithm used in engine (1). The choice between algorithms (3) and (4) has to be based on the expected quality of the data and observing conditions for a particular purpose, and on the computing performances required.

7 Conclusion and Perspectives

We have presented:

- A symbolic and structured shape description model which can be used for efficient indexing and retrieval of 2-D or 3-D generic object shapes.

- A dynamic hierarchical indexing method for hierarchical structures, that allows efficient update (used for learning in our case) and is data-driven retrieval oriented.

- A partial matching incremental retrieval method from which efficient algorithms allowing to process noisy, incomplete or totally new shapes can be implemented.

The combination of these features in a recognition system allows performances both in terms of quality and efficiency that are very promising for the development of higher level vision-based reasoning systems.

As previously mentioned, the algorithms we have developed represent the core of a fully integrated high-level recognition system, which leaves interesting work to do. On the input side, the system has to be directly connected with the low-level description process in order to create a single data flow from the image to the recognition hypothesis produced by the retrieval. After a straightforward adaptation of the description model, it can also be connected with a 3-D objects description engine.

Once the closest shape(s) in the database are retrieved, they are used to infer a classification (or identification information) for the new shape. The interpretation of the similarities and differences between the proposed shape and the retrieved closest known shapes (which in our case is an analysis of the correspondence hierarchy) allows to build an answer that can be justified (which is not possible with Neural Networks). This intelligent integration of the data retrieved requires knowledge. In shape recognition, there are two domains of knowledge: the knowledge related to the objects whose shapes are observed, and the meta-knowledge related to the observation process (image processing, description model, etc.). The former type is not to be considered in a generic shape recognition system (such knowledge could be added to customize the system for a specific application), but the latter can be used, for instance as a set of inference rules, without losing genericity (Rule-Based reasoning appears here at a meta-level). In our current implementation, the system simply gives the classe(s) of the closest shape(s) and the decision is left to the human operator (the number of cases proposed for the interpretation is a parameter of the retrieval process).

The CBR paradigm includes validation of the solution proposed, rectification of the solution and of the interpretation process in case of failure, and a possible update of the database (automatic learning). We have not addressed them in the context of shape recognition yet, but a few research directions are currently being investigated. For instance, more elaborate context information, such as functional cross-indexing, could be added to the system and used for validation.

Learning in the sense of storing shapes in the memory is at the core of our system. If we consider automatic learning, the strategy used for database update will affect the parameters of the index (like the branching factor and the load of the index nodes) and therefore should be determined according to the optimum efficiency of the index. For example, storing occluded shapes when a complete instance is already in the case-base is redundant and degrades the performances of the system (more index nodes in deeper levels but no useful discrimination on the first level). On the contrary, replacing several partial descriptions by one complete description, actually observed, would enhance the performances of the system. This however requires a (high) step towards higher-level intelligence in the management of the database (what makes the system discard a set of stored shapes and replace them by a more general occurrence of the same shape? Then why not allow the system to infer 3-D models from the known shapes and use the models for recognition? This path leads to intelligent learning). Once the recognition itself is complete, the engine can be integrated into a more complex reasoning system to handle complex scenes, for instance using a higher level cross-index describing the relations between the objects in scenes. An identification module could also use the recognition hypothesis for efficient selection of pertinent low-level information.

Finally, we want to point out that the problems addressed in this work and the solutions proposed to solve them are not specific to *our* recognition paradigm. For instance, the hierarchical indexing method together with the hierarchical description method allow to build an exact matching recognition system with the best efficiency that can possibly be expected (constant).

References

1. R. Bareiss. *Exemplar-based knowledge acquisition: a unified approach to concept representation, classification and learning.* Academic Press, 1989.

2. I. Biederman. Human image understanding: recent research and a theory. In *Computer vision, graphics and image understanding,* vol. 32, no. 1, pp. 29-73, October 1985.

3. R.C. Bolles and R.A. Cain. Recognizing and locating partially visible objects: the local-feature-focus method. In *International Journal of Robotics Research,* vol. 1, no. 3, pp. 57-82, 1982.

4. A. Califano and R. Mohan. Multidimensional indexing for recognizing visual shapes. In *Proc. IEEE Computer Vision and Pattern Recognition,* pp. 28-34, Maui, Hawaii, June 1991.

5. G. J. Ettinger. Large hierarchical object recognition using libraries of parametrized model subparts. In *Proc. IEEE Computer Vision and Pattern Recognition,* pp. 32-41, Ann Arbor, Michigan, June 1988.

6. W. E. L. Grimson. *Object recognition by computer - The role of geometric constraints.* MIT Press, Cambridge, Massachusetts, 1990.

7. W. E. L. Grimson and T. Lozano-Perez. Model-based recognition and localization from sparse range or tactile data. In *International Journal of Robotics Research,* vol. 3, no. 3, pp. 3-35, 1984.

8. W. E. L. Grimson and T. Lozano-Perez. Localizing overlapping parts by searching the interpretation tree. In *IEEE Transactions on Pattern Analysis and Machine Intelligence,* vol. 9, no. 4, pp. 469-482, 1987.

9. P. Havaldar, G. Medioni and F. Stein. Extraction of groups for recognition. In *Proc. European Conference on Computer Vision,* vol. 1, pp. 251-261, Stockholm, Sweden, May 1994.

10. A. Kalvin, E. Schonberg, J.T. Schwartz and M. Sharir. Two-dimensional, model-based, boundary matching using footprints. In *International Journal of Robotics Research,* vol. 5, no. 4, pp. 38-55, 1986.

11. J. Kolodner. An introduction to Case-Based Reasoning. In *Artificial Intelligence review,* vol. 6, pp. 3-34, 1992.

12. Y. Lamdan and H. J. Wolfson. Geometric hashing: a general and efficient model-based recognition scheme. In *Proceedings of IEEE International Conference on Computer Vision,* pp. 218-249, Tampa, Florida, december 1988.

13. H. Murase and S. K. Nayar. Visual learning and recognition of 3-D objects from appearance. In *International Journal of Computer Vision,* vol. 14, no. 1, pp. 5-24, January 1995.

14. R. Nevatia and Th. O. Binford. Description and recognition of curved objects. In *Artificial Intelligence*, vol. 8, no. 1, pp. 77-98, February 1977.

15. G. Provan, P. Langley and Th. O. Binford. Probabilistic learning of three-dimensional object models. In *Proc. Image Understanding Workshop*, pp. 1403-1413, Palm Springs, California, February 1996.

16. K. Rao. *Shape description from sparse and imperfect data*. PhD Thesis. University of Southern California, December 1988. IRIS Technical Report 250.

17. H. Rom and G. Medioni. Hierarchical decomposition and axial shape description. In *IEEE Transactions on Pattern Analysis and Machine Intelligence*. vol. 15, no. 10, pp. 973-981, October 1993.

18. S. Slade. Case-Based Reasoning: a research paradigm. In *AI Mag.*, pp. 42-55, Spring 1991.

19. F. Stein and G. Medioni. Structural indexing: efficient two dimensional object recognition. In *IEEE Transactions on Pattern Analysis and Machine Intelligence*, pp. 1198-1204, February 1992.

20. M. Zerroug and R. Nevatia. From an intensity image to 3-D segmented descriptions. In *Proc. IEEE International Conference on Pattern Recognition*, vol. 1, pp. 108-113, Jerusalem, Israel, October 1994.

A Hybrid Approach to 3D Representation

Nigel Ayoung-Chee, Gregory Dudek and Frank P. Ferrie

Center for Intelligent Machines, McGill University
Montréal, Québec, Canada H3A 2A7
e-mail: {ayoung, dudek, ferrie}@cim.mcgill.ca

Abstract. This paper deals with generic 3D shape modelling for the purposes of object recognition. Common problems with many existing methods are that they either capture insufficient detailed structure or fail to provide sufficiently abstract descriptions (global vs. local representation). As a result, they tend have a limited field of application. The approach presented here attempts to address this problem by building a composite representation of the data in terms of a superquadric augmented with multi-scale surface models.

This is illustrated experimentally using laser range data. The superquadric that results in the best possible fit is expressed in terms of its position, size, shape and pose parameters. The residual of the fit is then modelled at several scales using multiple surface patches with uniform mean and Gaussian curvature. A hierarchical ranking of these patches is used to describe the residual based on geometric properties. These geometric properties are ranked according to criteria expressing their stability and utility. The most stable patches are selected as the description of the residual. The resulting representation can then be used for both pose estimation and object recognition.

1   Introduction

In this paper, we consider 3D modelling with particular emphasis on recognition, but keeping in mind that generic models should be useful for other tasks such as grasping. Common approaches to the problem include the use of volumetric models such as superquadrics or deformable solids. While superquadrics are effective for capturing the global geometry of an object they lack the ability to model small scale features. Their use in recognition tends to be limited with respect to certain canonical shapes [2]. Deformable solids are able to *model* a wider class of shapes, however their many degrees of freedom make it difficult to extract stable representations of objects for recognition. In short, they do not provide a convenient abstraction mechanism for robust indexing.

An alternative involves modelling objects as a collection of surface patches or features. Typical examples involve the use of specialized edge information [20] or curvature extrema [10, 6, 8]. For object representation and recognition, this often involves fitting patches or membranes to the raw surface data. While the resulting description is often rich enough to accurately describe the surface of an object, the fitting and recognition process can be computationally expensive.

Such descriptions also do not capture the generic shape of objects which is needed for tasks such as grasping.

The solution presented in this paper involves constructing a model composed of two complementary representations: a volumetric primitive for simpler tasks such as obtaining the overall shape of an object (classification), and a more detailed model based on a surface decomposition for complex tasks such as recognition. This allows us to scale the complexity of the model to the task. In particular, we chose superquadrics to model the basic shape of objects. Rather than construct a separate surface representation of the original data, surface patches of uniform mean H and Gaussian K curvature are used to model the error in the superquadric fit to the data. These patches are then ranked according to their stability and salience. The most stable patches are used in conjunction with superquadrics in more complex tasks.

In short, we describe the residual surface that expresses the aspects of the data not captured by the superquadric model. This gives us a measure of the quality of the fit of the superquadric model to the original object and provides a relationship that is more meaningful than two independent modelling processes. It allows us to express features explicitly in terms of their geometry and provides us with a means by which the data can be reconstructed by adding the representation of the residual to the superquadric fit. This modelling approach is illustrated experimentally with range data.

1.1 Background

Barr [4] popularized the use of superquadric primitives and angle preserving transformations to extend the geometric primitives, quadric surfaces and parametric patches, to allow complex solids and surfaces to be constructed. Subsequently, Pentland [16] and also Solina and Bajcsy [17] used superquadrics for the recovery of compact volumetric models for shape representation from 3 dimensional range points. Model recovery was formulated as a least squares minimization of a cost function whose value depended on distance of the data points from the model's surface and the global parameters of the model (size, position and orientation). However, the class of objects that could be represented accurately are limited to single-part convex shapes.

Ferrie et al. [9] approached the problem of multi-part objects in a purely bottom-up approach which segments the range data based on curvature consistency. Each part of the segmented range image is then modelled with superquadrics. This form of modelling has been implemented in an active recognition system in which the data acquisition is guided by feedback from the recognition strategy [19]. In this way, as much data as possible is collected to ensure an accurate model of the object. However, even with this approach, superquadrics may be inadequate to represent the segmented parts uniquely. This can lead to ambiguity in the recognition process.

Metaxas and Terzopoulos [18] developed dynamic models based on global and local deformation properties from superquadric ellipsoids and membrane splines. However the parameters for local deformation are only used in the reconstruction process and only the gross shape information is used for recognition. Pentland and Sclaroff [16] propose a method for recovering a 3D model based on finite element methods (FEM) and parametric solid modelling using implicit functions. One issue with their technique is whether the selected modes capture the surface characteristics of interest.

Several methods for describing surfaces using local properties or hierarchies have been proposed [14]. We are particularly interested in methods based on perceptually significant cues. The significance of curvature in feature detection and object identification has been recognized for some time [3, 12] and is well documented in psychophysics. Besl and Jain [5] implemented a surface reconstruction strategy which involved segmenting the surface into regions of arbitrary shape and approximating the image data with bivariate functions. In their work, coarse segmentation is performed based on surface curvature. This is then refined using an iterative region growing method based on variable-order surface fitting. However experimental results focused on image segmentation and surface reconstruction and did not explore its applicability to recognition.

Various authors have used curvature information for the representation of curves or surfaces [13, 15]. An issue with curvature extraction is the selection of the correct scale. Most curvature measurement techniques assume that there is a unique curvature that can be measured at each point. Some work has considered the combination of scale information with curvature information using methods such as Gaussian scale-space or curvature-tuned smoothing. Mokhtarian [15] presents results that support the use of a multiple scale space representation for representing curves. However its use in the recognition process is not explored. Dudek and Tsotsos [8] use a multi-scale representation of curvature to perform recognition. A multi-scale hierachy of surface models is produced using "tuned regularization". This is used for recognition of 2D or 3D objects. Their approach, however, fails to produce or exploit a volumetric model which is useful for tasks such as grasping.

2 Theory

2.1 Fitting Superquadrics to Range Data

The form of superquadrics are the same as that used by Ferrie et al.[9]. The superquadric is described by 11 parameters: size (a_x, a_y, a_z), shape (ϵ_1, ϵ_2), position (t_x, t_y, t_z) and orientation (r_x, r_y, r_z) [4, 17]. The surface is defined by the following 3D vector:

$$v = \begin{pmatrix} a_x \cos^{\epsilon 1}(\eta) \cos^{\epsilon 2}(\omega) \\ a_y \cos^{\epsilon 1}(\eta) \sin^{\epsilon 2}(\omega) \\ a_z \sin^{\epsilon 1}(\eta) \end{pmatrix} \tag{1}$$

$$-\pi/2 \leq \eta \leq \pi/2$$
$$-\pi \leq \omega \leq \pi$$

where η and ω correspond to latitude and longitude angles of vector v expressed in spherical coordinates. The vector v originates in the coordinate centre and sweeps out a closed surface in space when the two independent parameters, η and ω change in the given intervals. Parallelepipeds are produced when ϵ_1 and ϵ_2 are $\ll 1$. When ϵ_1 and ϵ_2 are 1, the shape is more rounded. As ϵ_1 and ϵ_2 become greater than 2 the model becomes progressively more pinched.

The fitting of a superquadric in general position and orientation to range data is based on a least squares minimization of the superquadric inside-outside function [17]. There are 2 stages to the fitting process, an initialization stage followed by a stage in which the initial estimate is refined. The initial estimate for the model uses the moments of inertia of the data points to estimate the orientation of 3 three orthogonal axes of an ellipsoid. Once the orientation is obtained the data is projected onto the axes and the maximum projected value on each axis is used as the axis length (a_x, a_y, a_z) for that axis. The Levenberg-Marquardt method for nonlinear least squares minimization is then used to iteratively refine the initial estimate until the average error from one stage to the other is within some threshold.

2.2 Calculating the Residual Surface

Because superquadrics can only represent a restricted class of shapes, the resulting representation often fails to capture local or small scale (relative to the size of the object) features such as bumps or concavities. In addition, incomplete information such as that obtained from a single viewpoint can lead to gross errors in the estimation of the shape of the object.

To determine how well the object is modelled by a superquadric, the error in the fit is calculated. Computing the residual by a parallel (to the global coordinate system) projection from the range data to the surface of the superquadric was initially considered, but this resulted in unstable results for a superquadric in general position and orientation. In extreme cases, discontinuities in the residual surface can result. To overcome these problems the residual surface was coupled to the local coordinate system of the superquadric. This was done by defining the residual as the distance from the range data to the surface of the superquadric measured along a radial projection from the centre of the superquadric (Figure 1).

The residual data was then abstracted using surface patches with uniform *mean* (H) and *Gaussian* (K) curvatures. To address issues in selecting the correct scale, the patches were extracted at multiple scales and ranked based on their geometric properties [1].

2.3 Extracting Surface Patches

At a given point on a surface a curve is formed by the intersection of the surface and the normal plane in a given tangent direction. The curvature of this planar curve is called the *normal curvature* κ_n at that point. The maximum and minimum curvatures at a point define the principal curvatures κ_{max} and κ_{min}. The

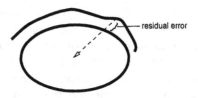

Fig. 1. Calculation of the residual surface using a radial projection.

Gaussian curvature at a point is defined as the product $\kappa_{max}\kappa_{min}$ and the mean curvature is defined as $(\kappa_{max} + \kappa_{max})/2$.

The procedure for extracting a multi-scale surface description of the residual surface is illustrated in Figure 2. First, an initial coarse segmentation of the surface based on H and K mapping is performed. The H and K curvature was chosen because of their invariance to rotation, translation and re-parametrization.

To calculate differential measurements it is necessary to filter the surface to smooth out local fluctuations in the surface due to quantization errors and measurement noise. This was done using a triangular approximation to a Gaussian kernel (Figure 3). This linear filter was chosen for its simplicity and because it is localized in the frequency domain. Therefore it is less likely to produce features that were not part of the original image.

The H and K curvatures are the basis for the segmentation of the surface into patches of constant curvature. A label or surface type is then assigned to every pixel in the image based on the signs of the H and K curvatures according to the following equation,

$$T(i,j)=1+3(1+sgn_{\gamma_H}(H(i,j)))+(1-sgn_{\gamma_K}(K(i,j))) \qquad (2)$$

$$\text{where } sgn_\gamma(x) = \begin{cases} +1 & \text{if } x > \gamma \\ 0 & \text{if } |x| \leq \gamma \\ +1 & \text{if } x < \gamma \end{cases}$$

where γ is pre-selected.

Coarse initial segmentation is obtained by finding four-connected regions. The *coherence* property of piecewise smooth surfaces makes pixels with the same label cluster together. After the range data has been segmented into regions of uniform curvature, a refinement to the initial coarse segmentation is performed by merging segments that appear to be part of the same surface primitive but are labelled as distinct due to noise, quantization error, or other factors. The resulting regions of constant curvature are approximated with biquadric polynomial surfaces. Initial coarse segmentation patches of size **N** pixels or higher are used as seed patches for refinement of the segmentation. Quadric surface patches are then fit to each patch and the fit error is calculated. Patches with a

RESIDUAL SURFACE

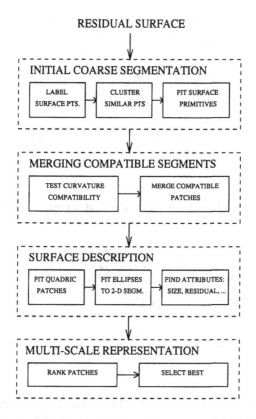

Fig. 2. Calculation of multi-scale surface description

low residual fit are selected for "growing". The patches are extrapolated to its surrounding regions and merged with a neighbouring patch if the residual of the extrapolation fit is low.

The shape of the surface patches are then abstracted, as consistent with our desire for a low-dimensional (stable) representation. The geometric size and shape of each patch is determined by fitting ellipses to the patches using a *best eigenvector fit*[7] (Figure 4).

In addition, the following descriptive properties are calculated:

- $p_{type}(S_i)$: *surface type*
- $p_{size}(S_i)$: *size*, area of the silhouette of a patch
- $p_{compact}(S_i)$: *compactness*, a measure of how close the silhouette is to a circle.
- $p_{max}(S_i)$: *maximum radius*, variance along the major axis of ellipse.
- $p_{min}(S_i)$: *minimum radius*, variance along the minor axis of the ellipse.
- $p_{elongation}(S_i)$: *elongation*, $(p_{max}(S_i)/p_{min}(S_i))$
- $p_{fit}(S_i)$: *goodness − of − fit* of patch to data.

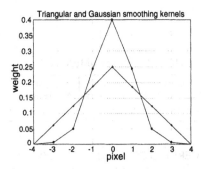

Fig. 3. Example of linear filters at a particular scale

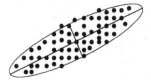

Fig. 4. Size of patch is estimated by fitting an ellipse

These properties are then used in a linear function to rank the utility and stability of patches. The most stable patches are used to describe an object of interest.

3 Experimental results

In this section we intend to demonstrate that the superquadric model is sufficient for coarsely classifying objects based on their global geometry. We will also show that a meaningful curvature-based description can be extracted from the fit residual, and that surface patches provide sufficient information to distinguish between objects in the same class.

For this paper, range data was used to test our approach. The size of the range images were 256x256 pixels and the objects were scanned from a single viewpoint using a laser range-finder attached to a Puma robot arm. Although this was adequate for these particular examples, generally speaking multiple viewpoints are desirable to ensure the stable extraction of the superquadric model. The objects were separated from the background by clipping the range data corresponding to the supporting platform. Figure 5 shows the range data used for the experiments. Three pairs of objects were used. One pair is composed of 2 faces, the second pair of jars while the third are two completely different models, a distributor cap and the top of a dome. The jars and dome provide fairly simple surfaces while the faces and distributor cap present more comples features for modelling.

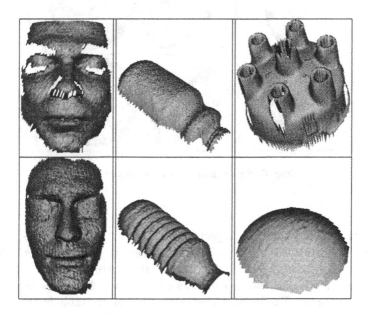

Fig. 5. Original range data of 6 individual models

Superquadrics were then fitted to the data as described in section 2.1. Figure 6 shows the range data superimposed over the fitted models. The resulting superquadric models for the faces have the same general shape but differ from the superquadric models for the jars. Table 1 lists the parameters (P_{sq}) for the superquadrics that were fitted to the data. The results show that the parameters for each pair of models, particularly the shape parameters, are similar enough that small perturbations in the data can make it difficult to distinguish between two objects belonging to the same class of models [2]. Figure 7 shows the residuals of the superquadric fit.

The surface R was then modelled with surface patches at multiple scales. Patches with a size greater that α and a mean squared error lower than ϵ were then ranked in terms of their utility and stability based on a weighted sum of the characteristics outlined in section 2.3. Specifically, this *reliability* measure is defined as follows:

$$reliability(S_i) = \omega_1 p_{scale}(S_i) + \omega_2 (p_{size}(S_i)/A)$$
$$+ \omega_3 p_{compact}(S_i) + \omega_4 p_{fit(S_i)}$$

where ω_i is a weighting vector, and α, ϵ and ω_i are experimentally determined and maintained fixed. This measure favours large, compact, low fit error patches

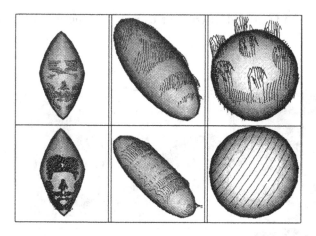

Fig. 6. Superquadric fitted to range data.

at high scale for describing an object.

Figure 8 illustrates the result of extracting surface patches at multiple scales (small to large) from the residual surface of the bottom jar in Figure 7. At a small scale, the smaller features such as the ridges and grooves encircling the bottle are apparent. The larger scales extract the more stable features and less noise but they also fail to detect the finer details. By combining these multiscalar descriptions, we can obtain a detailed description of the object. Figure 8 also demonstrates that certain features are apparent across several scales (different filter size). In this particular example the neck of the jar is extracted at all scales. Features such as these are considered valuable for performing recognition. Figure 9 shows an example of the significant patches that were extracted from the residuals shown in Figure 7 using a medium scale linear filter. At the representative scale the size, orientation and relative locations of the patches extracted differ enough to allow the objects to be readily distinguished from each other, in contrast with the superquadric alone. In the case of the faces and the jars, the surface patches allow models that belong to the same class to be

P_{sq}	Face 1	Face 2	Jar 1	Jar 2	Dist. Cap	Dome
a_x	179.8425	181.3416	133.1806	120.2001	48.9489	31.2853
a_y	92.4372	93.4064	51.9548	41.4134	48.9600	30.1849
a_z	88.6287	85.1957	46.3033	37.9726	48.5379	33.9946
ϵ_1	1.2022	1.2602	0.9820	0.9697	1.0512	0.9474
ϵ_2	1.4496	1.4951	1.0819	0.8463	1.1688	0.9986

Table 1. Parameters for fitted superquadrics

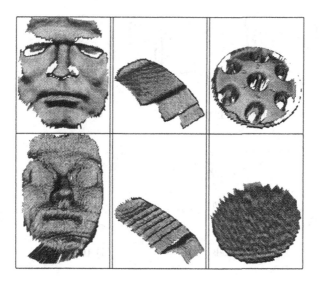

Fig. 7. Residual surface of superquadric fit.

distinguished from each other. However, in the case of the distributor cap and the dome, the surface patches prevent the two from being erroneously classified as the same object.

4 Discussion

As figure 6 shows, the superquadric models capture the canonical shape of the object. However it does not provide us with enough information for tasks such as recognition. The model is used, however, to stabilize a surfaced-based model. The patches shown in figure 9 extract enough information to differentiate one object from the other. Key characteristics of the patches that could be used for recognition are the size, orientation and location of the patches *relative* to each other since our goal is a stable description that is viewpoint independent.

We propose a recognition scheme in which superquadrics would be used to determine the class of shapes to which an object belongs. A surface description is then be used to refine the recognition process. This step would initially be performed at a coarse scale and progress to finer scales until the identity of the object is determined. Such a comparison can be performed by an algorithm such as an Interpretation Tree search [11]. In fact, our selection of significant surface patches is similar to the preprocessing used by Grimson and Lozano-Pérez to reduce the depth of the tree search.

The synthesis of superquadric and surface-based representations can be exploited in various ways. We have proposed using the superquadric as a primary indexing primitive. This, of course, presupposes that superquadric estimation

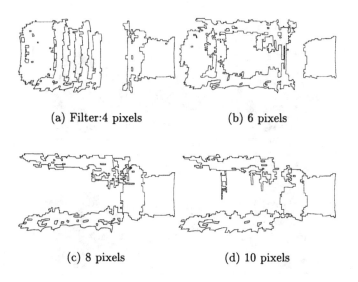

(a) Filter:4 pixels   (b) 6 pixels

(c) 8 pixels   (d) 10 pixels

Fig. 8. Surface patches extracted from jar

can be accomplished robustly. This can be achieved by acquiring multiple views of the object. An additional consideration is that the superquadric can be used for grasping or collision avoidance while a surface description may be used for tasks such as recognition.

References

1. Wassim Alami and Gregory Dudek. Multiscale object representation using surface patches. In *Proc. of SPIE - The International Society for Optical Engineering*, volume 2353, pages 108–119, Bellingham, WA USA, 1994.
2. T. Arbel, F. P. Ferrie, and P. Whaite. Recognizing volumetric objects in the presence of uncertainty. In *International Conference on Pattern Recognition*, volume 1, pages 470–476, Piscataway, NJ, USA, 1994. IEEE.
3. F. Attneave. Some informational aspects of visual perception. *Psychological Review*, (61):183–193, 1954.
4. Alan H. Barr. Superquadrics and angle preserving transformations. *IEEE Computer Graphics and Applications*, 1:11–23, Jan. 1981.
5. P. J. Besl and R. C. Jain. Segmentation through variable-order surface fitting. *IEEE Trans. on Pattern Anal. Machine Intell.*, 10(2):167–192, March 1988.
6. Paul Besl. Invariant surface characteristics for three-dimensional object recognition in range images. *IEEE Trans. Pattern Anal. Machine Intell.*, 33(1):33–80, January 1986.
7. Richard O. Duda and Peter E. Hart. *Pattern Classification and Scene Analysis*. John WIley & Sons, New York, N.Y., 1973.

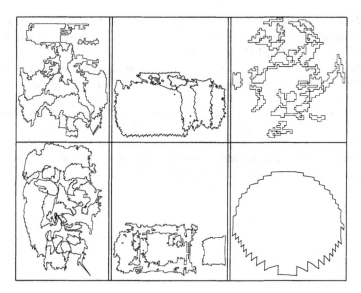

Fig. 9. Surface patches extracted from residual at a single scale

8. G. Dudek and J. K. Tsotsos. Shape representation and recognition from curvature. In *Proc. of the 1991 Conference on Computer Vision and Pattern Recognition*, pages 35–41, Maui, Hawaii, June 1991. IEEE.

9. F. P. Ferrie, J. Lagarde, and P. Whaite. Darboux frames, snakes and super-quadrics: Geometry from the bottom up. *IEEE Trans. Pattern Anal. Machine Intell.*, 15(8):771–783, Aug. 1993.

10. Dmitry B. Goldgof, Thomas S. Huang, and Hua Lee. Curvature-based approach to terrain recognition. *IEEE Trans. Pattern Anal. Machine Intell.*, 11(11):1213–1217, Nov 1989.

11. W. Eric L. Grimson and Tomás Lozano-Pérez. Localizing overlapping parts by searching the interpretation tree. *IEEE Trans. Pattern Anal. Machine Intell.*, 9(4):469–482, July 1987.

12. D. G. Lowe. *Perceptual organization and visual recognition*. Kluwer Academic Publishers, Boston, Mass., 1985.

13. David H. Marimont. A representation for image curves. *AAAI*, pages 237–242, 1984.

14. Art Matheny and Dmitry B. Goldgof. The use of three and four dimensional surface harmonics for rigid and nonrigid shape recovery and representation. *IEEE Trans. Pattern Anal. Machine Intell.*, 17(10):967–981, Oct. 1995.

15. F Mokhtarian. Evolution properties of space curves. In *Interlnational Conf. on Computer Vision*, pages 100–105, Tarpon Springs, Fla., Dec. 1988. IEEE 2nd International Conf. on Computer Vision.

16. A. P. Pentland and S. Sclaroff. Closed-form solutions for physically based shape modelling and recognition. *IEEE Trans. Pattern Anal. Machine Intell.*, 13(7):715–729, July 1991.

333

17. F. Solina and R. Bajcsy. Recovery of parametric models from range images: The case for superquadrics with global deformations. *IEEE Trans. Pattern Anal. Machine Intell.*, 12(2):131–147, Feb. 1990.

18. D. Terzopoulos and D. Metaxas. Dynamic 3d models with local and global deformations: Deformable superquadrics. *IEEE Trans. Pattern Anal. Machine Intell.*, 13(7):703–714, July 1991.

19. Peter Whaite and Frank P. Ferrie. From uncertainty to visual exploration. *IEEE Trans. Pattern Anal. Machine Intell.*, 13(10):1038–1049, Oct 1991.

20. D. Wilkes and J. Tsotsos. Active object recognition. In *Proc. of IEEE Conference on Computer Vision and Pattern Recognition*, pages 136–141, Urbana, Illinois, June 1992. IEEE Computer Society Press.

Finding Pictures of Objects in Large Collections of Images

David A. Forsyth[1], Jitendra Malik[1], Margaret M. Fleck[2], Hayit Greenspan[1,3],
Thomas Leung[1], Serge Belongie[1], Chad Carson[1], Chris Bregler[1]

[1] Computer Science Division, University of California at Berkeley, Berkeley CA 94720
[2] Dept. of Computer Science, University of Iowa, Iowa City, IA 52240
[3] Dept. of Electrical Engineering, Caltech, Pasadena CA 91125

Abstract. Retrieving images from very large collections, using image content as a key, is becoming an important problem. Users prefer to ask for pictures using notions of content that are strongly oriented to the presence of abstractly defined objects. Computer programs that implement these queries automatically are desirable, but are hard to build because conventional object recognition techniques from computer vision cannot recognize very general objects in very general contexts.

This paper describes our approach to object recognition, which is structured around a sequence of increasingly specialized grouping activities that assemble coherent regions of image that can be shown to satisfy increasingly stringent constraints. The constraints that are satisfied provide a form of object classification in quite general contexts.

This view of recognition is distinguished by: far richer involvement of early visual primitives, including color and texture; hierarchical grouping and learning strategies in the classification process; the ability to deal with rather general objects in uncontrolled configurations and contexts. We illustrate these properties with four case-studies: one demonstrating the use of color and texture descriptors; one showing how trees can be described by fusing texture and geometric properties; one learning scenery concepts using grouped features; and one showing how this view of recognition yields a program that can tell, quite accurately, whether a picture contains naked people or not.

1 Introduction

Very large collections of images are becoming common, and users have a clear preference for accessing images in these databases based on the objects that are present in them. Creating indices for these collections by hand is unlikely to be successful, because these databases can be gigantic. Furthermore, it can be very difficult to impose order on these collections. For example, the California Department of Water Resources collection contains of the order of half-a-million images; a subset of this collection can be searched at **http://elib.cs.berkeley.edu**. Another example is the collection of images available on the Internet, which is notoriously large and disorderly. This lack of structure makes it hard to rely on textual annotations in indexing - computer programs that could automatically assess image content are a more practical alternative (Sclaroff, 1995).

Another reason that manual indexing is difficult is that it can be hard to predict future content queries; for example, local political figures may reach national importance long after an image has been indexed. In a very large collection, the subsequent reindexing process becomes onerous.

Classical object recognition techniques from computer vision cannot help with this problem. Recent techniques can identify specific objects drawn from a small (of the order of 100) collection, but no present technique is effective at telling, say, people from cows, a problem usually known as *classification*. This paper presents case studies illustrating an approach to determining image content that is capable of object classification. The approach is based on constructing rich image descriptions that fuse color, texture and shape information to determine the identity of objects in the image.

1.1 Materials and Objects - "Stuff" vs "Things"

Many notions of image content have been used to organize collections of images (e.g. Layne, 1994). Relevant here are notions centered on objects; the distinction between materials - "stuff" - and objects - "things" - is particularly important. A material (e.g. skin) is defined by a homogeneous or repetitive pattern of fine-scale properties, but has no specific or distinctive spatial extent or shape. An object (e.g. a ring) has a specific size and shape. This distinction[4] and a similar distinction for actions, are well-known in linguistics and philosophy (dating back at least to Whorf, 1941) where they are used to predict differences in the behavior of nouns and verbs (e.g. Taylor, 1977;Tenney, 1987;Fleck, 1996).

To a first approximation, 3D materials appear as distinctive colors and textures in 2D images, whereas objects appear as regions with distinctive shapes. Therefore, one might attempt to identify materials using low-level image properties, and identify objects by analyzing the shape of and the relationships between 2D regions. Indeed, materials with particularly distinctive color or texture (e.g. sky) can be successfully recognized with little or no shape analysis, and objects with particularly distinctive shapes (e.g. telephones) can be recognized using only shape information.

In general, however, too much information is lost in the projection onto the 2D image for strategies that ignore useful information to be successful. The typical material, and so the typical color and texture of an object, is often helpful in separating the object from other image regions, and in recognizing it. Equally, the shapes into which it is typically formed can be useful cues in recognizing a material. For example, a number of other materials have the same color and texture as human skin, at typical image resolutions. Distinguishing these materials from skin requires using the fact that human skin typically occurs in human form.

---

[4] In computer vision, Ted Adelson has emphasized the role of filtering techniques in early vision for measuring stuff properties.

1.2 Object Recognition

Current object recognition systems represent models either as a collection of geometric measurements—typically a CAD or CAD-like model—or as a collection of images of an object. This information is then compared with image information to obtain a match. Comparisons can be scored by using a feature correspondences to backproject object features into an image. Appropriate feature correspondences can be obtained by various forms of search (for example, Huttenlocher and Ullman, 1986; Grimson and Lozano-Pérez, 1987; Lowe, 1987). A variant of this approach, due to Ullman and Basri (1991), uses correspondences to determine a new view of the object, which is defined by a series of images, and overlay that new view on the image to evaluate the comparison. Alternatively, one can define equivalence classes of features, each large enough to have distinctive properties (invariants) preserved under the imaging transformation. These invariants can then be used as an index for a model library (examples of various combinations of geometry, imaging transformations, and indexing strategies include Lamdan *et al.*, 1988; Weiss, 1988; Forsyth *et al.*, 1991; Rothwell *et al.*, 1992; Stein and Medioni, 1992; Taubin and Cooper, 1992; Liu *et al*, 1993; Kriegman and Ponce, 1994).

Each case described so far models object geometry exactly. An alternative approach, usually known as *appearance matching*, models objects by collections of images of the object in various positions and orientations and under various lighting conditions. These images are compressed, and feature vectors are obtained from the compressed images. Matches are obtained by computing a feature vector from a compressed version of the original image and then applying a minimum distance classifier (e.g. Sirovich and Kirby, 1987; Turk and Pentland, 1991; Murase and Nayar, 1995).

All of the approaches described rely heavily on specific, detailed geometry, known (or easily determined) correspondences, and either the existence of a single object on a uniform, known background (in the case of Murase and Nayar, 1995) or the prospect of relatively clear segmentation. None is competent to perform abstract classification; this emphasis appears to be related to the underlying notion of model, rather than to the relative difficulty of the classification vs. identification. Notable exceptions appear in Nevatia and Binford, 1977; Brooks, 1981; Connell, 1987; Zerroug and Nevatia, 1994, all of which attempt to code relationships between various forms of volumetric primitive, where the description is in terms of the nature of the primitives involved and of their geometric relationship.

1.3 Content Based Retrieval from Image Databases

Algorithms for retrieving information from image databases have concentrated on material-oriented queries, and have implemented these queries primarily using low-level image properties such as color and texture. Object-oriented queries search for images that contain particular objects; such queries can be seen either as constructs on material queries (Picard and Minka, 1995) as essentially textual matters (Price *et al.*, 1992), or as the proper domain of object recognition.

The best-known image database system is QBIC (Niblack *et al.*, 1993) which allows an operator to specify various properties of a desired image. The system then displays a selection of potential matches to those criteria, sorted by a score of the appropriateness of the match. The operator can adjust the scoring function. Region segmentation is largely manual, but the most recent versions of QBIC (Ashley *et al.*, 1995) contain simple automated segmentation facilities. The representations constructed are a hierarchy of oriented rectangles of fixed internal color and a set of tiles on a fixed grid, which are described by internal color and texture properties. However, neither representation allows reasoning about the shape of individual regions, about the relative positioning of regions of given colors or about the cogency of geometric coocurrence information, and so there is little reason to believe that either representation can support object queries.

Photobook (Pentland *et al.*, 1993) largely shares QBIC's model of an image as a collage of flat, homogeneous frontally presented regions, but incorporates more sophisticated representations of texture and a degree of automatic segmentation. A version of Photobook (Pentland *et al.*, 1993; p. 10) incorporates a simple notion of object queries, using plane object matching by an energy minimization strategy. However, the approach does not adequately address the range of variation in object shape and appears to require images that depict single objects on a uniform background. Further examples of systems that identify materials using low-level image properties include Virage (home page at http://www.virage.com/), Candid (home page at http://www.c3.lanl.gov/ kelly/CANDID/main.shtml and Kelly *et al.*, 1995) and Chabot (Ogle and Stonebraker, 1995). None of these systems code spatial organization in a way that supports object queries.

Variations on photobook (Picard and Minka, 1995; Minka, 1995) use a form of supervised learning known in the information retrieval community as "relevance feedback" to adjust segmentation and classification parameters for various forms of a textured region. When a user is available to tune queries, supervised learning algorithms can clearly improve performance given appropriate object and image representations. In some applications of our algorithms, however, users are unlikely to want to tune queries.

More significantly, the representations used in these supervised learning algorithms do not code spatial relationships. Thus, these algorithms are unlikely to be able to construct a broad range of effective object queries. To achieve an object-oriented query system there is a need to go to higher levels of the representation hierarchy and to encode spatial relationships using higher-level grouping features. Finally, there is a query mode that looks for images that are near iconic matches of a given image (for example, Jacobs *et al.*, 1995). This matching strategy cannot find images based on the objects present, because it is sensitive to such details as the position of the objects in the image, the composition of the background, and the configuration of the objects - for example, it could not match a front and a side view of a horse.

2 A Grouping Based Framework for Object Recognition

Our approach to object recognition is to construct a sequence of successively abstract descriptors, at an increasingly high level, through a hierarchy of grouping and learning processes. At the lowest level, grouping is based on spatiotemporal coherence of local image descriptors–color, texture, disparity, motion–with contours and junctions extracted simultaneously to organize these groupings. There is an implicit assumption in this process, that coherence of these image descriptors is correlated with the associated scene entities being part of the same surface in the scene. At the next stage, the assumptions that need to be invoked are more global (in terms of size of image region) as well as more class-specific. For example, a group that is skin-colored, has an extended bilateral image symmetry and has near parallel sides should imply a search for another such group, nearby, because it is likely to be a limb.

This approach leads to a notion of classification where object class is increasingly constrained as the recognition process proceeds. Classes need not be defined as purely geometric categories. For instance in a scene expected to contain faces, prior knowledge of the spatial configuration of eyes, mouth etc can be used to group together what might otherwise be regarded as separate entities. As a result, the grouper's activities become increasingly specialized as the object's identity emerges; constraints at higher levels are evoked by the completion of earlier stages in grouping. The particular attractions of this view are:

- that the primary activity is classification rather than identification;
- that it presents a coherent view of combining bottom-up with top-down information flow that is richer than brute search;
- and that if grouping fails at some point, it is still possible to make statements about an object's identity.

Slogans characterizing this approach are: *grouping proceeds from the local to the global*; and *grouping proceeds from invoking generic assumptions to more specific ones*. The most similar ideas in computer vision are those of a body of collaborators usually seen as centered around Binford and Nevatia (see, for example Nevatia and Binford, 1977; Brooks, 1981; Connell, 1987; Zerroug and Nevatia, 1994), and the work of Zisserman *et al.*, 1995. Where we differ is in:

1. offering a richer view of early vision, which must offer more than contours extracted by an edge detector (an approach that patently fails when one considers objects like sweaters, brick walls, or trees).
2. attributing much less importance to the recovery of generalized cylinders as the unifying theme for the recognition process.
3. attempting to combine learning with the hierarchical grouping processes.

A central notion in grouping is that of coherence, which is hard to define well but captures the idea that regions should (in some sense) "look" similar internally. Examples of coherent regions include regions of fixed color, tartan regions, and regions that are the projection of a vase. We see four major issues:

1. **Segmenting images into coherent regions based on integrated region and contour descriptors:** An important stage in identifying objects is deciding which image regions come from particular objects. This is simple when objects are made of stuff of a single, fixed color; however, most objects are covered with textured stuff, where the spatial relationships between colored patches are an important part of any description of the stuff. The content-based retrieval literature cited above contains a wide variety of examples of the usefulness of quite simple descriptions in describing images and objects. Color histograms are a particularly popular example; however, color histograms lack spatial cues, and so must confuse, for example, the English and the French flags. In what follows (Sec. 3), we show three important cases: in the first, features extracted from the orientation-histogram of the image are used for the extraction of coherent texture regions. This allows distinctions between uniform background and textured objects, and leads to higher-level information which can guide the recognition task. In the second, the observation that a region of stuff is due to the periodic repetition of a simple tile yields information about the original tile, and the repetition process. Such periodic textures are common in real pictures, and the spatial structure of the texture is important in describing them. Finally, measurements of the size and number of small blobs of color yield information about stuff regions - such as fields of flowers - that cannot be obtained from color histograms alone.

2. **Fusing color, texture and shape information to describe primitives:** Once regions that are composed of internally coherent stuff have been identified, 2D and 3D shape properties of the regions need to be incorporated into the region description. In many cases, objects either belong to constrained classes of 3D shapes - for example, many trees can be modeled as surfaces of revolution - or consist of assemblies of such classes - for example, people and many animals can be modeled as assemblies of cylinders. It is often possible to tell from region properties alone whether the region is likely to have come from a constrained class of shapes (eg Zisserman *et al.*, 1995); knowing the class of shape from which a region came allows other inferences. As we show in Sec. 4, knowing that a tree can be modeled as a surface of revolution simplifies marking the boundary of the tree, and makes it possible to compute an axis and a description of the tree.

3. **Learning as a methodology for developing the relationship between object classes and color, texture and shape descriptors** Given the color, texture and shape descriptors for a set of labeled objects, one can use machine learning techniques to train a classifier. In section 5, we show results obtained using a decision tree classifier that was trained to distinguish among a number of visual concepts that are common in our image database. A novel aspect of this work is the use of grouping as part of the process of constructing the descriptors, instead of using simple pixel-level feature vectors. Interestingly, the output of a classifier can itself be used to guide higher level grouping. While this work is preliminary, it does suggest a way

to make less tedious the processes of acquiring object models and developing class-based grouping strategies.

4. **Classifying objects based on primitive descriptions and relationships between primitives:** Once regions have been described as primitives, the relationships between primitives become important. For example, finding people or animals in images is essentially a process of finding regions corresponding to segments and then assembling those segments into limbs and girdles. This process involves exploring incidence relationships, and is constrained by the kinematics of humans and animals. We have demonstrated the power of this constraint-based representation by building a system that can tell quite reliably whether an image contains naked people or not, which is briefly described in Sec. 6.

3   Case Study 1: Color and Texture Properties of Regions

Color and texture are two important low-level features in the initial representation of the input image; as such they form the initial phase of the grouping framework. Texture is a well-researched property of image regions, and many texture descriptors have been proposed, including multi-orientation filter banks (e.g. Malik and Perona, 1990; Greenspan *et al.*,1994), the second-moment matrix (Förstner, 1993; Gårding and Lindeberg, 1995), and orientation histograms (Freeman and Roth, 1995). We will not elaborate here on some of the more classical approaches to texture segmentation and classification- both of which are challenging and well-studied tasks. Rather, we want to introduce several new perspectives related to texture descriptors and texture grouping which were motivated from the content-based retrieval task; and which we believe present new problems in the field.

The first task that we present is that of identifying regions of uniform intensity vs. regions that are textured. This categorization enables the extraction of foreground vs. background regions in the image, guiding the search for objects in the scene. In addition, distinguishing among texture patterns which are singly-oriented, multiply-oriented, or which are stochastic in nature, can allow for further categorization of the scene and for the extraction of higher-level features to aid the recognition process (e.g. single-oriented flow is a strong characteristic of water waves, grass is stochastic etc). Finally, boundaries between a textured region and the background, or between differing texture segments, are an additional important feature which can facilitate the extraction of contour descriptors.

A view unifying region finding with contour extraction can be facilitated by extracting informative features from the orientation histogram of the gradient image. One such feature is a 180° normalized cross-correlation measure of the orientation histogram. An *edge*, which separates two uniform-intensity regions, is characterized by a single dominant orientation in the gradient image. Its cross-correlation figure will correspondingly be close to zero. A *bar*, on the other hand, can be thought of as the basic texture unit, and is characterized by its gradient

Background   Textured region

Fig. 1. *An example of non-textured vs. textured region categorization. The categorization is based on orientation histogram analysis of overlapping local windows. A non-textured window is characterized by a low DC component of the histogram. A window is labeled as textured based on a strong response to a 180°-shift cross-correlation of the orientation histogram. The distinction into the two categories in this case allows for an important subdivision of the input image into the non-textured sky region and the textured city.*

map having two dominant orientations which are 180° phase shift apart. The normalized correlation figure is correspondingly close to one. Fitting the cyclic orientation histograms enables the extraction of additional informative features, such as relative peak energy as well as the corresponding peak angular localization. Finally, a frequency-domain analysis of the histogram harmonics can provide further region characterization.

In the following, features are extracted, characterizing the orientation histograms of two image resolutions ($8 * 8$ and $4 * 4$ overlapping windows). The combination of these features provide us with feature-vectors from which the desired categorization is enabled. Figs. 1 and 2 display preliminary results of the textured-region analysis.

A second problem of interest is the detection of periodic repetition of a basic tile, as a means for region grouping (Leung and Malik, 1996). Such regions can be described by a representation which characterizes the individual basic element, and then represents the spatial relationships between these elements. Spatial relationships are represented by a graph where nodes correspond to individual elements and arcs join spatially neighboring elements. With each arc r_{ij} is associated an affine map A_{ij} that best transforms the image patch $I(x_i)$ to $I(x_j)$. This affine transform implicitly defines a correspondence between points on the image patches at x_i and x_j.

Regions of periodic texture can be detected and described by:

- detecting "interesting" elements in the image;
- matching elements with their neighbors and estimating the affine transform between them;

– growing the element to form a more distinctive unit;

– and grouping the elements.

The approach is analogous to tracking in video sequences; an element is "tracked" to spatially neighboring locations in one image, rather than from frame to frame. Interesting elements are detected by breaking an image into overlapping windows and computing the second moment matrix (as in Förstner, 1993; Gårding and Lindeberg, 1995), which indicates whether there is much spatial variation in a window, and whether that variation is intrinsically one- or two-dimensional. By summing along the dominant direction, "flow" regions, such as vertical stripes along a shirt, can be distinguished from edges. Once regions have been classified, they can be matched to regions of the same type.

An affine transform is estimated to bring potential matches into registration, and the matches are scored by an estimate of the relative difference in intensity of the registered patches. The output of this procedure is a list of elements which form units for repeating structures in the image. Associated with each element is the neighboring patches which match well with the element, together with the affine transform relating them. These affine transforms contain shape cues, as well as grouping cues (Malik and Rosenholtz, 1994).

The final step is to group the elements together by a region growing technique. For each of the 8 windows neighboring an element, the patch which matches the element best, and the affine transform between them is computed. Two patches are grouped together by comparing the error between an element and its neighboring patch with the variation in the element. Of course, as the growth procedure propagates outward, the size and shape of the basic element in the image will change because of the slanting of the surface. An example of repetitive tile grouping is presented in Fig. 3. A more elaborate description of this work can be found in (Leung and Malik, 1996).

Of-course, texture need not be studied purely as a gray-scale phenomenon. Many interesting textures, such as fields of flowers, consist of a representative spatial distribution of *colored* elements. Color is yet another important cue in extracting information from images. Color histograms have proven a useful stuff query, but are poor at, for example, distinguishing between fields of flowers and a single large flower, because they lack information as to how the color is distributed spatially. This example indicates the importance of fusing color with textural properties. The size and spatial distribution of blobs of color is a natural first step in such a fusion. It is also a natural stuff description - and hence, query - which is particularly useful for outdoor scenes in the case of hues ranging from red to yellow. We achieve this query by the following method:

– forming hue, saturation, and value (HSV) channels;

– coarsely requantizing these channels for various colors to form color maps, where an orange color map would reflect those pixels which fall within a certain range around orange in HSV space;

– forming a Gaussian pyramid (after Burt & Adelson, 1983) for each color map;

- filtering each level of the pyramid with a center-surround "dot" filter and several oriented "line" filters (all zero-mean);
- thresholding the filter outputs and counting the number of distinct responses to a particular filter.

Responses at a coarse scale indicate large blobs of the appropriate color; responses at finer scales indicate smaller blobs. The number of blobs at each scale and orientation for each color is returned. As Figs. 4 and 5 show, queries composed of a combination of this information with textual cues, or with an estimate of a horizon, correlate extremely strongly with content in the present Cypress database. This query engine is available on the World Wide Web, at http://elib.cs.berkeley.edu/cypress.

Fig. 2. *Detection of non-textured regions, textured regions and boundary elements on a Cheeta image. Local windows, of size 4 * 4 are categorized into the 3 classes. Boundary elements correspond to both intensity-edges as well as textured-edges. Windows are labeled as boundary if they have a low histogram cross-correlation figure in both resolutions of analysis. We note the importance of detecting the no-texture region as a step which enables to focus further attention to the more-interesting, textured regions of the image - in this example, focusing the attention on the animal figure. The textured region extracts the entire cheeta, as well as the grass. Further, more detailed investigation on the extracted textured regions (using blob-finding or repetitive-pattern region-growing, for example, both schemes which will be described below), will enable a refined distinction between the cheeta and its surrounding. We note that the extraction of boundary information unified with textured-region information, helps eliminate the confusion between the true animal boundary and the many edges which exist within the cheeta's textured body (classic edge-finding schemes will detect all the circular edges as well).*

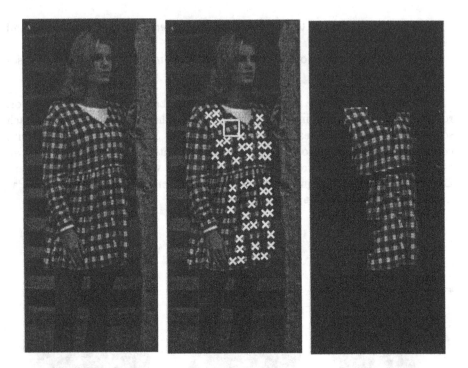

Fig. 3. *A textile image. The original image is shown on the left, and the center image shows the initial patches found. The crosses are the locations of units grouped together. The image on the right shows the segmented region is displayed. Notice that the rectangle includes two units in the actual pattern. This is due to the inherent ambiguity in defining a repeating unit - 2 tiles together still repeat to form a pattern.*

Fig. 4. *Querying the Cypress database for images that contain a large proportion of yellow pixels produces a collection of responses that is eclectic in content; there is little connection between the response to this query and particular objects. While these queries can be useful, particularly when combined with text information, they are not really concept or "thing" queries.*

Fig. 5. *Querying the Cypress database for images that contain a large number of small yellow blobs and a horizon yields scenic views of fields of flowers. The horizon is obtained by searching in from each boundary of the image for a blue region, extending to the boundary, that does not curve very sharply. In this case, the combination of spatial and color queries yields a query that encapsulates content surprisingly well. While the correlation between object type and query is fortuitous, and relevant only in the context of the particular database, it is clear that the combination of spatial and chromatic information in the query yields a more powerful content query than color alone. In particular, the language of blobs is a powerful and useful early cue to content. Note that the ordering of the images in response to the query (as presented in this figure) is arbitrary. No relative ranking is performed.*

4 Case Study 2: Fusing Texture and Geometry to Represent Trees

Generic grouping as studied in the previous subsection can only go so far; approaches can be made more powerful by considering classes of objects. We study trees as an interesting class. Recognizing individual trees makes no sense; instead it is necessary to define a representation with the following properties:

- It should not change significantly over the likely views of the tree.
- It should make visual similarities and visual differences between trees apparent. In particular, it should be possible to classify trees into intuitively meaningful types using this representation.
- It should be possible to determine that a tree is present in an image, segment it, and recover the representation without knowing what tree is present.

Trees can then be classified according to whether the representations are similar or not.

Branch length and orientation appear to be significant components of such a representation. Since trees are typically viewed frontally, with their trunks aligned with the image edges, and at a sufficient distance for a scaled affine viewing model to be satisfactory, it is tempting to model a tree as a plane texture. There are two reasons not to do so: considering a tree as a surface of revolution provides grouping cues; and there is a reasonable chance of estimating parameters of the distribution of branches in 3D. Instead, we model a tree as a volume with a rotational symmetry with branches and leaves embedded in it. Because of the viewing conditions, the image of a tree corresponding to this model will have a bilateral symmetry about a vertical axis, a special case of the planar harmonic homology of (Mukherjee *et al.*, 1995). This axis provides part of a coordinate system in which the representation can be computed. The other is provided by the outline of the tree, which establishes scale and translation along the axis and scale perpendicular to the axis. A representation computed in this coordinate system will be viewpoint stable for the viewpoints described.

Assuming that the axis and outline have been marked, the orientation representation is obtained by forming the response of filters tuned to a range of orientations. These response strengths are summed along the axis at each orientation and for a range of steps in distance perpendicular to the axis, relative to width. The representation resulting from this process (which is illustrated in Fig. 6) consists of a map of summed strength of response relative to orientation and distance from the axis. As the figure shows, this representation makes a range of important differences between trees explicit. Trees that have a strongly preferred branch orientation (such as the pine trees) show a strong narrow peak in the representation at the appropriate orientation; trees, such as monkey puzzle trees, which have a relatively broad range of orientations of branches show broader peaks in the representation. Furthermore, the representation distinguishes effectively between trees that are relatively translucent - such as the monkey puzzle - and those that are relatively opaque.

An axis and an outline are important to forming the representation. Both can be found by exploiting the viewing assumptions, known constraints on the geometry of volumetric primitives, and the assumed textural coherence of the branches. Figure 7 illustrates the axis finding procedure, and figure 8 shows how the outline follows from the axis.

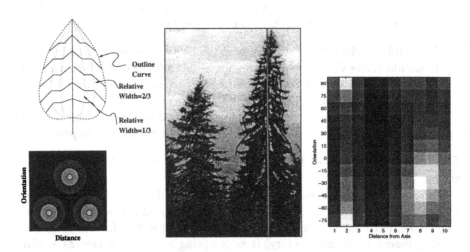

Fig. 6. *The orientation representation is obtained by computing the strength of response at various orientations with respect to the axis, at a range of perpendicular distances to the axis. These distances are measured relative to the width of the outline at that point, and so are viewpoint stable. Responses at a particular orientation and a particular distance are summed along the height of the outline. The figure on the left illustrates the process; the representation has three clear peaks corresponding to the three branch orientations taken by the (bizarre!) illustrative tree. The image on the extreme right shows the representation extracted for the tree in the center image. In our display of the orientation representation, brighter pixels correspond to stronger responses; the horizontal direction is distance perpendicular to the tree axis relative to the width of the tree at the relevant point, with points on the tree axis at the extreme left; the vertical direction is orientation (which wraps around). In the given case, there is a sharp peak in response close to the axis and oriented vertically, which indicates that the trunk of the tree is largely visible. A second peak oriented at about 30^0 and some distance out indicates a preferred direction for the tree branches.*

5   Case Study 3: Learning Scenery Concepts Using Grouped Features

The previous case study demonstrated the power of a hand-crafted grouping strategy for an important class of objects– trees. However a legitimate concern might be that to generalize this approach would be quite cumbersome– does one hand-craft groupers for trees, buildings, roads, chairs? Our view is that given the

Fig. 7. *The viewing assumptions mean that trees have vertical axes and a reflectional symmetry about the axis. This symmetry can be employed to determine the axis by voting on its horizontal translation using locally symmetric pairs of orientation responses and a Hough transform. Left: The symmetry axis superimposed on a typical image, showing also the regions that vote for the symmetry axis depicted. Right: In this image, there are several false axes generated by symmetric arrangements of trees; these could be pruned by noticing that the orientation response close to the axis is small.*

appropriate set of visual primitives, a suite of grouping strategies and classified examples, it should be possible to use machine learning techniques to aid this process.

As a demonstration of this idea, we have implemented a simple system for automatic image annotation using images from the DWR image database. The system is capable of detecting concepts such as sky, water and man-made structure in color images. Our approach begins with an early-visual processing step to obtain color and texture information and then proceeds with a number of parallel grouping strategies. The grouping strategies seek to combine the results of the first processing step in a manner which lends itself to the task of classification. For example, a region of an image which is (1) coherent with respect to its light blue color and lack of texture, (2) is located in the upper half of the image and (3) is elongated horizontally suggests the presence of sky. The classification of concepts based on grouped features is accomplished by means of a decision tree, which was learned using C4.5 (Quinlan, 1993).

Figure 9 illustrates the first step of our approach, wherein the image is decomposed into a number of binary color/texture 'separations.' The separation images are then fed to three parallel grouping strategies. Each grouping strategy attempts to specify regions in a binary image which are coherent with respect to

Fig. 8. *Once the axis is known, the outline can be constructed by taking a canonical horizontal cross-section and scaling other cross-sections to find the width that yields a cross-section that is most similar. Left: An outline and axis superimposed on a typical image. Center: The cross-sections that make up the outline, superimposed on an image of the tree. Right: The strategy fails for trees that are poorly represented by orientations alone, as in this case, as the comparisons between horizontal slices are inaccurate. Representing this tree accurately requires using filters that respond to blobs as well; such a representation would also generate an improved segmentation.*

one of the following rules: (1) solid contiguity, (2) similarity in local orientation and (3) similarity in diffuseness. The results of the three grouping strategies applied to the yellow, green, light blue and 'rough' separation images are shown in Figure 10.

Each blob is represented by a feature vector containing its area, coordinates in the image, eccentricity, principle orientation, mean saturation and mean intensity, as well as the color/texture separation and grouping strategy which gave rise to it. These feature vectors are the input to a decision tree classifier. The decision tree attempts to assign a label to each blob according to these characteristics. The class confusion matrix obtained in our experiments is shown in Figure 12. The performance of the system as summarized in the confusion matrix was obtained using 10-fold cross validation. For the most part, the confusion matrix contains large diagonal entries, indicating a tendency to classify concepts correctly. Notice that when errors do occur, the incorrectly chosen class tends to share some salient characteristics with the correct class (e.g. tree and vegetation).

While the current system's use of shape (i.e. area and principle axes) is somewhat primitive, there is a natural way to proceed within the same general framework. The additional use of symmetry features and repeating patterns, for example, promises to extend the capabilities of the system beyond simple blob-

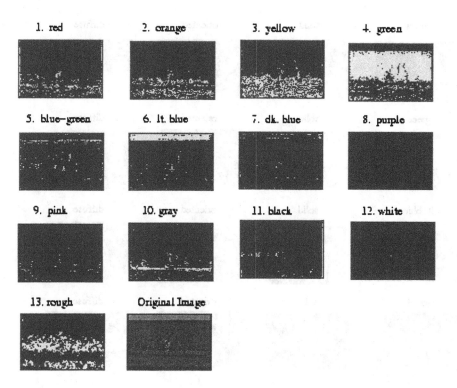

Fig. 9. *Illustrating the color/texture separations for a test image. Separations 1-9 were formed strictly by gating the hue; e.g., a hue in the interval [.07, .1) is labeled as 'orange.' Separations 10-12 each made use of saturation and/or intensity. Lastly, separation 13, 'rough,' made use of the eigenvalues of the windowed-image second moment matrix computed for the intensity component of the original image.*

like scenery concept detection. An interesting question to address is, given the component features needed to detect a tree in an image, can the performance of the hand-crafted tree recognizer of case study 2 be matched by a special instance of a general grouping-based concept learner?

In future work, we intend to answer this question by augmenting the toolbox of early vision descriptors and by adding more grouping strategies. We also intend to investigate additional learning strategies such as Bayes' nets and decision graphs for improved handling of spatial relationships between simpler concepts.

6 Case Study 4: Fusing Color, Texture and Geometry to Find People and Animals

A variety of systems have been developed specifically for recognizing people or human faces. There are several domain specific constraints in recognizing humans and animals: humans and (many!) animals are made out of parts whose shape is relatively simple; there are few ways to assemble these parts; the kinematics of

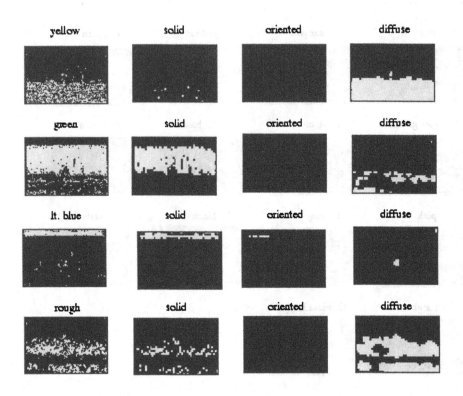

Fig. 10. *The results of the three grouping strategies for four of the color/texture bin images from the preceding figure. For example, the top row indicates that nearly all of the pixels in the 'yellow' separation were accounted for as a 'diffuse' region. Since there was no strongly oriented structure in the original image underlying the 'yellow' separation, no pixels were labeled as 'oriented.'*

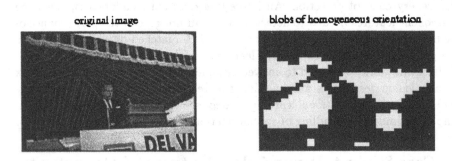

Fig. 11. *The homogeneously oriented regions in the striped canopy in the image on the left show up as distinct blobs in the result of the second grouping strategy, shown on the right. (The color/texture separation was equal to one everywhere, thus defining the entire image as a potential area of interest.)*

354

(a)	(b)	(c)	(d)	(e)	(f)	(g)	(h)	(i)	(j)	(k)	(l)	<-classified as
5				1	2							(a) cloud
	9	1		1			2					(b) dirt
		12						1				(c) flower
										8		(d) lawn
	1			27			2		1	1		(e) manmade
1					28				1	1		(f) sky
						6		1		1		(g) snow
							13					(h) tan-ground
		1						8		2	1	(i) tarmac
1	1								11	8		(j) tree
		1								36		(k) vegetation
									1	1	10	(l) water

Fig. 12. *The class confusion matrix. Off-diagonal entries correspond to misclassifications. Notice that all 8 of the lawn examples were misclassified as vegetation. Similarly, 8 out of 21 trees were misclassified as vegetation. This suggests that our features were not rich enough to discriminate these two classes. Notice also that 2 out of 8 cloud examples were misclassified as sky, probably because of noise in the training data.*

the assembly ensures that many configurations of these parts are impossible; and, when one can measure motion, the dynamics of these parts are limited, too. Most previous work on finding people emphasizes motion, but face finding from static images is an established problem. The main features on a human face appear in much the same form in most images, enabling techniques based on principal component analysis or neural networks proposed by, for example, Pentland *et al.*, 1994; Sung and Poggio, 1994; Rowley *et al.*, 1996; Burel and Carel, 1994. Face finding based on affine covariant geometric constraints is presented by Leung *et al.*, 1995.

However, segmentation remains a problem; clothed people are hard to segment, because clothing is often marked with complex colored patterns, and most animals are textured in a way that is intended to confound segmentation. Attempting to classify images based on whether they contain naked people or not provides a useful special case that emphasizes the structural representation over segmentation, because naked people display a very limited range of colors and are untextured. Our system (Fleck *et al.*, 1996) for telling whether an image contains naked people:

– first locates images containing large areas of skin-colored region;
– then, within these areas, finds elongated regions and groups them into possible human limbs and connected groups of limbs.

Images containing sufficiently large skin-colored groups of possible limbs are reported as potentially containing naked people. No pose estimation, back-projection or verification is performed.

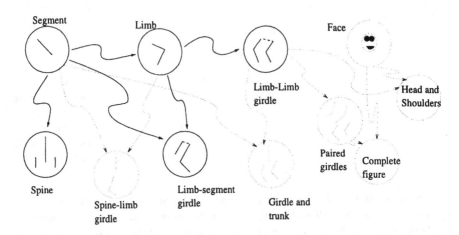

Fig. 13. *The grouping rules (arrows) specify how to assemble simple groups (e.g. body segments) into complex groups (e.g. limb-segment girdles). These rules incorporate constraints on the relative positions of 2D features, induced by geometric and kinematic constraints on 3D body parts. Dashed lines indicate grouping rules that are not yet implemented. Notice that this representation of human structure emphasizes grouping and assembly, but can be comprehensive.*

Marking skin involves stuff processing; skin regions lack texture, and have a limited range of hues and saturations. To render processing invariant to changes in overall light level, images are transformed into a log-opponent representation, and smoothed texture and color planes are extracted. To compute texture amplitude, the intensity image is smoothed with a median filter; the result is subtracted from the original image, and the absolute values of these differences are run through a second median filter. The texture amplitude and the smoothed $R-G$ and $B-Y$ values are used to mark as probably skin all pixels whose texture amplitude is no larger than a threshold, and whose hue and saturation lie in a fixed region. The skin regions are cleaned up and enlarged slightly, to accommodate possible desaturated regions adjacent to the marked regions. If the marked regions cover at least 30% of the image area, the image will be referred for geometric processing.

The input to the geometric grouping algorithm is a set of images, in which the skin filter has marked areas identified as human skin. Sheffield's implementation of Canny's (1986) edge detector, with relatively high smoothing and contrast thresholds, is applied to these skin areas to obtain a set of connected edge curves. Pairs of edge points with a near-parallel local symmetry (as in Brady and Asada, 1984) and no other edges between them are found by a straightforward algorithm. Sets of points forming regions with roughly straight axes ("ribbons"; Brooks, 1981) are found using a Hough transformation.

Grouping proceeds by first identifying potential segment outlines, where a segment outline is a ribbon with a straight axis and relatively small variation in average width. Pairs of ribbons whose ends lie close together, and whose cross-sections are similar in length, are grouped together to make limbs. The grouper then proceeds to assemble limbs and segments into putative girdles. It has grouping procedures for two classes of girdle; one formed by two limbs, and one formed by one limb, and a segment. The latter case is important when one limb segment is hidden by occlusion or by cropping. The constraints associated with these girdles use the same form of interval-based reasoning as used for assembling limbs. Finally, the grouper can form spine-thigh groups from two segments serving as upper thighs, and a third, which serves as a trunk.

In its primary configuration, the system uses the presence of either form of girdle group or of a spine-thigh group to assert that a naked human figure is present in the image. This yields a system that is surprisingly accurate for so abstract a query; as figure 14 shows, using different groups as a diagnostic for the presence of a person indicates a significant trend. The selectivity of the system increases, and the recall decreases, as the geometric complexity of the groups required to identify a person increases, suggesting that our representation used in the present implementation omits a number of important geometric structures and that the presence of a sufficiently complex geometric group is an excellent guide to the presence of an object.

7 Conclusion

Object models quite different from those commonly used in computer vision offer the prospect of effective recognition systems that can work in quite general environments. The primary focus is on *classification* instead of *identification*. The central process is that of hierarchical grouping. Initially, the grouping is based on quite local (short range in the image) measurements of color and texture coherence; as it proceeds more global and more specific models, e.g. surfaces of revolution, are invoked. In this approach, the object database is modeled as a loosely coordinated collection of detection and grouping rules. An object is recognized if a suitable group can be built. Grouping rules incorporate both surface properties (color and texture) and shape information. This type of model gracefully handles objects whose precise geometry is extremely variable, where the identification of the object depends heavily on non-geometrical cues (e.g. color and texture) and on the interrelationships between parts. Learning can be incorporated into the framework as a convenient way of associating object class labels with color, texture and shape descriptors.

We demonstrated the paradigm with four case studies that are prototype implementations of modules of such a grouping based recognition system.

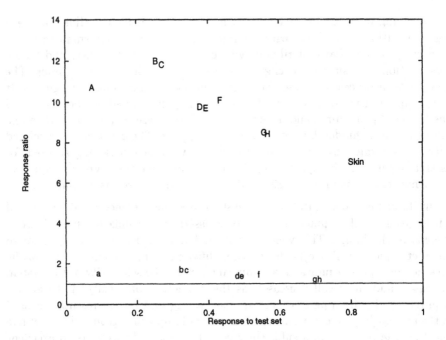

Fig. 14. *The response ratio, (percent incoming test images marked/percent incoming control images marked), plotted against the percentage of test images marked, for various configurations of the naked people finder. Labels "A" through "H" indicate the performance of the entire system of skin filter and geometrical grouper together, where "F" is the primary configuration of the grouper. The label "skin" shows the performance of the skin filter alone. The labels "a" through "h" indicate the response ratio for the corresponding configurations of the grouper, where "f" is again the primary configuration of the grouper; because this number is always greater than one, the grouper always increases the selectivity of the overall system. The cases differ by the type of group required to assert that a naked person is present. The horizontal line shows response ratio one, which would be achieved by chance. While the grouper's selectivity is less than that of the skin filter, it improves the selectivity of the system considerably. There is an important trend here; the response ratio increases, and the recall decreases, as the geometric complexity of the groups required to identify a person increases. This suggests (1) that the presence of a sufficiently complex geometric group is an excellent guided to the presence of an object (2) that our representation used in the present implementation omits a number of important geometric structures. **Key:** A: limb-limb girdles; B: limb-segment girdles; C: limb-limb girdles or limb-segment girdles; D: spines; E: limb-limb girdles or spines; F: (two cases) limb-segment girdles or spines and limb-limb girdles, limb-segment girdles or spines; G, H each represent four cases, where a human is declared present if a limb group or some other group is found.*

Acknowledgments

We would like to thank R. Blasi and K. Murphy who collaborated with S. Belongie in the work on learning decision trees for visual concept classification. We thank Joe Mundy for suggesting that the response of a grouper may indicate the presence of an object. Aspects of this research were supported by the National Science Foundation under grants IRI-9209728, IRI-9420716, IRI-9501493, an NSF Young Investigator award, an NSF Digital Library award IRI-9411334, an instrumentation award CDA-9121985, and by a Berkeley Fellowship.

References

1. Ashley, J., Barber, R., Flickner, M.D., Hafner, J.L., Lee, D., Niblack, W. and Petkovich, D. (1995) "Automatic and semiautomatic methods for image annotation and retrieval in QBIC," *SPIE Proc. Storage and Retrieval for Image and Video Databases III*, 24-35.

2. Belongie, S., Blasi, R., and Murphy, K. (1996) "Grouping of Color and Texture Features for Automated Image Annotation," Technical Report for CS280, University of California Berkeley.

3. Brady, J.M. and Asada, H. (1984) "Smoothed Local Symmetries and Their Implementation," *Int. J. Robotics Res.* 3/3, 36-61.

4. Brooks, R.A. (1981) "Symbolic Reasoning among 3-D Models and 2-D Images," *Artificial Intelligence* 17, pp. 285-348.

5. Burel, G., and Carel, D. (1994) "Detecting and localization of face on digital images" *Pattern Recognition Letters 15* pp 963-967.

6. Burt, P.J., and Adelson, E.H., (1983) "The Laplacian Pyramid as a Compact Image Code," *IEEE Trans. on Communications*, vol. com-31, no. 4.

7. Canny, J.F. (1986) "A Computational Approach to Edge Detection," *IEEE Patt. Anal. Mach. Int.* 8/6, pp. 679-698.

8. Connell, J.H., and Brady, J.M. (1987) "Generating and Generalizing Models of Visual Objects," *Artificial Intelligence*, **31**, 2, 159-183.

9. Fleck, Margaret M. (1996) "The Topology of Boundaries," in press, *Artificial Intelligence*.

10. Fleck, M.M., Forsyth, D.A., and Bregler, C. (1996) "Finding Naked People," *Fourth European Conference on Computer Vision*, Cambridge, UK, Vol 2, pp. 593-602.

11. Förstner, W. (1993) Chapter 16, in Haralick, R. and Shapiro, L. *Computer and Robot Vision*, Addison-Wesley.

12. Forsyth, D.A., Mundy, J.L., Zisserman, A.P., Heller, A., Coehlo, C., and Rothwell, C.A. (1991) "Invariant Descriptors for 3D Recognition and Pose," *IEEE Trans. Patt. Anal. and Mach. Intelligence*, **13**, 10.

13. Freeman, W., and Roth, M. (1995) Orientation histograms for hand gesture recognition. *International Workshop on Automatic Face- and Gesture-Recognition*.

14. Garding, J., and Lindeberg, T. (1996) Direct computation of shape cues using scale-adapted spatial derivative operators. *Int. J. of Computer Vision*, 17, February 1996.

15. Greenspan, H., Goodman R., Chellappa, R., and Anderson, S. (1994) "Learning Texture Discrimination Rules in a Multiresolution System," in the special issue on "Learning in Computer Vision" of the *IEEE Transactions on Pattern Analysis and Machine Intelligence (PAMI)*, Vol. 16, No. 9, 894-901.

16. Greenspan, H., Belongie, S., Perona, P., and Goodman, R. (1994) "Rotation Invariant Texture Recognition Using a Steerable Pyramid," *12th International Conference on Pattern Recognition (ICPR)*, Jerusalem, Israel.

17. Grimson, W.E.L. and Lozano-Pérez, T. (1987) "Localising overlapping parts by searching the interpretation tree", *PAMI*, **9**, 469-482.

18. Huttenlocher, D.P. and Ullman, S. (1986) "Object recognition using alignment," *Proc. ICCV-1*, 102-111.

19. Jacobs, C.E., Finkelstein, A., and Salesin, D.H. (1995) "Fast Multiresolution Image Querying," *Proc SIGGRAPH-95*, 277-285.

20. Kelly, P.M., Cannon, M., Hush, D.R. (1995) "Query by image example: the comparison algorithm for navigating digital image databases (CANDID) approach," *SPIE Proc. Storage and Retrieval for Image and Video Databases III*, 238-249.

21. Kriegman, D. and Ponce, J. (1994) "Representations for recognising complex curved 3D objects," *Proc. International NSF-ARPA workshop on object representation in computer vision*, LNCS-994, 89-100.

22. Lamdan, Y., Schwartz, J.T. and Wolfson, H.J. (1988) "Object Recognition by Affine Invariant Matching," Proceedings CVPR, p.335-344.

23. Layne, S.S. (1994) "Some issues in the indexing of images," *J. Am. Soc. Information Science*, **45**, 8, 583-588.

24. Leung, T.K., Burl, M.C., Perona, P. (1995) "Finding faces in cluttered scenes using random labelled graph matching, " *International Conference on Computer Vision* pp 637-644.

25. Leung, T.K., and Malik, J., "Detecting, localizing and grouping repeated scene elements from an image," (1996) *Fourth European Conference on Computer Vision*, Cambridge, UK, Vol 1, pp. 546-555.

26. Liu, J., Mundy, J.L., Forsyth, D.A., Zisserman, A.P., and Rothwell, C.A. (1993) "Efficient Recognition of rotationally symmetric surfaces and straight homogenous generalized cylinders," *IEEE Conference on Computer Vision and Pattern Recognition '93*.

27. Lowe, David G. (1987) "The Viewpoint Consistency Constraint," *Intern. J. of Comp. Vis*, 1/1, pp. 57-72.

28. Malik, J., and Perona, P. (1990) "Preattentive texture discrimination with early vision mechanisms," *J. Opt. Soc. Am. A*, 7(5):923-932.

29. Malik, J., and Rosenholtz, R. (1994) "Recovering surface curvature and orientation from texture distortion: a least squares algorithm and sensitivity analysis," *Proc. of Third European Conf. on Computer Vision*, Stockholm, published as J.O. Eklundh (ed.) LNCS **800**, Springer Verlag, pp. 353-364.

30. Minka, T. (1995) "An image database browser that learns from user interaction," MIT media lab TR 365.

31. Mukherjee, D.P., Zisserman, A., and Brady, J.M. (1995) "Shape from symmetry - detecting and exploiting symmetry in affine images," *Proc. Roy. Soc.*, **351**, 77-106.

32. Murase, H. and Nayar, S.K. (1995) "Visual learning and recognition of 3D objects from appearance," *Int. J. Computer Vision*, **14**, 1, 5-24.

33. Nevatia, R. and Binford, T.O. (1977) "Description and recognition of curved objects," *Artificial Intelligence*, **8**, 77-98, 1977

34. Niblack, W., Barber, R, Equitz, W., Flickner, M., Glasman, E., Petkovic, D., and Yanker, P. (1993) "The QBIC project: querying images by content using colour, texture and shape," *IS and T/SPIE 1993 Intern. Symp. Electr. Imaging: Science and Technology, Conference 1908, Storage and Retrieval for Image and Video Databases*.

35. Ogle, Virginia E. and Michael Stonebraker (1995) "Chabot: Retrieval from a Relational Database of Images," *Computer* 28/9, pp. 40–48.

36. Pentland A., Moghaddam, B., Starner T., (1994) "View-based and modular eigenspaces for face recognition," *Computer Vision and Pattern Recognition*, pp 84-91.

37. Pentland, A., Picard, R.W., and Sclaroff, S. (1993) "Photobook: content-based manipulation of image databases," MIT Media Lab Perceptual Computing TR No. 255.

38. Picard, R.W. and Minka, T. (1995) "Vision texture for annotation," *J. Multimedia systems*, **3**, 3-14.

39. Polana, R., Nelon, R. (1993) "Detecting Activities" *Computer Vision and Pattern Recognition* pp 2-13.

40. Price, R., Chua, T.-S., Al-Hawamdeh, S. (1992) "Applying relevance feedback to a photo-archival system," *J. Information Sci.*, **18**, 203-215.

41. J. R. Quinlan, *C4.5 Programs for Machine Learning*, Morgan Kauffman, 1993.

42. Rothwell, C.A., Zisserman, A., Mundy, J.L., and Forsyth, D.A. (1992) "Efficient Model Library Access by Projectively Invariant Indexing Functions," *Computer Vision and Pattern Recognition 92*, 109-114.

43. Rowley, H., Baluja, S., Kanade, T. (1996) "Human Face Detection in Visual Scenes" *NIPS*, volume 8, 1996.

44. Sclaroff, S. (1995) "World wide web image search engines," Boston University Computer Science Dept TR95-016.

45. Sirovitch, L. and Kirby, M., "Low-dimensional procedure for the characterization of human faces," *J. Opt. Soc. America A*, **2**, 519-524, 1987.

46. Stein, F. and Medioni, G. (1992) "Structural indexing: efficient 3D object recognition," *PAMI-14*, 125-145.

47. Sung, K.K, Poggio, T., (1994) "Example-based Learning from View-based Human Face Detection" MIT A.I. Lab Memo No. 1521.

48. Taubin, G. and Cooper, D.B. (1992) "Object recognition based on moment (or algebraic) invariants," in J.L. Mundy and A.P. Zisserman (ed.s) *Geometric Invariance in Computer Vision*, MIT Press.

49. Taylor, B., (1977) "Tense and Continuity" Linguistics and Philosophy 1 199–220.

50. Tenny, C.L. (1987) "Grammaticalizing Aspect and Affectedness," Ph.D. thesis, Linguistics and Philosophy, Massachusetts Inst. of Techn.

51. Turk, M. and Pentland, A., "Eigenfaces for recognition," *J. Cognitive Neuroscience*, **3**, 1, 1991.

52. Ullman, S. and Basri, R. (1991) "Recognition by linear combination of models," *IEEE PAMI*, **13**, 10, 992-1007.

53. Weiss, I. (1988) "Projective Invariants of Shapes," Proceeding DARPA Image Understanding Workshop, p.1125-1134.

54. Whorf, B.L. (1941) "The Relation of Habitual Thought and Behavior to Language," in Leslie Spier, ed., *Language, culture, and personality, essays in memory of Edward Sapir*, Sapir Memorial Publication Fund, Menasha, WI.

55. Zerroug, M. and Nevatia, R. (1994) "From an intensity image to 3D segmented descriptions," *Proc 12'th ICPR*, 108-113.

56. Zisserman, A., Mundy, J.L., Forsyth, D.A., Liu, J.S., Pillow, N., Rothwell, C.A. and Utcke, S. (1995) "Class-based grouping in perspective images", *Intern. Conf. on Comp. Vis.*

Beyond the Hough Transform: Further Properties of the $R\theta$ Mapping and Their Applications [1]

M. Wright   A. Fitzgibbon   P. J. Giblin [2]   R. B. Fisher

Department of Artificial Intelligence
University of Edinburgh
Edinburgh, Scotland EH1 2QL
`markwr`, `andrewfg` or `rbf@aifh.ed.ac.uk`

Abstract. The Hough transform is a standard technique for finding features such as lines in images. Typically, points or edgels are mapped into a partitioned parameter or Hough space as individual votes where peaks denote the feature of interest. The standard mapping used for line detection is the $R\theta$ mapping and the key property the Hough transform exploits is that lines in the image map to points in Hough space. In this paper we introduce and explore three further properties of the $R\theta$ mapping and suggest applications for them. Firstly, we show that points in Hough space with maximal R for any value of θ are on the convex hull of the object in image space. It is shown that approximate hulls of 2D and 3D hulls of objects can be constructed in linear time using this approach. Secondly, it is shown that a simple relationship exists between the occluding contour of an object and the $R\theta$ mapping and that this could in principle be used to generate approximate aspect graphs of objects whose geometry was known. Thirdly, it is shown that antipodal points on object boundaries, (which are optimal robot grasp points), can be found by translation and reflection of the $R\theta$ representation. In addition we show the relationship between the $R\theta$ mapping used in the Hough transform and the classical mathematical theory of duals. We use this analysis to prove formally stated properties of the $R\theta$ mapping from image space to Hough space and in particular the relationship to the convex hull.

---

[1] This work was funded by EPSRC grant number GR/J44018
[2] Department of Pure Mathematics, University of Liverpool

1 Introduction

The Hough transform [5] is a standard technique in computer vision having many applications for finding lines, circles and other features in images [6]. Typically edgels or other features are mapped into a partitioned parameter or Hough space as individual votes. The target image features are detected as peaks in the Hough space.

One particular variant of the Hough transform for line detection is shown in Figure 1. This uses the $R\theta$ parameterization and makes use of edgel orientation to reduce computation [7]. For each edgel a perpendicular is dropped from the origin in image space to a line passing through the edgel having the same orientation. This is then mapped to a space with two parameters R and θ where R is the length of perpendicular and θ is the angle the line makes with the x axis. As orientation information is available then each edgel maps to a single point in Hough space. If the edgels lie on a line then all will map to a common point in the Hough space and this peak in the Hough space or accumulator signifies a linear feature in image space. If no orientation information is available then each edge point maps to a sine wave in Hough space corresponding to the set of all lines passing through that point. The intersection of these sine waves signifies the parameters of the line that passes through all the points.

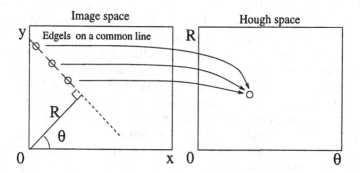

Fig. 1. On the left is a set of edgels in image space with the same orientation and aligned such that they lie on the same line. On the right we can see that they map to a single point in Hough space and so the line they create can be detected as a peak at this point.

The specific property of the R, θ mapping which is used in the Hough transform is that lines in images space map to points in Hough space. In this paper we explore further properties of the R, θ mapping and suggest applications for them. Previously [15], we have shown the relationship of the R, θ mapping to the convex hull of objects and the occluding contour. In this paper we provide further examples of convex hulls computed using our approach, we show timings and code listings to demonstrate the simplicity of the algorithm. We show how a variant of the R, θ mapping can be used to detect optimal grasps for a robot gripper. We show the relationship of the R, θ mapping and the classical theory of duals. We also provide a proof of our assertion that maximal R for any θ in Hough space corresponds to the convex hull in image space.

2 The R, θ Mapping of the Shape Boundary

In the Hough transform algorithm described above individual edgels are mapped into the Hough space and the peaks detected. Those edgels not contributing to peaks in the accumulator are ignored. We will now consider what happens to these other edgels and what this can tell us about the geometry of the curve. Consider the smooth image curve in Figure 2a. We can sample the boundary of this curve and at each point determine the orientation of the tangent to the curve. As we traverse the curve in image space a curve is also traced out in R, θ space. The mapping to Hough space for our smooth curve is given in Figure 2b. This curve is remarkably structured and we will now describe how its geometric features relate to the image space curve.

The main features of the curve are self intersections, cusps, maxima, minima and inflexions. The self intersections of the R, θ curve at C and F in Figure 2b are due to two points on the curve boundary at C and F in Figure 2a mapping to the same point in Hough space. This signifies where there is a bi-tangent line in the image *i.e.* a line tangent to the boundary at two distinct points. The loop formed by the self intersection corresponds in this case to the indentation in the curve boundary. The cusps within this loop at D and E in Figure 2b correspond to points of inflexion at D and E on the image space curve in Figure 2a. The maxima (B, G, I), minima (H) and inflexion (A) points on the Hough space curve in Figure 2a all correspond to points on the shape boundary where the normal at that point passes through the origin as shown in Figure 2c. It is a minima if the distance along the normal measured from the shape boundary to the origin is less than the radius of curvature of the curve at that point. It is a maxima if the normal length is greater than the radius of curvature and it is an inflexion if they are equal. The locus of centres of curvature defines a curve called the evolute which is depicted over the shape boundary as a dotted line in Figure 2d. The evolute consists of separate smooth sections running between cusp which correspond to curvature extrema on the image curve boundary. So the minima of the Hough space curve correspond to normals in image space which reach the origin before they cross their corresponding section of evolute. Conversely, maxima correspond to normals which do cross their section of evolute before reaching the origin. Finally, inflexions of the Hough space curve occur when the origin lies on the corresponding section of evolute. See §6 for proofs of the assertions made in this section.

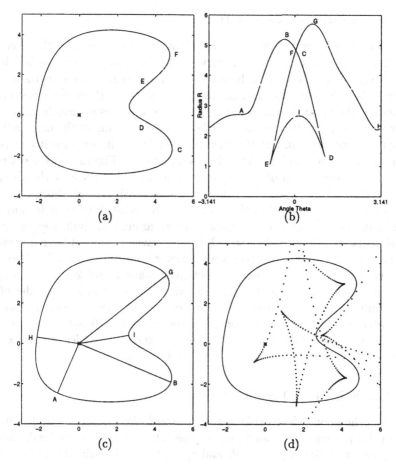

Figure 2a shows a smooth image curve. Figure 2b shows the mapping of the image curve to R, θ or Hough space. Features on the Hough space curve have been labeled to show corresponding features in image space depicted in Figures 2a and c (see text). Figure 2d shows the evolute of the image curve. The relationship between the evolute (dotted line) and origin (astrix) determine the type of turning points that exist on the Hough space curve (see text).

Fig. 2. The mapping from image space to R, θ space

3 The R, θ Mapping and Convex Hull

The relationship between the R, θ mapping and the convex hull is as follows. **If we take for every value of θ that portion of the Hough space curve that has maximum R i.e. the upper envelope of the R, θ space graph, then the corresponding points in image space are those points on the convex hull.** (See §6 for proof).

3.1 Dealing with Discontinuity

It would appear at first sight that this might only work for smooth curves because if the derivative of the boundary is discontinuous then there is a break in the Hough space curve. It is however a simple matter to join up these gaps. We remember that, for the original Hough transform, if there is no orientation information then points in image space map to sine waves in Hough space. At a corner we merely join the end of one edge to the beginning of the next with the appropriate sine wave. In fact this idea can be taken to its extreme if we wish to construct the convex hull of sets of isolated points. In Figure 3a we see a point set in image space. Each of these points maps to a sine wave in Hough space which are labeled in Figure 3b. The convex hull consists of all those points for which their corresponding Hough space sine wave is maximal in R for any particular value of θ. [3] Furthermore their ordering around the hull is given simply by their adjacency in Hough space. Only sine waves $1, 2$ and 3 have a maximal R and are on the convex hull. Sine wave 5 never approaches a maximal position but wave 4 comes close at the intersection of wave 1 and 2. This is because point 4 in image space is nearly collinear with points 1 and 2. The sides of the convex hull correspond to the points in Hough space where the maximal sine waves intersect. In Figure 4 we see the convex hull of 7870 isolated points randomly distributed within a unit circle. Figure 5 shows examples of convex hulls computed for various objects using the Hough transform.

3.2 Extending to 3D Hulls

Extending the approach to 3D is straight forward. We now drop a perpendicular to a tangent plane rather than a tangent line. The parameters are the distance R of the plane from the origin and θ_1 and θ_2 which are the azimuth and elevation of the perpendicular. The Hough space now has three dimensions rather than two. (In the general case we drop the perpendicular to a hyperplane and the Hough space becomes n-dimensional.) Figure 6 shows the Hough space for a small three dimensional point set which is not shown. This is the upper envelope of the sinusoids generated by all points and each one corresponds to a point on the convex hull; compare this with Figure 3b.

Figure 7 is an example of a 3D convex hull generated using the R, θ mapping. The object on the left is Figure 7 is a Volkswagen Beetle body shell and the object on the right is its convex hull.

3.3 Algorithms

Algorithm 1: Sampling In Hough space the upper envelope of all sine waves defines the convex hull, in that each sine wave segment on this envelope corresponds to a single point in the hull. We now describe an algorithm to identify these segments. The array *hull* is workspace.

---

[3] The fact that the upper *and* lower envelope of sine waves in the accumulator corresponds to the convex hull was first noted by Murakami *et al* [10] (see discussion).

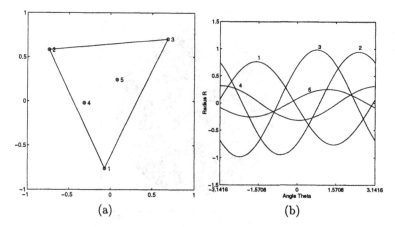

(a)   (b)

Figure 3a shows a set of 5 image points and their convex hull. In Figure 3b we can see their corresponding sine waves in Hough space. Only those waves with maximal R for some value of θ contribute to the convex hull, in this case waves 1, 2 and 3.

Fig. 3. The relationship between the convex hull and the Hough transform for isolated points

Step 1 Let $p_1(\theta), \ldots, p_n(\theta)$ be the sine waves in Hough space corresponding to the n points in the image space.

Step 2 Set index k to 1; for $\theta = 0$ to 2π in steps of Δ_θ (see discussion below) fill an array $hull[k]$ with indices as follows:

Step 2a Find the greatest sine at this θ: Set $hull[k] := i$ where $p_i(\theta) \geq p_j(\theta)$ for all $1 \leq j \leq n$.

Step 2b Increment index k by 1.

Step 3 The distinct indices in $hull$ are those points in the image plane in the convex hull. The order of the indices within the array determines the order of the points around the hull *i.e.* if we draw a line from one point to the next using these indices we will draw the convex hull in 2D (For higher dimensions than 2D the adjacency of indices in the accumulator encodes adjacency of points on the hull).

(Termination is trivial since $\Delta_\theta > 0$.) An attractive aspect of the algorithm is its simplicity. Figure 8 shows the actual matlab code used to generate the 2D hulls, apart from these six lines, no other code is required.

An important issue is the granularity of the Hough space from the viewpoint of the complexity of computing the convex hull by this algorithm. Consider three collinear points a, b and c, with a and b in the convex hull, and c lying between them. Without loss of generality we assume that c lies at the origin of the image plane and a and b on either side of it along the x-axis, at $-x_a$ and x_b. We will consider the behaviour in Hough space as c is perturbed along the y-axis by δc, and derive from this the required sampling interval of the Hough space so as to

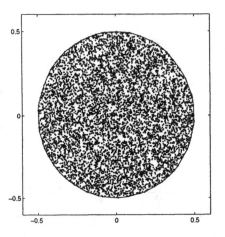

Fig. 4. The convex hull of 7870 points generated using the Hough transform

detect that c is now part of the hull.

Perturbing c in this way contributes $\delta c. \sin \theta$ to Hough space, and we see that the interval Δ_θ must be such that

$$\Delta_\theta < \tan^{-1}(\delta c/x_a) + \tan^{-1}(\delta c/x_b)$$

for c to be seen to be in the hull. This gives an approximate complexity

$$O(\frac{n}{\tan^{-1}(1/p)})$$

Where p measures the greatest distance between any two points a and b. Choosing p *a priori* to be the image diagonal renders it independent from (a bounded) n, and overall complexity is $O(n)$; however, constants are much greater than in the Graham's scan algorithm [4] [11].

For example, for a 512×512 image we have $512.4\pi n$ for n points with a twice-over sampling interval.

Algorithm 2: Non-Sampling We can make a significant (see below) average-case improvement in speed of the algorithm, with the following approach, which has complexity $O(n^2)$. (We adopt the same notation as above.)

Step 1 Determine i such that $p_i(0) > p_j(0)$ for all $1 \leq j \leq n$. We use this hull point to start.

Step 2 Record that the point i is on the hull. If $i = 2\pi$ then we have completed the hull, so stop.

Step 3 Compute the next point i' known to be on the hull as follows. Determine the smallest $\theta_{i'}$ such that $p_i(\theta_{i'}) = p_{i'}(\theta_{i'})$ subject to $\theta_{i'} > \theta_i$. i' is the 'nearest' sine wave which intersects with since wave i.

Set i to $'i$ and go to Step 2.

(We omit proofs of termination.)

This algorithm has quadratic complexity since there are n computations in Step 3 (finding the next intersection point), and we do this for each point in the hull, which is $O(n)$ (think of all n points arranged as as circle). Hence $O(n^2)$. However, we anticipate much better that this worst-case performance in practice, since typically considerably fewer points are in the hull in comparison with n.

Simple timing trials for algorithm 1 were carried out for both the 2D and 3D case. Algorithms were implemented in matlab [14] and run on a Sparc 10 workstation. The algorithms were tested on points randomly distributed within the unit square for 2D and unit cube for the 3D case. The number of input points was varied from 1000 to 10,000 in increments of 1000. For any fixed number of points the algorithms were run 100 times and then mean run times and standard deviations were calculated. Figure 9a shows the results for the 2D case and Figure 9b. In both cases the solid line indicates the mean run time and the error bars are +/- the standard deviation for each trial.

4 The R, θ Mapping and Aspect Graph

In the last section we computed the convex hull of 2D and 3D objects by using the R, θ mapping of object boundaries with a stationary origin in image space. In this section we consider what happens to the same mapping as we move the origin in image space.

If the origin is outside the shape then we can consider it as a vantage point from which we look at the object. Using this perspective the first striking observation is that the occluding contour of the object, *i.e.* those point where the local tangent line or plane lies along the line of sight, is simply those points where $R = 0$. Furthermore those points with $R > 0$ have normals pointing toward the origin or vantage point and those with $R < 0$ have normals points away. We suggest that these observations can form the basis of an algorithm to build aspect graphs [9] of arbitrary 2D and 3D shapes. We can move the origin about the viewing sphere and keep track of the self intersections and cusps of the Hough space curve. As these features cross the θ axis *i.e.* $R = 0$ these constitute event boundaries in the aspect graph. To illustrate this principle we have devised a graphics program which animates in real time the mapping from image space to Hough space as the origin is moved or the shape boundary is deformed. Figure 10 shows a few different frames of the animation. As the vantage point (cross) is moved so the concavity comes into view the self intersection and cusps of the curve in Hough space move below the θ axis. We suggest that tracking of these events could be used to trigger the building of new branches and nodes in the aspect graph.

5   Detection of Robot Grasp Points

In this section we show how a variant of the R, θ mapping can be used to detect optimal robot grasp points. Blake [1], has developed a theory of planar grasp where optimal grasps are defined using the criterion of least friction required to obtain force closure. Antipodal grasps, *i.e.* where the boundary normals precisely oppose one another, are identified as being particularly good as they requiring nominal friction for force closure. Normally, these grasp points would be detected using an n^2 search. We now show how these antipodal grasps can be detected using a variant of the R, θ mapping.

Consider the ellipse in figure 11. We compute an R, θ curve for the boundary with respect to the arbitrary origin at O. However, instead of the using the tangent to construct the curve we use the normal, *i.e.* we drop a perpendicular from the origin to the inward normal *not* to the tangent line.

The solid curve of figure 12 is the R, θ curve for the ellipse constructed using the normals instead of the tangents. Antipodal points can be found by a simple manipulation of the R, θ curve. Firstly, the curve is phase shifted with respect to itself by an angle of π; this can be visualised as a translation along the θ axis by π. Secondly, R values of the phase shifted curve are negated; this can be visualised as a reflection in the θ axis. The intersections thus resulting with the original R, θ curve correspond to the antipodal grasp points. The dotted curve in figure 12 is the result of applying the translation and reflection operations to the original solid curve. The intersection points A and C in figure 12 correspond to antipodal grasp points A and C on the ellipse in figure 11. Similarly, the intersection points B and D in figure 12 correspond to antipodal grasp points B and D on the ellipse in figure 11. Notice in figure 12 that antipodal pairs are separated by exactly π radians along the θ axis.

6   Duals and the Hough Transform

In this section, we relate the Hough transform to the classical mathematical theory of duals, and prove the assertions made in §2 above.

The *dual* of a smooth curve in the plane is a model of its tangent lines (and, for a surface, a model of its tangent planes). There is an extensive theory of duals, going back to the work of Plücker [12, 13] in the early 19th century; some modern references are [2, 3]. Consider a plane curve C given parametrically by $\mathbf{r}(t)$, where t lies in some interval of real values, which could be all of the real numbers. The curve comes then with an *orientation*, given by t going from 'left to right', i.e. increasing. We usually take the orientation of a *closed* curve C without self-crossings to be such that the interior of the curve lies on the *left* as we traverse the curve.

Assuming that the velocity vector $\mathbf{r}'(t)$ is never zero, there is a well-defined (oriented) tangent line at each point: it passes through $\mathbf{r}(t)$ and has the direction of the velocity vector $\mathbf{r}'(t)$. The *oriented dual* of C is a model for the set of these tangent lines, and there are many ways of setting up such a model. If we forget the orientation and regard the tangent simply as a line in the plane, then a model for the set of these tangent lines is the *dual* of C. The oriented dual is easier to describe and we concentrate on that here. (The reason is that the set of all oriented lines in the plane is conveniently parameterized by the points of a cylinder, as we show below, but the unoriented lines are parameterized by the points of a Möbius band, which is a little harder to handle.)

Given an oriented line l in the plane (a line with an arrow on it) we can describe l completely by an angle θ and a real number R. Let $\mathbf{v}$ be a unit vector obtained as follows: start with a unit vector along the line l, i.e. pointing with the arrow, and turn this clockwise through $\frac{1}{2}\pi$. (This will later be along the outward normal to our curve C.) Let $\mathbf{x}_0$ be any point of l. Then R, θ are defined by

$$\mathbf{v} = (\cos\theta, \sin\theta), \quad R = \mathbf{x}_0.\mathbf{v},$$

the dot being scalar product. If a different point $\mathbf{x}_1$ of l is selected then the scalar product is unaltered since $\mathbf{x}_1 - \mathbf{x}_0$ is perpendicular to $\mathbf{v}$. Geometrically, R is, up to sign, the distance from the origin to the line l and θ is the anticlockwise angle turned from the positive x-axis in the plane to the vector $\mathbf{v}$. Thus l determines R, θ; conversely R, θ determine a line with equation

$$\mathbf{x}.\mathbf{v} = R$$

and this line, which is perpendicular to $\mathbf{v}$, is oriented by turning $\mathbf{v}$ anticlockwise through $\frac{1}{2}\pi$. (This line will later be the oriented tangent line to C.) In this way, all oriented lines in the plane are parameterized by an angle and a real number, that is by the points of a cylinder, which we conventionally represent by a rectangle in R, θ space with $\theta = \pi$ identified with $\theta = -\pi$.

This 'R, θ representation' is one of many classical ways of describing lines in the plane, and is sometimes referred to as the 'Hesse normal form' (1861); for this form and some others, see [3, pp.123-127]. Returning now to our curve C with parameterization $\mathbf{r}(t)$, the oriented tangent is given by the velocity vector $\mathbf{r}'(t)$, and conventionally the oriented normal $\mathbf{N}(t)$ is given by turning $\mathbf{r}'(t)$ anticlockwise through $\frac{1}{2}\pi$. Thus in the above discussion we take $\mathbf{v} = -\mathbf{N}(t)$. If $\mathbf{r}(t) = (X(t), Y(t))$ then, making $\mathbf{N}$ a unit vector, we have (using ' for d/dt)

$$\mathbf{N}(t) = \frac{(-Y', X')}{\sqrt{X'^2 + Y'^2}},$$

and associated with the oriented tangent at $\mathbf{r}(t)$ we have the R, θ values given by

$$\mathbf{N}(t) = -(\cos\theta, \sin\theta), \quad R = -\mathbf{r}(t).\mathbf{N}(t). \tag{1}$$

As t varies, the point R, θ describes a parameterization of the oriented dual of C. It is a model for the set of oriented tangent lines to C, and is the Hough transform as used in the text of this paper. As it is a model for the oriented tangent lines, it is evident that self-crossings of the R, θ curve correspond to oriented lines tangent to C twice, that is, to *bi-tangent lines* to C.

We can use the above parameterization to deduce geometrical properties of the Hough transform (or oriented dual). Here is a brief indication of how this is done. The curve parameterized $\theta(t), R(t)$ can be considered as a 'graph', that is as R a function of θ, provided the tangent line is not 'vertical', that is provided $\theta' \neq 0$. We assume that t, the parameter on C, is arclength. This does not affect the results but shortens the calculations, since the Serret formulae then take the following simple form (see [8, p.170], where $\tau = 0$ since our curves are planar)

$$\mathbf{r}' = \mathbf{T}, \quad \mathbf{T}' = \kappa\mathbf{N}, \quad \mathbf{N}' = -\kappa\mathbf{T}. \tag{2}$$

Here $\mathbf{T}$ is the unit tangent and κ is curvature. It follows from (1) and (2) that

$$\mathbf{T} = (-\sin\theta, \cos\theta),$$
$$-\kappa\mathbf{T} = -\theta'(-\sin\theta, \cos\theta) = -\theta'\mathbf{T}, \quad \text{so} \quad \theta' = \kappa \tag{3}$$
$$R' = -\mathbf{T}.\mathbf{N} + \mathbf{r}.\kappa\mathbf{T} = \kappa\mathbf{r}.\mathbf{T} \tag{4}$$

The following are now easy deductions.

1. $\theta' = 0$ if and only if $\kappa = 0$, that is the curve C has an *inflexion*. When $\kappa = 0$ we also have $R' = 0$, which means that the R, θ curve is actually *singular* (generally has a cusp). Thus inflexions of C correspond to singular points (cusps) of the R, θ curve, and, between cusps, the R, θ curve is a graph, that is R is a function of θ. For a given θ, there may be several values of R (as in the example of Figure 2b in §2). On a section between two cusps there are no vertical tangents, so locally R is a single valued function of θ; we denote the slope of this curve by $dR/d\theta$.

2. Suppose $\kappa \neq 0$. We have $dR/d\theta = 0$ if and only if $R' = 0$, which from (4) means $\mathbf{r}.\mathbf{T} = 0$, that is to say the normal to C passes through the origin (the equation of the normal is $(\mathbf{x} - \mathbf{r}(t)).\mathbf{T}(t) = 0$). So the R, θ curve has a maximum, minimum or inflexion if and only if the normal to C passes through the origin.

3. From (4) we can find the second derivative:

$$R'' = \kappa'\mathbf{r}.\mathbf{T} + \kappa\mathbf{T}.\mathbf{T} + \kappa\mathbf{r}.\kappa\mathbf{N} = \kappa^2\left(\frac{1}{\kappa} - R\right)$$

when $R' = 0$, that is when $\mathbf{r}.\mathbf{T} = 0$. From this it follows that, when $\mathbf{r}.\mathbf{T} = 0$, R has, as a function of t or of θ,

  – a maximum if $R > \rho$, where $\rho = 1/\kappa$ is the radius of curvature of C,
  – a minimum if $R < \rho$,
  – generally an inflexion with horizontal tangent if $R = \rho$. (The condition for a non-horizontal inflexion of the R, θ curve, namely $R'' = 0$ without $R' = 0$, does not appear to have a simple geometrical interpretation.)

(*Remark* The process of taking duals is in a sense 'involutory', that is the dual of the dual of C is the curve C again. In that case, as inflexions of C give cusps on the dual, it might be expected that inflexions on the dual could only come from cusps on C. But whether or not a plane curve, such as the dual, has inflexions, depends on the choice of coordinate system used to describe it. In another standard representation of the dual of a nonsingular curve C (not the Hough transform, see [3, p.126]) there are no inflexions. Cusps, on the other hand, are present irrespective of the coordinate system, so in any representation of the dual, inflexions of C (using the ordinary coordinate system in the plane) give cusps on the dual.)

4. Consider a particular value of θ, corresponding to a unit vector

$$\mathbf{v} = (\cos\theta, \sin\theta)$$

Turning $\mathbf{v}$ anticlockwise through $\frac{1}{2}\pi$ to give $\mathbf{T}$, the curve C may have several oriented tangents parallel to $\mathbf{T}$. Using the convention on orienting C mentioned above, the interior of C is, near the point of contact, always to the left of such an oriented tangent. Going in the direction of $\mathbf{v}$, the *last* such oriented tangent encountered must have the whole curve to its left, that is it is a 'support line' of C. Wherever the origin O is situated, this tangent will have the *largest* value of R for that θ (this R is > 0 if the oriented tangent is 'anticlockwise about O' and < 0 otherwise). So the support lines of C, for each θ, correspond to the maximum value of R for that θ. Since the support lines are tangents to the convex hull, this verifies the connexion with convex hulls in §3.

Finally, a few words about how these ideas extend to 3 dimensions. For a surface M in 3-space, it is possible to parameterize the oriented tangent planes by pairs $(\mathbf{v}, R)$ where $\mathbf{v}$ is a unit vector in 3-space, hence a point of the unit sphere S^2, and R is a real number. In fact $\mathbf{v}$ can be chosen as the oriented (outward) normal of M and $R = \mathbf{r}.\mathbf{v}$ where $\mathbf{r}$ is the point of contact of the tangent plane. Unlike the circle, there is no overall parameterization of the sphere S^2; for example azimuth and elevation break down at the poles, where azimuth is undefined. Away from the poles, we can locally use a 'Hough space' given by azimuth, elevation and R, or we can do the job globally by using the space $S^2 \times \mathcal{R}$ where $\mathcal{R}$ is the real numbers. The oriented dual of M is a locus in this space, and it has many properties analogous to those for a curve above. For example a parabolic point of M (Gauss curvature zero) corresponds to a singular point (generally an 'edge of cusps') on the dual. See [2].

7 Discussion

The convex hull algorithms presented above has similarities to Graham's algorithm [4] where points are sorted according to angle. [4] The complexity of Graham's algorithm is $O(n\ln(n))$. However our Hough based approach (algorithm 1) has a high constant term due to the sampling of Hough space. Algorithm 2 has much poorer worse case complexity but in the average case we anticipate better performance than algorithm 1. Neither algorithm can compete with Graham's algorithm in performance as a practical convex hull algorithm. At first

---

[4] Graham's algorithm cannot be applied to problems of dimension more than two.

sight it might appear that algorithm 1 breaks the lower bound for convex hull computation but we must remember that we are computing an *approximate* convex hull in *discrete* space. It is interesting to note that an equivalence has been proved between convex hull computation and the operation of sorting and therefore interesting comparison can be made between convex hull algorithms and sorting algorithms. There are *linear* time algorithms for sorting and perhaps one of the simplest is bucket sort. Both bucket sort and the R, θ approach to convex hull computation work in linear time, both discretize space and both work approximately for arbitrary input, giving exact correct output for restricted input. (Bucket sort requires that the minimum distance between numbers is greater than the bucket interval, and the convex hull algorithm presented here requires that the angle between facets on the hull is greater than sampling interval in Hough space.

We believe that sampling may also be the basis for producing approximate convex hulls (and aspect graphs) in a principled way which has been identified as an important problem. Unlike Graham's algorithm there is a global representation of the hull. We could in principle use this to assess the effect of leaving out a particular point. This could be on the principle of the area of the sine wave for which it is maximal in R. We have also experimented with approximate hulls in the following way; rather than take the strictly maximal R for any θ, we take all points within a threshold of R to be on the hull. This detects all points which are within this threshold from the hull. This might be important if we require to extract all of a curved convex boundary outline but points would be missed if they were not exactly on the hull due to noise or quantization. This technique also registers degenerate cases such as colinear points which are implicitly merged using strictly maximal R.

Similar work on convex hull computation has been carried out by Murakami *et al* [10]. Murakami *et al* were the first to note that the upper and lower envelope of sine waves in the Hough accumulator, using the R, Θ mapping, correspond to the convex hull in image space. The contribution of our work in the area of convex hull computation is as follows:

- **Simple algorithm using much less memory and fewer operations.**
  Murakami *et al* used the full 2D accumulator for the 2D case only. Sine waves are input to the accumulator as a string of single votes. The accumulator is scanned to find the envelope of sine waves and intersections of sine waves on the envelope which correspond to the parameters of sides of the convex hull. The Murakami *et al* algorithm requires separate accumulation, envelope detection and envelope segmentation stages. If we define the accumulator to be of size L rows for the R axis and K columns for the Θ axis, this gives a principle memory requirement of $L \times K$. We use a single stage Z buffering technique which has a principle memory requirement of $2 \times K$. Resolution of R is a function of machine precision rather than segmentation of the R axis. Murakami *et al* extracts the convex hull from the Hough space representation alone and therefore has to use a rather complex method to solve for the x, y positions of hull points in image space from sine wave samples in Hough

space. We use image space information to label the image points and the fact that order of sine waves on the envelope across the accumulator gives the order of points around the hull to extract the hull.

In discussion Murakami *et al* does allude to a more efficient algorithm using image space but gives no details apart from the fact that this algorithm would have a memory requirement of "a few 1D arrays of size L" where L is the size of the R axis. As our algorithm segments the Θ axis, using two 1D arrays of size K we deduce that our algorithm is different to the one mentioned by Murakami *et al.*

- **Detailed complexity analysis.** We give a detailed complexity analysis for our algorithm. Murakami *et al* noted correctly that the complexity is O(N) but does not show the exact relationship to the quantization of the accumulator. We give an equation for the order and how to work out accumulator size for specified error.

- **Extension from 2D to 3D.** We demonstrate the extension of the approach from 2D to 3D with an example and describe how the algorithm could be applied in N dimensions.

- **Timings of our approach.** We show timing trials for 2D and 3D training sets as empirical evidence demonstrating the linear nature of the algorithm.

- **Oriented dual as theoretical model.** We demonstrate other applications of the R, Θ mapping and describe them all, including convex hull computation, using the classical theory of duals as a unifying theoretical framework.

The Hough approach to building the aspect graph has the problem that although we can tell if a face points towards the viewer we cannot tell if it is occluded by another part of the object. It is as if we are considering the singularities of a transparent object. We could solve this by filtering the Hough space output with some kind of visibility check. One such check would be to compute a different transform, say d, α where d is the length of a line from the boundary point to the origin and α is the angle this line makes with the x axis. The points in d, α space with minimal d for a particular α are those points which are unoccluded.

Currently we are investigating new applications of the R, θ mapping. We have found that we can extract radius and diameter functions of shapes using the mapping. A radius function is the locus of the centre of mass of an object as it is rolled along a straight edge. Up to a change of coordinates, the radius function is equivalent to the R, θ mapping of the shape taken with the origin at the centre of mass. The diameter function is the diameter of the object measured between two parallel lines plotted as a function of object orientation. The diameter function is easily extracted from the R, θ mapping by taking both maximal and minimal R for every θ and simply subtracting their difference. The application we intend to use these functions for is the orientation of objects by pushing with a robot manipulator or conveyor and fences.

8 Conclusions

The Hough transform uses a particular property of the R, θ, *i.e* that lines in image space map to points in Hough space, to detect linear features in images. In this paper we have presented and explored three further properties of the R, θ mapping and suggest useful applications for them. We have shown the relationship of the mapping to the convex hull and computed 2D and 3D hulls of objects. We have seen the relationship of the mapping to the occluding contour and pointed to the possibility of constructing aspect graphs of objects with known geometry. We have also shown how translations and reflections of the mapping can be used to detect optimal robot grasp points. We have shown the relationship of the Hough transform and R, θ mapping to the classical mathematical theory of duals. We hope that we have challenged the traditional view of the the R, θ mapping and Hough transform as purely for feature detection. The R, θ mapping has many interesting properties and uses and many more may yet be waiting to be discovered.

9 Acknowledgement

Mark Wright would like to thank Ian Green for helpful discussions concerning algorithm complexity. Part of this work has been submitted for review to the journal Image and Vision Computing.

References

1. A. Blake. A theory of planar grasp. OUEL Report 1958/92, Oxford University Engineering Department, October 1992.
2. J. W. Bruce. The duals of generic hypersurfaces. *Math. Scand*, 49:36–69, 1981.
3. J.W Bruce and P.J. Giblin. *Curves and singularities*. Cambridge University Press, second edition, 1992.
4. R. L. Graham. An efficient algorithm for determining the convex hull of a finite planar set. *Information processing letters*, 1:132–133, 1972.
5. P. V. C. Hough. *Method and means for recognising complex patterns*. US Patent 3069654, 1962.
6. J. Illingworth and J. Kittler. A survey of the Hough transform. *CVGIP*, 44:87–116, 1988.
7. C. Kimme, D. Ballard, and J. Sklansky. Finding circles by an array of accululators. *Comm ACM*, 18:120–122, 1975.
8. J. J. Koenderink. *Solid shape*. M.I.T Press, 1990.
9. J. J. Koenderink and A. J. van Doorn. The internal representations of solid shape with respect to vision. *Biological Cybernetics*, 32:211–216, 1979.
10. K. Murakami, H. Koshimizu, and K. Hasegawa. An algorithm to extract convex hull on $\theta - \rho$ Hough transform space. In *9th International Conference on Pattern Recognition*, 1988.
11. J. O'Rourke. *Computational geometry in C*. Cambridge University press, 1992.
12. J. Plucker. *System der analytischen geometrie*. Berlin, 1835.

13. J. Plucker. *Theorie der algebraischen kurven.* Bonn, 1839.
14. Math Works. *MATLAB user guide*, 1992.
15. M. W. Wright, A. Fitzgibbon, P. J. Giblin, and R. B. Fisher. Convex hulls, occluding contours, aspect graphs and the Hough transform. In *British Machine Vision Conference*, pages 493–502, 1995.

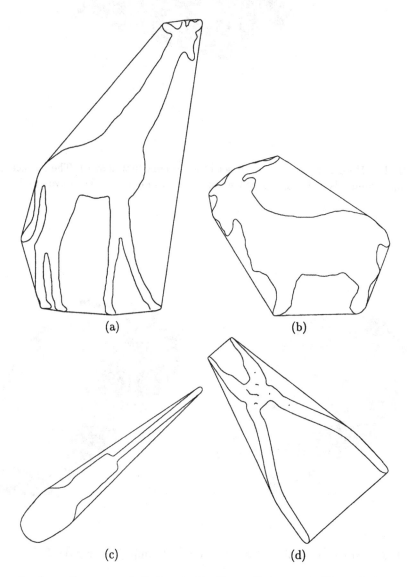

(a)

(b)

(c)

(d)

Figure 5a shows the convex hull of a giraffe, Figure 5b is a goat, Figure 5c is a screwdriver and Figure 5d is a pair of pliers.

Fig. 5. Examples of convex hulls computed using the Hough transform

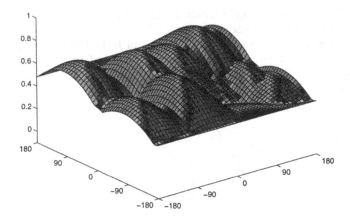

Fig. 6. The Hough space for a small set of 3D points (not shown). The surface is of the upper sinusoids in Hough space corresponding to points on the convex hull.

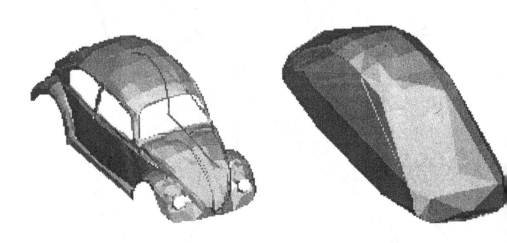

Fig. 7. A Beetle body shell and its 3D convex hull computed using the R, θ mapping.

```
function hull_indices = chull2(P, max_error)
% 1. Calculate accumulator size from max_error
nsamples = floor(1/atan(max_error))

% 2. Generate unit circle (2D unit sphere)
theta = linspace(0,2*pi, nsamples)';
Normals = [cos(theta) sin(theta)];

% 3. Calculate hull
[Rmax, hull_indices] = max(P * Normals');

% 4. Remove duplicate indices
hull_indices = hull_indices(find(diff([hull_indices Inf])));
```

Fig. 8. Demonstration of algorithm simplicity: Matlab code for 2D Hull

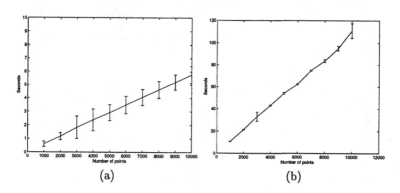

(a)   (b)

Figure 9a shows timing trials for the 2D case and Figure 9b shows the 3D case. In both cases the solid line is mean for 100 runs and it can be seen that it is linear. Error bars indicate +/- the standard deviation of each run.

Fig. 9. Timing trials for computation of the convex hull using the Hough transform

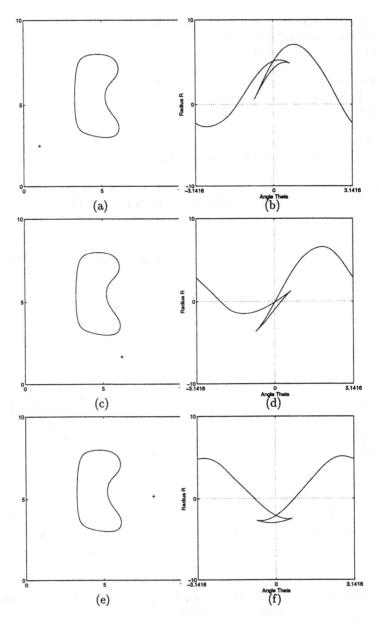

(a)

(b)

(c)

(d)

(e)

(f)

Figures 10a,c and e show a kidney bean shape in image space where different view points are marked with a cross. In Figure 10a no part of the concavity can be seen and the loop in Hough space in Figure 10b is above the x axis. In Figure 10c the viewpoint is precisely along the bi-tangent line, this is signified by the self intersection in Figure 10d crossing the x axis. In Figure 10e the concavity is in plain view and the loop in Figure 10f is below the x axis.

Fig. 10. The transitions in Hough space relating to changing view point.

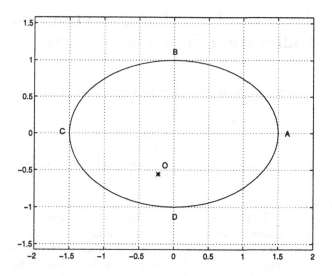

Fig. 11. An ellipse with arbitrary origin O and antipodal grasp pairs A, C and B, D

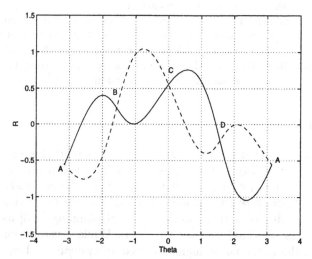

The smooth curve is the R, θ mapping of the ellipse in figure 11 obtained using the normals rather than the tangent lines. The dotted line is obtained by translating the original curve by π and reflecting it in the θ axis. Intersections thus formed correspond to antipodal grasps.

Fig. 12. Antipodal grasps identified using transformations of the R, θ mapping

Non-Euclidean Object Representations for Calibration-Free Video Overlay

Kiriakos N. Kutulakos and James R. Vallino

Computer Science Department
University of Rochester
Rochester, NY 14627-0226
Email: kyros@cs.rochester.edu, vallino@cs.rochester.edu

Abstract. We show that the overlay of 3D graphical objects onto live video taken by a mobile camera can be considerably simplified when the camera, the camera's environment and the graphical objects are represented in an affine frame of reference. The key feature of the approach is that it does not use any metric information about the calibration parameters of the camera, the position of the user interacting with the system, or the 3D locations and dimensions of the environment's objects. The only requirement is the ability to track across frames at least four features (points or lines) that are specified by the user at system initialization time and whose world coordinates are unknown. Our approach is based on the following observation: Given a set of four or more non-coplanar 3D points, the projection of all points in the set can be computed as a linear combination of the projections of just four of the points. We exploit this observation by (1) tracking lines and feature points at frame rate, and (2) representing graphical objects in an *affine* frame of reference that allows the projection of virtual objects to be computed as a linear combination of the projection of the feature points.

1 Introduction

Recent advances in display technology and graphics rendering techniques have opened up the possibility of mixing live video from a camera with computer-generated graphical objects registered in a user's three-dimensional environment. Applications of this powerful visualization tool include overlaying clinical 3D data with live video of patients during surgical planning [1–3] as well as developing three-dimensional user interfaces [4]. While several approaches have demonstrated the potential of augmented reality systems for human-computer interaction [5], the process of embedding three-dimensional "virtual" objects into a user's environment raises three issues unique to augmented reality:

- *Establishing 3D geometric relationships between physical and virtual objects:* The locations of virtual objects must be initialized in the user's environment before user interaction can take place.
- *Rendering virtual objects:* Realistic augmentation of a 3D environment can only be achieved if objects are continuously rendered in a manner consistent with their assigned location in 3D space and the camera's viewpoint.

Object-to-World

↓

World-to-Camera

↓

Camera-to-Image

Fig. 1. Calibration of a video-based augmented reality system requires specifying three transformations that relate the coordinate systems of the virtual objects, the environment, the camera, and the image it produces. This paper focuses on how specification of the top two transformations in the figure can be avoided.

– *Dynamically modifying 3D relationships between real and virtual objects:* Animation of virtual objects relative to the user's environment should be consistent with the objects' assigned physical properties (e.g., rigidity).

At the heart of these issues is the ability to describe the camera's motion, the user's environment and the embedded virtual objects in the same frame of reference (Figure 1). Typical approaches rely on 3D position tracking devices [6] and precise calibration [7] to ensure that the entire sequence of transformations between the internal reference frames of the virtual and physical objects, the camera tracking device, and the user's display is known exactly. In practice, camera calibration and position tracking are prone to errors which accumulate in the augmented display. Furthermore, initialization of virtual objects requires additional calibration stages [1,8], and camera calibration must be performed whenever its intrinsic parameters (e.g., focal length) change. Even though recent work has shown that mis-registration errors can be reduced by tracking known objects within the camera's environment [9–11], an important requirement is that the Euclidean shape of the tracked objects is accurately known.

This paper presents a new approach for embedding 3D virtual objects into live video. Its key feature is that it allows object initialization, real-time rendering and animation to be performed without any *a priori* information about the calibration parameters of the camera, the camera's motion, or the availability of Euclidean 3D models for objects physically present in the environment. The only requirement is the ability to track across frames at least four features (points or lines) that are specified by the user during system initialization and whose world coordinates are unknown.

Our approach is motivated by recent work on non-metric scene reconstruction [12,13], uncalibrated active vision [14], as well as recent human-computer interaction techniques [15,16]. We show that the task of overlaying 3D virtual objects onto live video becomes considerably simplified when the camera, the virtual objects, and the environment are represented in an *affine* frame of reference. In particular, initialization of virtual objects is reduced to the operation of *affine reconstruction* [13], visible-surface rendering is reduced to *re-projection* [17,18], and rigid animation is reduced to the problem of *synthesizing conjugate rotation transformations* [19].

Our approach is based on a well-known property of affine point representations [13]: Given a set of four or more non-coplanar 3D points represented in an affine reference frame, the projection of all points in the set can be computed as a linear combination of the projections of just four of the points. We exploit this property by (1) tracking linear and point features in the user's environment at frame rate, and (2) representing virtual objects in an affine reference frame in which the coordinates of points on a virtual object are expressed relative to a subset of the tracked features.

Affine object representations have been applied extensively to 3D reconstruction and recognition tasks [13, 20]. While our results draw heavily from this research, the use of affine 3D graphical models for interactive graphics and augmented reality applications has not been previously studied. Here we show that placement of affine virtual objects, visible-surface rendering, as well as animation can be performed using simple linear methods that operate at frame rate and exploit the ability of the augmented reality system to interact with its user.

Very little work has been published on augmented reality systems that reduce the effects of calibration errors through real-time processing of the live video stream [2, 6, 8, 9]. To our knowledge, only two systems have been reported [2, 8] that operate without specialized camera tracking devices and without relying on the assumption that the camera is always fixed [5] or perfectly calibrated. The system of Mellor [8] is capable of overlaying 3D medical data over live video of patients in a surgical environment. The system tracks circular features in a known 3D configuration to invert the object-to-image transformation using a linear method. Even though the camera does not need to be calibrated at all times, camera calibration is required at system initialization time and the exact 3D location of the tracked image features is recovered using a laser range finder. The most closely related work to our own is the work of Uenohara and Kanade [2]. Their system allows overlay of planar diagrams onto live video by tracking feature points in an unknown configuration that lie on the same plane as the diagram. Calibration is avoided by expressing diagram points as linear combinations of the feature points. Their study did not consider uncalibrated rendering, animation and interactive placement of 3D virtual objects.

Our approach both generalizes and extends previous approaches in three ways. First, we show that by representing virtual objects in an affine reference frame and by performing computer graphics operations such as projection and visible-surface determination directly on affine models, the entire video overlay process is described by a single 4×4 homogeneous *view transformation matrix* [21]. Furthermore, the elements of this matrix are simply the image x- and y-coordinates of feature points. This not only enables the efficient estimation of the view transformation matrix but also leads to the use of optimal estimators such as the Kalman filter [22] to both track the feature points and compute the matrix. Second, the use of affine models leads to a simple *through-the-lens method* [23] for interactively placing virtual objects within the user's 3D environment and for animating them relative to other physical or virtual objects. Third, efficient execution of computer graphics operations on affine virtual objects is achieved by implementing affine projection computations directly on dedicated graphics hardware.

2 Geometrical Foundations

Accurate projection of a virtual object requires knowing precisely the combined effect of the object-to-world, world-to-camera and camera-to-image transformations [21]. In homogeneous coordinates this projection is described by the equation

$$
\begin{bmatrix} u \\ v \\ h \end{bmatrix} = \mathbf{P}_{3\times4}\mathbf{C}_{4\times4}\mathbf{O}_{4\times4} \begin{bmatrix} x \\ y \\ z \\ w \end{bmatrix} \tag{1}
$$

where $[x\ y\ z\ w]^T$ is a point on the virtual object, $[u\ v\ h]^T$ is its projection, $\mathbf{O}_{4\times4}$ and $\mathbf{C}_{4\times4}$ are the matrices corresponding to the object-to-world and world-to-camera homogeneous transformations, respectively, and $\mathbf{P}_{3\times4}$ is the matrix modeling the object's projection onto the image plane.

Eq. (1) implicitly assumes that the 3D coordinate frames corresponding to the camera, the world, and the virtual object are not related to each other in any way. The main idea of our approach is to represent both the object and the camera in a single, *non-Euclidean* coordinate frame defined by feature points that can be tracked across frames in real time. This change of representations, which amounts to a 4×4 homogeneous transformation of the object and camera coordinate frames, has two effects:

- It simplifies the projection equation. In particular, Eq. (1) becomes

$$
\begin{bmatrix} u \\ v \\ h \end{bmatrix} = \mathit{\Pi}_{3\times4} \begin{bmatrix} x' \\ y' \\ z' \\ w' \end{bmatrix}, \tag{2}
$$

where $[x'\ y'\ z'\ w']^T$ are the transformed coordinates of point $[x\ y\ z\ w]^T$ and $\mathit{\Pi}_{3\times4}$ models the combined effects of the change in the object's representation as well as the object-to-world, world-to-camera and projection transformations.

- It allows the elements of the *projection matrix*, $\mathit{\Pi}_{3\times4}$, to simply be the image coordinates of the feature points. Hence, the image location of the feature points contains all the information needed to project the virtual object; the 3D position and calibration parameters of the camera as well as the Euclidean 3D configuration of the feature points can be unknown. Furthermore, the problem of determining the projection matrix corresponding to a given image becomes trivial.

Affine object representations become important because they can be constructed for any virtual object without requiring any information about the object-to-world, world-to-camera, or camera-to-image transformations. The only requirement is the ability to track across frames a few feature points, at least four of which are not coplanar. The basic principles behind these representations are briefly reviewed next. We assume in the following that the camera-to-image transformation can modeled using the *weak perspective projection* model [12].

2.1 Affine Point Representations

A basic operation in our method for computing the projection of a virtual object is that of *re-projection* [17, 18]: given the projection of a collection of 3D points at two positions of the camera, compute the projection of these points at a third camera position. Affine point representations allow us to re-project points without knowing the camera's position and without having any metric information about the points (e.g., 3D distances between them).

In particular, let $p_1, \ldots, p_n \in \Re^3, n \geq 4$, be a collection of points, at least four of which are not coplanar. An *affine representation* of those points is a representation that does not change if the same non-singular linear transformation (e.g., translation, rotation, scaling) is applied to all the points. Affine representations consist of three components: The *origin*, which is one of the points $p_1, \ldots, p_n$; the *affine basis points*, which are three points from the collection that are not coplanar with the origin; and the *affine coordinates* of the points $p_1, \ldots, p_n$, expressing the points $p_i, i = 1, \ldots, n$ in terms of the origin and affine basis points. We use the following two properties of affine point representations [13, 24, 25] (Figure 2):

Property 1 (Re-Projection Property) *When the projection of the origin and basis points is known in an image I_m, we can compute the projection of a point p from its affine coordinates:*

$$
\begin{bmatrix} u_p^m \\ v_p^m \\ 1 \end{bmatrix} = \begin{bmatrix} u_{b_1}^m & u_{b_2}^m & u_{b_3}^m & u_{p_o}^m \\ v_{b_1}^m & v_{b_2}^m & v_{b_3}^m & v_{p_o}^m \\ 0 & 0 & 0 & 1 \end{bmatrix} \begin{bmatrix} x \\ y \\ z \\ 1 \end{bmatrix} \tag{3}
$$

where $[u_p^m \; v_p^m \; 1]^T$ is the projection of p; b_1, b_2, b_3 are the basis points; $[u_{p_o}^m \; v_{p_o}^m \; 1]^T$ is the projection of the origin; and $[x \; y \; z \; 1]^T$ is the homogeneous vector of p's affine coordinates.

Property 1 tells us that the projection process for any camera position is completely determined by the matrix collecting the image coordinates of the affine basis points in Eq. (3). This equation, which makes precise Eq. (2), implies that if the affine coordinates of a virtual object are known, the object's projection can be trivially computed by tracking the affine basis points. The following property suggests that it is possible, in principle, to extract the affine coordinates

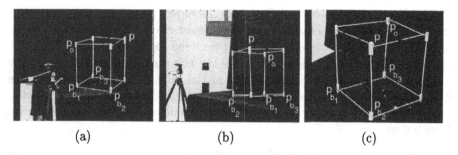

Fig. 2. Properties of affine point representations. Three views of a real wireframe object are shown. Points $p_o, p_{b_1}, p_{b_2}, p_{b_3}$ define an affine coordinate frame within which all world points can be represented: Point p_o is the origin, and points $p_{b_1}, p_{b_2}, p_{b_3}$ are the basis points. The affine coordinates of a fifth point, p, are computed from its projection in images (a) and (b) using Property 2. p's projection in image (c) can then be computed from the projections of the four basis points using Property 1.

of an object without having any 3D information about the position of the camera or the affine basis points:

Property 2 (Affine Reconstruction Property) *The affine coordinates of* $p_1, \ldots, p_n$ *can be computed using Eq. (3) when their projection along two viewing directions is known.*

Intuitively, Property 2 shows that this process can be inverted if at least four non-coplanar 3D points can be tracked across frames as the camera moves. More precisely, given two images I_1, I_2, the affine coordinates of a point p can be recovered by solving an over-determined system of equations

$$
\begin{bmatrix} u_p^1 \\ v_p^1 \\ u_p^2 \\ v_p^2 \end{bmatrix} = \begin{bmatrix} u_{b_1}^1 & u_{b_2}^1 & u_{b_3}^1 & u_{p_o}^1 \\ v_{b_1}^1 & v_{b_2}^1 & v_{b_3}^1 & v_{p_o}^1 \\ u_{b_1}^2 & u_{b_2}^2 & u_{b_3}^2 & u_{p_o}^2 \\ v_{b_1}^2 & v_{b_2}^2 & v_{b_3}^2 & v_{p_o}^2 \end{bmatrix} \begin{bmatrix} x \\ y \\ z \\ 1 \end{bmatrix}. \tag{4}
$$

In Section 5 we consider how this property can be exploited to interactively "position" a virtual object within an environment in which four feature points can be identified and tracked.

3   Affine Augmented Reality

The previous section suggests that once the affine coordinates of points on a virtual object are determined relative to four features in the environment, the points' projection becomes trivial to compute. The central idea in our approach is to ignore the original representation of the object altogether and perform all graphics operations with the new, affine representation of the object. This representation is related to the original object-centered representation by a homogeneous transformation: if p_1, p_2, p_3, p_4 are the coordinates of four non-coplanar points on the virtual object expressed in the object's coordinate frame and p_1', p_2', p_3', p_4' are their corresponding coordinates in the affine frame, the two frames are related by an invertible, homogeneous *object-to-affine* transformation **A** such that

$$[p_1'\ p_2'\ p_3'\ p_4'] = \mathbf{A}[p_1\ p_2\ p_3\ p_4]. \tag{5}$$

The affine representation of virtual objects is both powerful and weak: it allows us to compute an object's projection without requiring information about the camera's position or calibration. On the other hand, this representation captures only properties of the virtual object that are maintained under affine transformations—metric information such as the distance between an object's vertices and the angle between object normals is not captured by the affine model. Nevertheless, our purpose is to show that the information that *is* maintained is sufficient for correctly rendering the virtual object. The augmented reality system we are developing based on this principle currently supports the following operations:

- *Object rendering:* Efficient and realistic rendering of virtual objects requires that operations such as point projection and z-buffering can be performed accurately and can exploit graphics rendering hardware when available. This is made possible by describing the entire projection process with a view transformation matrix expressed directly in terms of measurements in the image (Section 4).
- *Interactive placement of virtual objects:* This operation allows virtual objects to be "placed" in the environment with a simple through-the-lens interaction of the user with the system. The operation effectively determines the object-to-affine transformation that maps points on the virtual object into the affine frame of the feature points being tracked (Section 5).
- *Real-time affine basis tracking:* Determination of the projection matrix for rendering virtual objects requires tracking the affine basis points reliably and efficiently across frames. Our system tracks lines and polygonal regions in real time and uses line intersections to define the basis points (Section 6).
- *Object animation:* Because virtual objects are represented in terms of objects physically present in the environment, their motion relative to such objects becomes particularly easy to specify without any metric information about their 3D locations or about the camera (Section 7).

4 Visible Surface Rendering

The projection of points on an affinely-represented virtual object is completely determined by the location of the feature points defining the affine frame. One of

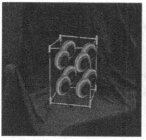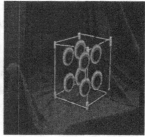

Fig. 3. Real time visible-surface rendering of texture-mapped affine virtual objects. The objects were represented in OpenInventor. Affine basis points were defined by the intersections of lines on the wireframe object which were tracked in real time. The virtual objects were defined with respect to those points (see Section 5). The wireframe was rotated clockwise between frames. Note that hidden-surface elimination occurs only between virtual objects; correct occlusion resolution between real and virtual objects requires information about the real objects' 3D structure [6].

the key aspects of affine object representations is that even though they are non-Euclidean, they nevertheless allow rendering operations such as z-buffering and clipping [21] to be performed accurately. This is because both depth order as well as the intersection of lines and planes is preserved under affine transformations.

More specifically, z-buffering relies on the ability to order in depth two object points that project to the same pixel in the image. Typically, this operation is performed by assigning to each object point a z-value which orders the points along the optical axis of the (graphics) camera. The observation we use to render affine objects is that the actual z-value assigned to each point is irrelevant as long as the correct ordering of points is maintained. To achieve such an ordering we represent the camera's optical axis in the affine frame defined by the feature points being tracked, and we order object points along this axis.

The optical axis of the camera can be defined as the 3D line whose points project to a single pixel in the image. This is expressed mathematically by representing the optical axis of the camera with the homogeneous vector $[\zeta^T \ 0]^T$ where ζ is given by the cross product

$$\zeta = \begin{bmatrix} u_{b_1} - u_{p_o} \\ u_{b_2} - u_{p_o} \\ u_{b_3} - u_{p_o} \end{bmatrix} \times \begin{bmatrix} v_{b_1} - v_{p_o} \\ v_{b_2} - v_{p_o} \\ v_{b_3} - v_{p_o} \end{bmatrix} \qquad (6)$$

and $[u_{b_i} \ v_{b_i} \ 1]^T, i = 1, \ldots, 3$ are the image locations of the affine basis points. To order points along the optical axis we assign to each point p on the model a z-value equal to the dot product $p \cdot [\zeta^T \ 0]^T$. Hence, the entire projection process is described by a single 4×4 homogeneous matrix.

Observation 1 (Projection Equation) *Visible surface rendering of a point p on an affine object can be achieved by applying the following transformation to p:*

$$
\begin{bmatrix} u \\ v \\ w \\ 1 \end{bmatrix} = \begin{bmatrix} u_{b_1} & u_{b_2} & u_{b_3} & u_{p_o} \\ v_{b_1} & v_{b_2} & v_{b_3} & v_{p_o} \\ & \zeta^T & & z_o \\ 0 & 0 & 0 & 1 \end{bmatrix} \begin{bmatrix} x \\ y \\ z \\ 1 \end{bmatrix} \tag{7}
$$

where u and v are the image coordinates of p's projection and w is p's assigned z-value.

The matrix in Eq. (7) is an affine generalization of the *view transformation matrix*, which is commonly used in computer graphics for describing arbitrary orthographic and perspective projections of Euclidean objects and for specifying clipping planes. A key practical consequence of the similarity between the Euclidean and affine view transformation matrices is that graphics operations on affine objects can be performed using existing hardware engines for real-time projection, clipping and z-buffering. In our experimental system the matrix of Eq. (7) is input directly to a Silicon Graphics RealityEngine2 for implementing these operations efficiently (Figure 3).

5   Interactive Object Placement

Before virtual objects can be overlaid with images of a three-dimensional environment, the geometrical relationship between these objects and the environment must be established. Our approach for placing virtual objects in the 3D environment borrows from a few simple results in stereo vision [26]: given a point in space, its 3D location is uniquely determined by the point's projection in two images taken at different positions of the camera (Figure 4(a)). Rather than specifying the virtual objects' affine coordinates explicitly, our system allows the user to interactively specify what the objects should "look like" in two images of the environment. In practice, this involves specifying the projection of points on the virtual object in two images in which the affine basis points are also visible. The main questions here are (1) how many point projections need to be specified in the two images, (2) how does the user specify the projection of these points, and (3) how do these projections determine the objects' affine representation?

The number of point correspondences required to determine the position and shape of a virtual object is equal to the number of points that uniquely determine the object-to-affine transformation. This transformation is uniquely determined by specifying the 3D location of four non-coplanar points on the virtual object that are selected interactively (Eq. (5)).

To fix the location of a selected point p on the virtual object, the point's projection in two images taken at distinct camera positions is specified interactively using a mouse. The process is akin to stereo triangulation: by selecting interactively the projections, q^L, q^R, of p in two images in which the projection

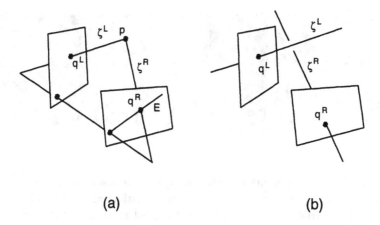

(a)   (b)

Fig. 4. Positioning virtual objects in a 3D environment. (a) Any 3D point is uniquely specified by its projection in two images along distinct lines of sight. The point is the intersection of the two visual rays, ζ^L, ζ^R that emanate from the camera's center of projection and pass through the point's projections. (b) In general, a pair of arbitrary points in two images does not specify two intersecting visual rays. A sufficient condition is to require the point in the second image to lie on the *epipolar line*, i.e., on the projection of the first visual ray in the second image. This line is determined by the projection matrix.

of the affine basis points is known, p's affine coordinates can be recovered using the Affine Reconstruction Property.

The two projections of point p cannot be selected in an arbitrary fashion. In general, the correspondence induced by q^L and q^R may not define a physical point in space. Once p's projection is specified in one image, its projection in the second image must lie on a line satisfying the *epipolar constraint* [12]. This line is computed automatically and is used to constrain the user's selection of q^R in the second image. In particular, if Π^L, Π^R are the projection matrices associated with the first and second image, respectively, and ζ^L, ζ^R are the corresponding optical axes defined by Eq. (6), the epipolar line can be parameterized by the set [27]

$$\{\Pi^R[(\Pi^L)^{-1}q^L + t\zeta^L] \mid t \in \Re\}. \tag{8}$$

By taking the epipolar constraint into account, we can ensure that the points specified by the user provide a physically valid placement of the virtual object.

Once the projections of a point on a virtual object are specified in the two images, the points' affine coordinates can be determined by solving the linear system of equations in Eq. (4). This solves the placement problem for virtual objects. The entire process is shown in Figure 5.

Affine object representations lead naturally to a through-the-lens method for constraining further the user degrees of freedom during the interactive placement

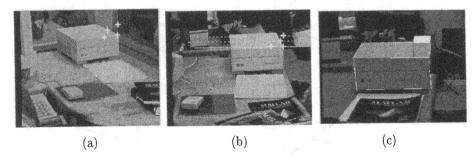

(a)   (b)   (c)

Fig. 5. Steps in placing a virtual cube on top of a workstation. (a) The mouse is used to select four points in the image at the position where four of the cube's vertices should project. In this example, the goal is to align the cube's corner with the right corner of the workstation. (b) The camera is moved to a new position and the epipolar line corresponding to each of the points selected in the first image is computed automatically. The epipolar line corresponding to the lower right corner of the cube is drawn solid. Crosses represent the points selected by the user. (c) View of the virtual cube from a new position of the camera, overlaid with live video. Also overlaid are the lines whose intersections are used to define the affine frame. These lines are tracked at frame rate by the system. No information about the camera's position or the Euclidean shape of the workstation is used in the above steps.

of virtual objects. Such constraints have been used in vision-based pointing interfaces [15]. An example of such a constraint is planarity: if r_1, r_2, r_3 are three points on a physical plane in the environment and p is a point on the virtual object, the constraint that p lies on the plane of r_1, r_2, r_3 implies that p's projection is a linear combination of the projections of r_1, r_2, r_3. This constraint completely determines the projection of p in a second image in which r_1, r_2, r_3 are visible. The planarity constraint allows virtual objects to be "snapped" to physical objects and "dragged" over their surface by forcing one or more points on the virtual object to lie on planes in the environment that are selected interactively (Figure 6).

6   Affine Basis Tracking

The ability to track the projection of 3D points undergoing rigid transformations with respect to the camera becomes crucial in any method that relies on image information to represent the position and orientation of the camera [2, 8, 9, 16]. Real-time tracking of image features has been the subject of extensive research (e.g., see [28–31]). Below we describe a simple approach that exploits the existence of more than the minimum number of feature points to increase robustness,

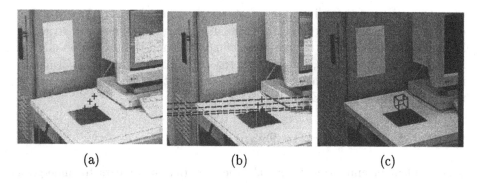

(a)　　　　　　　　(b)　　　　　　　　(c)

Fig. 6. Aligning a virtual quadrangle with a mousepad. The affine frame is defined by the vertices of the white planar region on the wall and the mousepad. Crosses show the points selected in each image. Dotted lines in (b) show the epipolars associated with the points selected in (a). The constraints provided by the epipolars, the planar contact of the cube with the table, as well as the parallelism of the cubes' sides with the side of the workstation allows points on the virtual object to be specified interactively even though no image feature points exist at any four of the object's vertices. (c) Real time overlay of a virtual cube.

tracks lines rather than points, and automatically provides an updated projection matrix used for rendering virtual objects.

The approach is based on the following observation: Suppose that the affine coordinates of a collection of N non-coplanar feature points is known. Then, the changes in the projection matrix caused by a change in the camera's position, orientation, or calibration parameters can be modeled by the equation

$$\Delta I = \Delta \Pi_{3 \times 4} \, M \qquad (9)$$

where ΔI is the change in the image position of the feature points, $\Delta \Pi$ is the change in the projection matrix, and M is the matrix holding the affine coordinates of the feature points.[1] Eq. (9) leads directly to a Kalman filter based method for both tracking the feature points and for continuously updating the projection matrix. We use two independent constant acceleration Kalman filters [22] whose states consist of the first and second rows of the projection matrix $\Pi_{3 \times 4}$, respectively, as well as their time derivatives. The filter's measurement equation is given by Eq. (9).

Feature points and their affine coordinates are determined at system initialization time. Tracking is bootstrapped by interactively specifying groups of coplanar linear features in the initial image of the environment. Lines are tracked using an algorithm similar to that of Harris [32]. Feature points are defined to be

---

[1] When only four feature points are available the matrix M degenerates to a unit 4×4 matrix.

the intersections of these lines. Once feature points have been tracked over several frames their affine coordinates are computed using the Affine Reconstruction Property. Eq. (9) is used to update the affine basis and the projection matrix. During the tracking phase, the first two rows of the affine view transformation matrix are contained in the state of the Kalman filters. The third row of the matrix is computed from Eq. (6).

7   Rigid Animation By Example

The problem of animating a virtual object so that it appears to undergo a rigid transformation is particularly challenging when the camera's position and calibration parameters are unknown because metric information is lost in the projection process. In general, one cannot distinguish, based solely on an object's affine coordinates, between arbitrary homogeneous 4×4 transformations that cause shear and scaling of a virtual object from those that transform the object rigidly.

Our approach for rigidly animating affine virtual objects is twofold. First we note that pure translations of a virtual object do correspond to rigid translations in the world coordinate frame. Second, to realize arbitrary rigid rotations we exploit the user's ability to interact with the augmented reality system: the user effectively "demonstrates" to the system which transformations to use to generate rigid rotations.

Suppose that all feature points defining the affine basis lie on a single rigid object and that this object is viewed simultaneously by two cameras. To specify a valid rigid rotation, the user simply rotates the object containing the affine basis in front of the two cameras. The image positions of the feature points at the end of the rotation define four points in 3D space whose affine coordinates with respect to the original affine basis can be determined using the Affine Re-Projection Property. Eq. (5) tells us that these coordinates define a 4×4 homogeneous transformation $\mathbf{T}$ that corresponds to a rigid rotation since the object itself was rotated rigidly. By repeating this process three times, the user provides enough information to span the entire space of rotations.

More precisely, we use the following two theorems which exploit the special structure of matrix $\mathbf{T}$. This matrix is a *conjugate rotation transformation* [19] that allows generation of all 4×4 transformations corresponding to rigid rotations about the axis of rotation of $\mathbf{T}$:

Theorem 1 (Rotation about the axis of T) *Transformations corresponding to rotation about the axis of* $\mathbf{T}$ *are of the form*

$$\mathbf{T}(\theta) = \mathbf{Q} \begin{bmatrix} \cos\theta + i\sin\theta & 0 & 0 \\ 0 & \cos\theta - i\sin\theta & 0 \\ 0 & 0 & 1 \end{bmatrix} \mathbf{Q}^{-1}, \tag{10}$$

where θ *is a continuous parameter in the interval* $[0, 2\pi)$ *and* $\mathbf{Q}\Lambda\mathbf{Q}^{-1}$ *is a diagonalization of* $\mathbf{T}$.

Theorem 2 (Rotation about an arbitrary axis α) *If $\mathbf{Q_k}\Lambda_\mathbf{k}\mathbf{Q_k^{-1}}$ is a diagonalization of $\mathbf{T_k}, k = 1, 2, 3$ and α is an arbitrary affine vector, a transformation corresponding to rotation about α is given by the product $\mathbf{T} = \mathbf{T_i}(\theta_i)\mathbf{T_j}(\theta_j)\mathbf{T_k}(\theta_k)$, where $\{i, j, k\} = \{1, 2, 3\}$ and $\theta_1, \theta_2, \theta_3$ satisfy*

$$\mathbf{T_i}(\theta_i)\mathbf{T_j}(\theta_j)\mathbf{T_k}(\theta_k)\alpha = \alpha. \tag{11}$$

Together, Theorems 1 and 2 completely solve the problem of continuously animating a virtual object about an arbitrary axis. Theorem 1 gives us a way to generate object rotations about a single axis when a conjugate rotation transformation corresponding to a rotation of unknown angle about that axis is known. Theorem 2 tells us how to synthesize a single conjugate rotation transformation about an arbitrary axis from three example transformations. Given a value for θ_i, Eq. (11) accepts an analytic solution for θ_j and θ_k. In practice, the value of θ_i is found numerically by minimizing the residual of Eq. (11) in the interval $[0, 2\pi)$.

Theorems 1 and 2 are closely related to recent techniques for Euclidean and affine camera calibration that exploit the special structure of conjugate rotation transformations [19, 33]. Intuitively, the problem of synthesizing conjugate rotation transformations is equivalent to the indirect computation of metric quantities (e.g., the Grammian matrix [25]). Rather than explicitly recovering Euclidean structure and propagating it through our entire system, we use this information indirectly and only when necessary, i.e., for implementing rigid animations. To exploit Theorems 1 and 2 for rigidly animating virtual objects we need to have at least three examples of conjugate rotation transformations. This is achieved interactively by rotating a rigid object in front of a pair of stereo cameras while the object and the affine basis points are being tracked by the system.

8 Experimental System

We have implemented a prototype augmented reality system consisting of a Silicon Graphics RealityEngine2 that handles all graphics operations using the OpenGL and OpenInventor graphics libraries, and a tracking subsystem implemented in C that runs on a SUN SPARCserver2000. Video input is provided by a Sony camcorder, a TRC BiSight stereo head and a Datacube MaxVideo 10 board used only for frame grabbing (Figure 7). The intrinsic and extrinsic parameters of the cameras were not computed.

Operation of the system involves three steps: (1) initialization of the affine basis, (2) virtual object placement, and (3) affine basis tracking and projection update. Initialization of the affine basis establishes the frame in which all virtual objects will be represented during a run of the system. Basis points are initialized as intersections of line segments that are selected interactively in the initial image. Virtual object initialization follows the sequence of steps shown in Figure 5. Once the affine coordinates of all points on a virtual object are computed, the

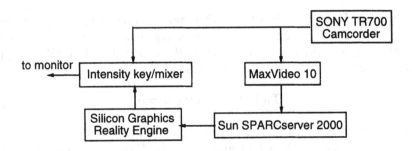

Fig. 7. Configuration of our augmented reality system.

affine object models are transmitted to the graphics subsystem where they are treated as if they were defined in a Euclidean frame of reference.

Upon initialization of the affine basis, the linear features defining the basis are tracked automatically. Line tracking runs on a single processor at rates between 30Hz and 60Hz for approximately 12 lines and provides updated Kalman filter estimates for the elements of the projection matrix [22]. Conceptually, the tracking subsystem can be thought of as an "affine camera position tracker" that returns the current affine projection matrix asynchronously upon request. For each new video frame, the rows of the projection matrix are used to build the view transformation matrix. This matrix is sent to the graphics subsystem. Figure 8 shows snapshots from example runs of our system. The image overlay was initialized using the placement method of Section 5 at two viewpoints close to the view in Figure 8(a). The objects were then rotated together through the sequence of views in Figure 8(b)-(e) while tracking was maintained on the two black squares.

The accuracy of the image overlays is limited by radial distortions of the camera [9] and the affine approximation to perspective projection. Radial distortions are not currently taken into account. In order to assess the limitations resulting from the affine approximation to perspective we computed mis-registration errors as follows. We used the image projection of vertices on a physical object in the environment (a box) to serve as ground truth and compared these projections at multiple camera positions to those predicted by their affine representation and computed by our system. The image points corresponding to the projection of the affine basis in each image were not tracked automatically but were hand-selected on four of the box's corners to establish a best-case tracking scenario for affine-based image overlay.[2] These points were used to define the affine view transformation matrix. The affine coordinates of the remaining vertices on the box were then computed using the Affine Reconstruction Property, and their projection was computed for roughly 50 positions of the camera. As the camera's distance to the object increased, the camera zoom was also increased

---

[2] As a result, mis-registration errors reported in Figure 9 include the effects of small inaccuracies due to manual corner localization.

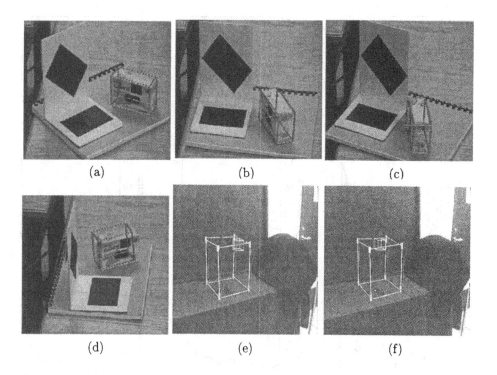

Fig. 8. Experimental runs of the system. (a) View from the position where the virtual object was interactively placed over the image of the box. The affine basis points were defined by tracking the outline of the two black polygonal regions. These regions were only used to simplify tracking; the Euclidean shape of the regions unknown and was not recovered by the system. (b),(c) Image overlays after a combined rotation of the box and the object defining the affine basis. (d) Limitations of the approach due to tracking errors. Since the only information used to determine the projection matrix comes from tracking the basis points, tracking errors inevitably lead to wrong overlays. In this example, the extreme foreshortening of the top square led to inaccurate tracking of the affine basis points. (f),(g) Two snapshots of an animated virtual wireframe cube.

in order to keep the object's size constant and the mis-registration errors comparable. Results are shown in Figures 9 and 10. While errors remain within 15 pixels for the range of motions we considered (in a 512×480 image), the results show that, as expected, the affine approximation to perspective leads to errors as the distance to the object decreases. These effects strongly suggest the utility of projectively-invariant representations for representing virtual objects in calibration-free video overlay.

The accuracy of the image overlays generated by our system was measured as follows. A collection of line trackers was used to track the outline of the two

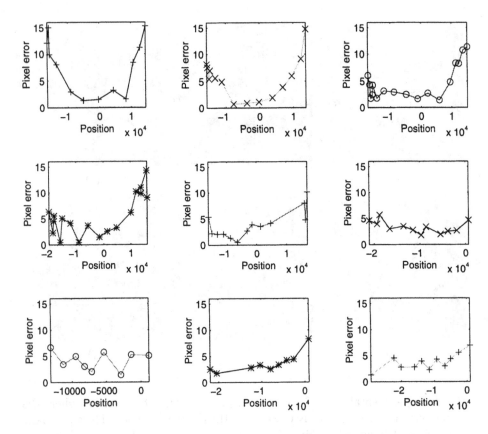

Fig. 9. Mis-registration errors. The errors are averaged over three vertices on a $15cm \times 15cm \times 15cm$ box that are not participating in the affine basis. The line style of the plots corresponds to the camera paths shown in Figure 10.

black squares on the object of Figure 11(a). The affine coordinates of an easily-identifiable 3D point that was rigidly attached to the object were then computed with respect to the basis defined by the two squares. These coordinates were sent to the graphics subsystem and used to display in real time a small dot at the predicted position of the 3D point as the affine basis points and the 3D point were rotated freely. Two correlation-based trackers were used to track in real-time the image projections of the 3D point and the synthetic dot thus providing an on-line estimate of the ground truth for image overlay. Plots of the $x-$ and $y-$ coordinates of the tracked and re-projected points are shown in Figure 11(b).

9   Concluding Remarks

Current limitations of our system are (1) the affine approximation to perspective which inevitably introduces errors in the re-projection process, and (2) the

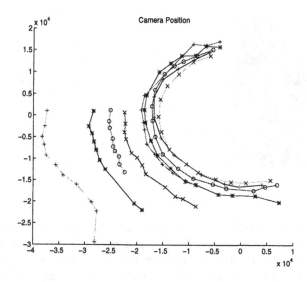

Fig. 10. Camera positions used for computing mis-registration errors. The plot shows the first and third components of the (unnormalized) vector ζ, corresponding to the computed optical axis of the affine camera. Because the camera was moved on a plane, the plot provides a qualitative indication of the path along which the camera was moved to quantify the mis-registration errors. The path followed a roughly circular course around the box at distances ranging from $0.5m$ to $5m$. The same four non-coplanar vertices of the box defined the affine frame throughout the measurements. The affine coordinates of all visible vertices of the box were computed from two views near position (-1.5,0.5).

existence of a 5-10 frame lag in the projection of virtual objects due to communication delays between the tracking and graphics subsystems. On a theoretical level, we are extending the basic approach by representing virtual objects in a projective reference frame and investigating the use of image synthesis techniques for shading non-Euclidean virtual objects. On a practical level, we are planning to use gray level feature trackers [2] to increase tracking accuracy and versatility.

Acknowledgements

The authors would like to thank Chris Brown for many helpful discussions and for his constant encouragement throughout the course of this work. The financial support of the National Science Foundation under Grant No. CDA-9503996, of the University of Maryland under Subcontract No. Z840902 and of Honeywell under Research Contract No. 304931455 is also gratefully acknowledged.

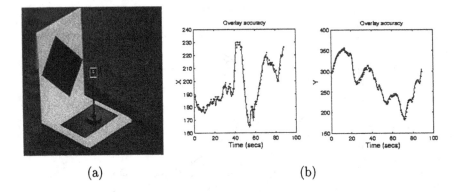

(a)   (b)

Fig. 11. (a) Experimental setup for measuring the accuracy of image overlays. Affine basis points were defined by the corners of the two black squares. The affine coordinates for the tip of a nail rigidly attached to the object were computed and were subsequently used to generate the overlay. Overlay accuracy was measured by independently tracking the nail tip and the generated overlay using correlation-based trackers. (b) Real-time measurement of overlay errors. Solid lines correspond to the tracked image point and dotted lines to the tracked overlay. The object in (a) was manually lifted from the table and freely rotated for approximately 90 seconds. The mean absolute overlay error in the $x-$ and $y-$ directions was 1.74 and 3.47 pixels, respectively.

References

1. W. Grimson *et al.*, "An automatic registration method for frameless stereotaxy, image guided surgery, and enhanced reality visualization," in *Proc. IEEE Conf. Computer Vision and Pattern Recognition*, pp. 430–436, 1994.
2. M. Uenohara and T. Kanade, "Vision-based object registration for real-time image overlay," in *Proc. CVRMED'95*, pp. 14–22, 1995.
3. M. Bajura, H. Fuchs, and R. Ohbuchi, "Merging virtual objects with the real world: Seeing ultrasound imagery within the patient," in *Proc. SIGGRAPH'92*, pp. 203–210, 1992.
4. S. Feiner, B. MacIntyre, and D. Soligmann, "Knowledge-based augmented reality," *Comm. of the ACM*, vol. 36, no. 7, pp. 53–62, 1993.
5. T. Darrell, P. Maes, B. Blumberg, and A. P. Pentland, "A novel environment for situated vision and action," in *IEEE Workshop on Visual Behaviors*, pp. 68–72, 1994.
6. M. M. Wloka and B. G. Anderson, "Resolving occlusion in augmented reality," in *Proc. Symposium on Interactive 3D Graphics*, pp. 5–12, 1995.
7. M. Tuceyran *et al.*, "Calibration requirements and procedures for a monitor-based augmented reality system," *IEEE Trans. Visualization and Computer Graphics*, vol. 1, no. 3, pp. 255–273, 1995.
8. J. Mellor, "Enhanced reality visualization in a surgical environment," Master's thesis, Massachusetts Institute of Technology, 1995.

9. M. Bajura and U. Neumann, "Dynamic registration correction in video-based augmented reality systems," *IEEE Computer Graphics and Applications*, vol. 15, no. 5, pp. 52–60, 1995.

10. D. G. Lowe, "Robust model-based tracking through the integration of search and estimation," *Int. J. Computer Vision*, vol. 8, no. 2, pp. 113–122, 1992.

11. S. Ravela, B. Draper, *et al.*, "Adaptive tracking and model registration across distinct aspects," in *Proc. 1995 IEEE/RSJ Int. Conf. Intelligent Robotics and Systems*, pp. 174–180, 1995.

12. L. S. Shapiro, A. Zisserman, and M. Brady, "3D motion recovery via affine epipolar geometry," *Int. J. Computer Vision*, vol. 16, no. 2, pp. 147–182, 1995.

13. J. J. Koenderink and A. J. van Doorn, "Affine structure from motion," *J. Opt. Soc. Am.*, vol. A, no. 2, pp. 377–385, 1991.

14. G. D. Hager, "Calibration-free visual control using projective invariance," in *Proc. 5th Int. Conf. Computer Vision*, 1995.

15. R. Cipolla, P. A. Hadfield, and N. J. Hollinghurst, "Uncalibrated stereo vision with pointing for a man-machine interface," in *Proc. IAPR Workshop on Machine Vision Applications*, 1994.

16. A. Azarbayejani, T. Starner, B. Horowitz, and A. Pentland, "Visually controlled graphics," *IEEE Trans. Pattern Anal. Machine Intell.*, vol. 15, no. 6, pp. 602–605, 1993.

17. A. Shashua, "A geometric invariant for visual recognition and 3D reconstruction from two perspective/orthographic views," in *Proc. IEEE Workshop on Qualitative Vision*, pp. 107–117, 1993.

18. E. B. Barrett, M. H. Brill, N. N. Haag, and P. M. Payton, "Invariant linear methods in photogrammetry and model-matching," in *Geometric Invariance in Computer Vision*, pp. 277–292, MIT Press, 1992.

19. P. A. Beardsley, I. D. Reid, A. Zisserman, and D. W. Murray, "Active visual navigation using non-metric structure," in *Proc. 5th Int. Conf. Computer Vision*, pp. 58–64, 1995.

20. Y. Lamdan, J. T. Schwartz, and H. J. Wolfson, "Object recognition by affine invariant matching," in *Proc. Computer Vision and Pattern Recognition*, pp. 335–344, 1988.

21. J. D. Foley, A. van Dam, S. K. Feiner, and J. F. Hughes, *Computer Graphics Principles and Practice*. Addison-Wesley Publishing Co., 1990.

22. Y. Bar-Shalom and T. E. Fortmann, *Tracking and Data Association*. Academic Press, 1988.

23. M. Gleicher and A. Witkin, "Through-the-lens camera control," in *Proc. SIGGRAPH'92*, pp. 331–340, 1992.

24. J. L. Mundy and A. Zisserman, eds., *Geometric Invariance in Computer Vision*. MIT Press, 1992.

25. D. Weinshall and C. Tomasi, "Linear and incremental acquisition of invariant shape models from image sequences," in *Proc. 4th Int. Conf. on Computer Vision*, pp. 675–682, 1993.

26. O. D. Faugeras, *Three-Dimensional Computer Vision: A Geometric Viewpoint*. MIT Press, 1993.

27. S. M. Seitz and C. R. Dyer, "Complete scene structure from four point correspondences," in *Proc. 5th Int. Conf. on Computer Vision*, pp. 330–337, 1995.

28. A. Blake and A. Yuille, eds., *Active Vision*. MIT Press, 1992.

29. C. M. Brown and D. Terzopoulos, eds., *Real-Time Computer Vision*. Cambridge University Press, 1994.

30. A. Blake and M. Isard, "3D position, attitude and shape input using video tracking of hands and lips," in *ACM SIGGRAPH'94*, pp. 185–192, 1994.

31. K. Toyama and G. D. Hager, "Incremental focus of attention for robust visual tracking," in *Proc. Computer Vision and Pattern Recognition*, 1996. To appear.

32. C. Harris, "Tracking with rigid models," in *Active Vision* (A. Blake and A. Yuille, eds.), pp. 21–38, MIT Press, 1992.

33. R. Horaud, F. Dornaika, B. Boufama, and R. Mohr, "Self calibration of a stereo head mounted onto a robot arm," in *Proc. 3rd European Conf. on Computer Vision*, pp. 455–462, 1994.

Author Index

Springer-Verlag
and the Environment

We at Springer-Verlag firmly believe that an international science publisher has a special obligation to the environment, and our corporate policies consistently reflect this conviction.

We also expect our business partners – paper mills, printers, packaging manufacturers, etc. – to commit themselves to using environmentally friendly materials and production processes.

The paper in this book is made from low- or no-chlorine pulp and is acid free, in conformance with international standards for paper permanency.

Lecture Notes in Computer Science

For information about Vols. 1–1075

please contact your bookseller or Springer-Verlag